Land Policies for Inclusive Growth

About the Editor

T. Haque is a well-known Agricultural Economist, having specialization in the field of Agricultural Development and Policy. He obtained his Ph.D. degree from Patna University and postdoctoral training from Iowa State University, Ames, U.S.A.

Dr. Haque is currently Director, Council for Social Development, New Delhi and Senior Adviser, Rural Development Institute, New Delhi. He served as Chairman, Commission for Agricultural Costs and Prices, Government of India, for six years, ICAR National Fellow at the National Centre for Agricultural Economics and Policy, New Delhi for five years and Director, Centre for Agrarian Studies, NIRD, Hyderabad for five years and Assistant Professor/Associate Professor at the Indian Agricultural Research Institute, New Delhi for about thirteen years.

Dr. Haque served as Consultant/Specialist in several international organizations, namely, Food and Agriculture Organization, International Labour Organization, the World Bank, Afro-Asian Rural Development Organization etc.

Dr. Haque is author of several books including (i) "Agrarian Reforms and Institutional Changes in India", (ii) "Impact of Land Reforms on Agriculture and Rural Development", (iii) "Transfer of Technology to Small Farmers", (iv) "Impact of Tenancy Reform on Agricultural Productivity and Socio-economic Status of Tenants", and (v) "Sustainability of Small Holder Agriculture". Besides, he has published about 100 research papers and reports. His e-mail address is: drt.haque@gmail.com.

Land Policies for Inclusive Growth

Edited by
T. Haque

in collaboration with
Council for Social Development, New De[illegible]
and
Rural Development Institute, [illegible] Delhi

by
CONCEPT PUBLISHING C[illegible]ANY PVT. LTD.
NEW DEL[illegible] 59

ISBN-13 : 978-81-8069-885-9

First Published 2012

Published and Printed by

Concept Since 1974

Concept Publishing Company Pvt. Ltd.
A/15-16, Commercial Block, Mohan Garden
New Delhi-110059 (India)
T : 25351460, 25351794, *F* : 091-11-25357109
E: publishing@conceptpub.com, W: www.conceptpub.com
Editorial Office: H-13, Bali Nagar, New Delhi-110 015, India.

Cataloging in Publication Data--*Courtesy:* D.K. Agencies (P) Ltd. <docinfo@dkagencies.com>

Land policies for inclusive growth / edited by T. Haque.
p. cm.
Includes bibliographical references and index.
ISBN 9788180698859

1. Land reform—Government policy—India—Congresses. I. Haque, T. (Tajamul), 1947-

DDC 333.3154 23

Introduction

T. Haque

Land continues to be of great economic and social value in India as the majority of the population continues to be dependent on agriculture for food and livelihood. Besides, land policies adopted by the government, from time to time, determine not only the pace and patterns of rural growth but also whether or not the poor and the marginalized people in the country effectively and equitably participate in the growth process. This volume analyses the emerging land policy issues in India in the context of land access and land security for India's rural poor, including landless labourers, marginal farmers and other vulnerable groups such as women and tribals. Such an exercise assumes an especial significance in the wake of rising complexities due to market-driven land use policy and diversified rural growth.

The contributors to the volume are well-known experts in the field of land reforms, including researchers, lawyers and policy administrators. The majority of the papers were presented in a national seminar jointly organized by the Council for Social Development and Rural Development Institute, New Delhi, on November 9-10, 2010; these have been supplemented by some papers especially commissioned for this volume. The nineteen papers included here are presented under five broad titles, namely (i) Redistributive Land Reform: Old and New Approaches, (ii) Tenancy Reform, (iii) Land Rights for Women, (iv) Tribal Land Issues, and (v) Land Acquisition and Inclusive Development.

Redistributive Land Reform

Since independence, several measures of redistributive land reform have been undertaken including abolition of intermediaries and conferring ownership/occupancy right on tenants, imposition of ceiling on land holdings and distribution of surplus land among landless and semi-landless families, distribution of government land, *bhoodan* land etc. However, these measures though well intentioned, have yielded only partial results.

Still, about 10 per cent of the rural households are landless and about 50 per cent of them do not even own homestead plots.

Marginalisation of landholding is rising and inequality in the ownership of land continues to be high (Haque: 2010). The situation undoubtedly demands accessibility to land by the landless poor and improvement in the condition of marginal and semi-marginal farmers. Roy L. Prosterman and Tim Hanstad in their paper entitled "Land Reform in the 21st Century: New Challenges, New Responses", point out how land reform can improve the lives of poor people in a number of ways. According to them, the experience of countries where land reform has been effectively implemented, shows that it can lead to increased crop production, improved nutrition of poor households and lay the foundation for (i) sustained and inclusive growth, (ii) grass-roots empowerment, (iii) movement towards more democratic societies, (iv) reduction in social unrest and instability, (v) reduced urban migration, (vi) better environmental stewardship, and (vii) creation of wealth. There is cogent international evidence to suggest that smaller holdings generally produce more and family-operated farms are more efficient than either collective or capitalist farms. Besides, a cultivator with ownership or secure long-term ownership rights like tenure has a greater incentive to make long-term as well as short-term investment for productivity enhancement. The most significant aspect of redistributive land reform in a situation of highly skewed land distribution is that it lays down the foundation for inclusive economic growth and grassroots democracy.

However, there is a limit beyond which various state governments in India may not go for enforcement of ceiling laws, mainly fearing backlash from the landed class. Also, under the impact of population pressure on land as well as lack of adequate off-farm and non-farm employment opportunities, the size of large farms also is on the decline, thereby leaving very limited scope for availability of ceiling surplus land. Therefore, it may be desirable to look for other options for improving the rural poor's access to land. T. Haque's paper on 'Land Policies for Social Inclusion in India', suggests that the government should consider: (a) purchasing land from the open market and allocating at least 10 to 15 cents of land to all landless and houseless families, (b) legalization of land leasing, (c) speedy settlement of all ceiling surplus cases pending in courts, and (d) regularization/allotment of available government and *bhoodan* lands to the poor, (ii) protecting the land and forest rights of tribals, following (a) special drive to complete records of rights, (b) banning transfer of tribal land to non-tribal without any exception, (c) complete ban on land acquisition for private industrialization in scheduled areas, (d) special drive to restore all alienated lands to tribals and effective implementation of PESA and Indian Forest Rights Act, 2006. Haque's paper also underscores the need for strengthening women's landownership right, land acquisition policies and records of land rights for social inclusion.

Mitchell and Hanstad in their paper entitled 'Small Homegarden Plots and Sustainable Livelihoods for the Poor', point out that allocation of small homegarden plots can be an useful strategy for promoting sustainable livelihood for the poor, including secure access to land and water, improved financial security, improved leverage in wage bargaining, improved nutrition and improved socio-political status. Based on international experience, they see women benefitting significantly from allocation of homestead plots to them, as sales of homegarden produce may be an independent source of income which can be used for overall family welfare such as improved education and nutrition of the children.

Srijit Mishra's paper on 'Access to Land: Some Issues', underscores the need for improved land access of the tiller from the point of view of both equity and agricultural efficiency. He suggests that the landlease market should be opened up in states where it is illegal so that tenancy does not go underground. In states where tenants are protected, as in West Bengal, there can be a provision for making them owners of part of the land, while giving up the tenancy rights for the remaining part. Besides, the government can consider buying land with the purpose of distribution. If market rates are paid, then landowners who do not have much stake in agriculture may agree to sell land which can be distributed to the landless and semi-landless poor. Another important suggestion made by Mishra is that women should be given independent or joint land rights.

Nair and Banerjee analyses the structural changes in land distribution and its implication for improving access to land in their paper entitled 'Structural Changes in Land Distribution and its Implications for Improved Access to Land'. The main findings of the paper are that the households in those states with higher level of land reform witnessed their investment and welfare improve more in comparison to households in states where land reform implementation remained lackluster. Since access to land has a positive impact on reduction in poverty and improvement in overall well-being of the household, the question that confronts the policy makers is the nature and direction of the reform needed to improve access to land by poor households, They argue that any process of reform contributing to removal in the restriction of lease market is likely to work more in favour of the larger holdings compared to the smaller ones, which is, in fact, contrary to the views held by the majority of the researchers on this subject in India.

The paper on 'Land Bank: An Institutional Mechanism for Correcting the Land Related Distortions' by B.K. Sinha points out that the introduction of a land bank scheme which maintains land resources for cultivation by groups of poor women may bring about a new social and economic equilibrium in the villages and will have a significant impact on rural poverty.

A.K. Singh's paper on 'Land Reforms in Uttar Pradesh' points out that consolidation of holdings has been one of the most successful land reform measures in the state. It also indicates that land leasing has helped mostly the small and marginal farmers in the state, while the beneficiaries of ceiling-surplus land although small in number have improved their income levels and social status.

Agricultural Tenancy Reform

There are eight papers on various aspects of tenancy reform. The paper on 'Agricultural Tenancy in India – Policy, Practice and Impact' by Haque analyses various positive and negative aspects of post-independence tenancy reform policy in India and suggests certain policy changes, keeping in view the emerging agrarian situation. He has made a strong case for legalization of land leasing in order to improve the rural poor's access to land through leasing and also for upward occupational mobility of the rural people. He argues that removal of various restrictions on land leasing will result in better utilization of land and labour, as landowners will not keep their land fallow for fear of losing land in case they lease out. Both landowners and tenants can take more interest in cultivation, provided there is land security for the owner and tenurial security for the tenant. The paper recommends that leasing-in and leasing-out of agricultural land by an agriculturist within the ceiling limit of each state should be legally permissible. The ceiling on operational holding would take care of any likely problem of increased concentration of land due to legalization of land leasing. It is also suggested that in order to encourage the tenants to undertake land improvement, it would be necessary to have at least three to five years of lease agreement with the condition that any capital investment made by the tenant on leased land should be duly compensated by the landowner on the expiry of the lease period.

Bhaumik's paper entitled 'Legalising Agricultural Land Leasing in India: An Assessment of Possible Consequences' analyses whether the mechanism of land leasing in rural India plays an equalizing role in the sense of helping the land-poor households to access more land for cultivation and whether legalization of land leasing would have a favourable impact on the status of landless, marginal and small farmers. The paper concludes by suggesting that the gains or losses for the poorer sections following liberalization of land leasing would not be uniform across states. It suggests that any effort towards liberalization of land leasing should be preceded by imposition of a ceiling on the size of the operational holding in order to combat its possible adverse distributional consequences. Other recommendations of the paper are (i) longer duration contracts to encourage the lessees to undertake investment on leased-in lands which would raise agricultural productivity, (ii) an upper limit of fixed rental to prevent exploitation of the lessees, and (iii) recognition of landowner's right of resumption of leased land after the expiry of the lease duration to motivate them to lease-out land.

The paper on 'Equity and Efficiency Impacts of Rural Land Lease Restrictions' by Deininger, Jin and Nagarajan reviews evidence on the impact of land rental market restrictions and analyses the varying impact of different models of land rental market participation. Deininger *et al.* conclude by saying that land rental markets help to improve not only productivity but also equity, as it helps the landless poor households to access land and improve their livelihood. The findings of the study indicate that restrictive tenancy laws might have led to significant social gains in the past, but are resulting in an increasingly potent obstacle to greater land access and productivity.

H.R. Sharma's paper on 'Changing Tenancy Relations in Rural India' puts forth the view that leasing-in and leasing-out of land should be legalized with adequate safeguards to protect the interests of small and marginal farmers. The tenancy laws should be suitably amended, incorporating provisions such as fixing the tenure of lease, recording of lease and allowing landowners to resume land for self-cultivation after the expiry of lease. Sharma points out that there is a need to separate the ownership rights from the use of land rights, as such a distinction would contribute towards broad-based rural development.

Naresh K. Sharma's paper on 'Land Leasing, Poverty and Inequality' points out the adverse impact of restriction of lease market on the poor, as it increases income inequality, reduces land access and economic opportunities for the poor and results in misallocation of land and other resources. According to him, a redistributive land reform, providing a minimum landholding to every rural labour household and a more open lease market should be the basis for pro-poor land reforms policy. To guard against the concentration of land in a few hands, land ceiling should be applicable to both ownership and operational holdings.

Padmaja Mishra's paper entitled 'Legalising Tenancy and Vitiating Agriculture: A study in Odisha' recommends tenancy be legalized to provide security of livelihood to the tenant farmers and to increase land productivity. However, care should be taken to check reverse tenancy and dominance of moneypower so that the interest of poor farmer can be protected in the lease market.

Pyare Lal's paper on 'Status and Relevance of Legalisation of Land Leasing in Bihar' points out that there is a strong case for amending tenancy legislation to legalise the lease market in the interest of different categories of farmers including landless, marginal and small households and improvement in agricultural production through efficient use of scarce land resources.

The paper on 'Land Leasing by Women in Andhra Pradesh' by Vakati *et al.* points out that women in Andhra Pradesh are using land leasing to escape some of the constraints imposed by caste, gender and economic circumstances. In most cases, the leasing relationships are mutually beneficial, although in some cases, landowners have more bargaining power than the tenants.

Land Rights of Women

The only paper on land rights of women by Hina Haque reveals how religious and customary laws as well as practices which govern property rights relating to inheritance, marriage and divorce are gender biased and work against women. She points out how the redistributive land reform policy is actually biased against women, and even land market do not favour women as dependent women cannot afford to purchase land in their names. She, therefore, proposes a national policy whereby every landless family is provided with at least ten cents of land in women's name. She also suggests a nationwide campaign and social mobilization for effective implementation of Hindu Succession Act, 1956, as amended in 2005.

Tribal Land Issues in North-East India

Non-codification and violation of customary land rights are said to be one of the main sources of conflict in the north-east and other tribal pockets of the country. Most tribes of the north-east region manage their civil affairs on the basis of custom and tradition, though individual-based land laws followed in the rest of the country have been superimposed on them in several places.

Fernandez's paper on 'Tribal Customary and Formal Law Interface in North-Eastern India', reveals how tribal and customary laws have changed over time in response to formal education and the new economic, religious and political relations and yet most tribes continue to treat their customary laws and the community ethos as intrinsic to their identity. A crucial issue is commercialization of land and forests and the resultant transition of most ethnic groups from subsistence to a commercial economy and of their land from livelihood to commodity. The government has supported the transition by insisting on individual *pattas* for subsidies and loans. Commercialization as such may not be bad, but the way it has happened for the benefit of a new class has caused harm to local communities. It has also resulted in ethnic conflicts. Related to this is the change-over from an egalitarian to a class society. The Sixth Schedule, which was meant to be a protective measure, has not always gone in favour of the community, as there is a change-over to individual ownership that goes against women and leads to class formation.

Land Acquisition and Inclusive Development

In the wake of rapid land acquisition for various development activities in the past few years, the people who are dependent on land for livelihood have shown resistance and agitation, not just for adequate compensation, but against the very principle of land acquisition for private benefit in the name of public purpose. The Land Acquisition Act, 1894 which vests the Government with absolute power is under criticism on grounds of its inconsistency with democratic governance and principles of equity and social justice.

Saxena's paper on 'Land Acquisition and Peasant Resistance' examines various grey areas in the land acquisition policy followed by central and state governments. It points out that a land policy for accelerated agricultural growth should aim at protecting agricultural land from diversion for non-agricultural purposes and check its degradation. On both these counts, the land policy has miserably failed. Under the proposed Land Acquisition Bill, 2011, if enacted, the government may not directly acquire prime agricultural land for diversion to non-agricultural uses, but acquisition of agricultural land by private parties would remain unchecked. There would be no let-up in acquisition of agricultural land, especially through negotiated procurement by companies. Also, there may be more intensified local resistance with its implication for governance and democracy. Indiresan's paper however, mentions that there is need to re-examine the entire process of rural development. The aim should not be, he stresses mere adequate compensation for displaced people, but the development of the entire population. Otherwise, both rural-urban and intra-village social and economic disparities will increase.

Emerging Issues and Conclusions

On the basis of the arguments presented in the different papers included in this volume as well as the views of experts expressed in the course of the seminar, the following issues are presented before the Indian government for consideration:

I. Keeping in view the existing socio-political situation in various states, there is not much scope for redistributive land reform through enforcement of ceilings. While ceiling on landholding has its own importance in preventing concentration of land in a few hands, the development and distribution of government land as well as purchase of land from the open market and their distribution among the landless poor would form the core of any non-ceiling land reform measures in future.

II. Leasing-in and leasing-out agricultural land for agriculture should be legalised within the ceiling limits of each state. This would help increase the landless and marginal farmers' access to land through leasing and also encourage much needed occupational mobility of rural people as well as result in efficient utilization of land labour and other resources.

III. The tenure of lease could be of three years or more, provided the landowner has a right of automatic resumption of land on the expiry of agreed lease period and also the clause of adverse possession of land in land reform laws of various states, is deleted.

IV. Land acquisition may be necessary for various non-agricultural development such as industrialization, infrastructure, school/ college buildings, hospitals, etc. However, unless unavoidable, fertile agricultural land should not be acquired as this may result in food crisis. Also, it would be essential to first rehabilitate and resettle those who have been displaced, before actual land acquisition takes place.

V. While the proposed Land Acquisition Bill, 2011 provides that irrigated land capable of producing multiple crops will not be acquired unless unavoidable as this may threaten the food security of the nation and livelihood security of the farmers, it would be necessary to extend this provision to all other categories of fertile land as well.

VI. The term 'public purpose' should be properly defined. The proposed Land Acquisition Bill includes even mining within the definition of infrastructure, so as to qualify for land acquisition. This is wrong. The necessary amendment needs to be carried out before the Proposed Land Acquisition Bill, 2011 is passed by the Parliament.

VII. Regarding payment of compensation, replacement value and not just the market value of land, should be considered. There is often a time lag between land acquisition and payment of compensation. Therefore, land price *at the time of payment* and not at the time of land acquisition should be the reference point.

VIII. Farmers and other affected people should get an equity share or some other kind of partnership in the proposed project for which land is acquired. In a housing project, adequate number of houses should be earmarked and allocated to the affected families.

IX. The project affected people who receive compensatory cash should be guided for proper, albeit productive utilization of the money received.

X. The size of land to be acquired by and for various projects should be properly, albeit contexually, examined as some projects acquire more land than what is required and use the extra land for speculative purposes.

XI. As far as possible, land for land in the nearby areas should be given as compensation, as farmers even with money do not have the skills to invest in non-farm enterprises. Also, they should be given exemption from capital gains tax in case of appreciation of land value. In hilly areas, there should be no land acquisition for any large industrial development, as it would destroy the ecology/environment of the area.

XII. The problem of indiscriminate land acquisition and project-related displacement is quite serious in tribal areas. This should be addressed on priority basis so as to prevent growth of extremism and lawlessness in tribal areas.

XIII. The state governments should sincerely implement PESA and Forest Rights Act so as to ensure effective ownership and utilization of land and forest resources by the tribals and other forest dwellers, for which massive legal aid and awareness building campaign among tribals have to be initiated, with support from various civil society groups and Panchayati Raj Institutions.

XIV. PESA provides for consultation with the Gram Sabha in all land acquisition and rehabilitation and resettlement cases. However, this is often understood as mere consultation, not consent. The amended law as such or additional rules should clarify that consent and not mere consultation is binding.

XV. There should be a comprehensive survey and settlement and updation of land records in both tribal and non-tribal areas for effective implementation of some existing well-meaning land laws.

Acknowledgements

I am extremely grateful to Prof. Muchkund Dubey, President, Council for Social Development and Mr. Tim Hanstad, President, Landesa, for their constant encouragement and help in completing this volume. I also thankfully acknowledge the help and co-operation received from Mr. Greogory Rake, Mr. K.B. Saxena, Ms. Jaya Nair and Mr. Gitesh Sinha. Besides, the help received from Ms. Reeta Khurana and Mr. Dev Dutt in typing the final manuscript, is thankfully acknowledged.

T. Haque (*Editor*)

Contents

Part II
TENANCY REFORMS

Part III
LAND RIGHTS FOR WOMEN AND TRIBAL

Part IV
LAND ACQUISITION AND INCLUSIVE DEVELOPMENT

Contributors

Ajit Kumar Singh is Director, Giri Institute of Development Studies, Lucknow. He has served as member of several planning and expert groups both central and state government. He is author of several books and research papers.

Arindam Banerjee is Consultant at the Research and Information System (RIS), New Delhi. His field of specialization is Agriculture, Food Security and Bio-fuels.

B.K. Sinha is Secretary to Government of India, in the Ministry of Rural Development. He also served as Director-General, NIRD, Hyderabad. He has deep interest and experience in research and pro-poor policy formulation and implementation in the fields of land reform, employment guarantee etc.

Derryl Vhugen is a senior researcher at Landesa, USA. He also served as Director of Rural Development Institute in India, up to 2008.

H.R. Sharma is Visiting Professor, Division of Food, Resource and Environmental Economics, Konkuk University, Korea and a regular Professor at the Himachal Pradesh Agriculture University, Palanpur.

Hari K. Nagarajan is a Senior economist at the National Council of Applied Economic Research, New Delhi. He has several research papers to his credit.

Hina Haque is Research Fellow at the Institute for Sustainable Development, New Delhi. She has published research paper on rural poverty in journal of international repute. Her areas of specialization are land rights of women and rural poverty.

K.B. Saxena is Professor of Governance and Social Justice, Council for Social Development, New Delhi. He is a retired civil servant, well-known for his outstanding contribution to pro-poor policy interventions. He served as Secretary, to Government of India in the Departments of Rural Development and Social Welfare. He has written several research papers and monographs on land reform and land acquisition policy in India.

K.N. Nair is Professor and Director, Centre for Development Studies, Thiruvananthapuram. His areas of specialization are land reform and natural resource management.

Karuna Vakati is sub-collector, District Warangal, Andhra Pradesh. Her contribution to the success of IKP Project, under the guidance of K. Raju is well-known. She has published several papers based on field research.

Klaus Deininger is a lead economist in the World Bank, Washington, DC. He has published several research papers on various land policy issues in journals of international repute. He has done extensive field research in several countries of Asia and Africa.

Naresh K. Sharma is Reader in the Department of Economics, University of Hyderabad. He has done extensive research on land reform and rural poverty and published several papers based on field research.

P.V. Indiresan is former Director, Indian Institute of Technology, Madras. He writes extensively in newspapers and journals on various issues of social relevance, concerning especially the poor and the marginalized.

Padmaja Mishra is Professor and Head, Department of Economics, Utkal University, Bhubaneswar. She has done extensive field research in the field of land reform, agriculture and rural development.

Pyare Lal is Director, Bihar Institute of Economic Studies, Patna. His areas of specialization are rural development and economic planning.

Robert Mitchell is Senior Land Tenure specialist, Landesa. He has been actively involved in research and policy advocacy in the field of land reform globally. He obtained his law degree from the University of Washington.

Roy L. Prosterman is Professor of Law at the University of Washington, Seattle, U.S.A. and founder President of Rural Development Institute, renamed Landesa, U.S.A. He has done extensive research and policy advocacy in the field of land reform in as many as 40 countries including China, India and former Soviet Union to name a few. He contributed significantly in shaping P.R. China's land reform policy in the past few years.

Sankar Kumar Bhaumik is Professor of Economics, at Calcutta University. His research in the field of tenancy reform is widely quoted. His area of specialization are land reform, rural development and macro-economic policy.

Songqing Jin is a Associate Professor, Michigan State University, U.S.A. His areas of specialization are land reform and rural development.

Srijit Mishra is Associate Professor at Indira Gandhi Institute of Development Research, Mumbai. His field of specialization is Agricultural Development Policy. He has authored a book on Agrarian Crisis in India and several research papers.

T. Haque is currently, Director, Council for Social Development. Prior to this, he was Chairman, Commission for Agricultural Costs and Prices, Government of India. He has done extensive research and Policy advocacy on small holders agriculture, land reform and food security. He has six books and about 100 research papers and reports, to his credit.

Tim Hanstad is President of Landesa, (formerly Rural Development Institute) and Assistant Professor of Law at the University of Washington. He has been actively involved in agrarian reform research and Policy advocacy globally and missioning to improve the rural poors' land access and land tenure security for poverty reduction globally.

Walter Fernandez is former Director, Indian Social Institute and Editor of "Social Action". He has done extensive research on tribal issues and published papers in journals of repute.

Part I

REDISTRIBUTIVE LAND REFORMS
Old and New Approaches

1

Land Reform in the 21st Century: New Challenges and New Responses

Roy L. Prosterman and *Tim Hanstad*

I. Land Reform: Scope of the Problem

Citizens of the industrialized world do not commonly recognize the continuing great importance of agriculture to the well-being of people in poorer countries. An estimated 45 per cent of the world's population still makes their living primarily from agriculture, depending directly on the land for their income, status, and security. In many traditional less developed countries, and a number of the economies in transition, these agricultural families still constitute a substantial majority of the population. However, great numbers of these agricultural families lack a stable and predictable relationship to the land they farm, and thus face serious economic and social insecurity.

How to improve and secure the relationship agricultural families have with land, which is the central question of "land reform", thus persists as a crucial issue in much of the world. In this report, we define land reform broadly to include reforms that increase the ability of the rural poor and other socially excluded groups to gain access to land and to exercise effective control over land.

Potential land reform beneficiaries in the developing world can generally be grouped into three categories: (1) rural households who lack ownership to the land they cultivate in traditional less developed countries; (2) rural households who live on state or collective farms in communist or formerly communist countries; and (3) rural households who occupy public lands, or indigenous groups lacking demarcated lands.

First, in traditional less developed countries, roughly 100 million farm families, comprising about 500 million people, lack ownership or owner-like rights to the land they cultivate. Most of these families earn their living as tenant farmers or agricultural labourers. The tenant farmers typically pay high rents and have little security of land possession from season to season. The agricultural labourers generally work for extremely low wages and often as itinerants.

These half-billion people are among the poorest on earth. They constitute majorities, or near-majorities, of the agricultural population in countries such as India, Bangladesh, Pakistan, Indonesia, the Philippines, South Africa, Brazil, Colombia, Guatemala and Honduras. In addition, they are a significant part of the agricultural population in a number of other countries, ranging from Zimbabwe and Egypt to Afghanistan, Nepal and Venezuela. In countries where they represent a large part of the agricultural population, this group forms a deep concentration of poverty and human suffering, as well as an impediment to the process of economic development and, in many settings, a potential threat to political stability.

Although the number of desperately poor tenant farmer and agricultural labourer families in the world today is very large, it would be much larger had it not been for a series of major land reforms carried out over the past century. In some of these land reforms, nearly all land reform beneficiaries ended up with individual family farms, held in ownership or owner-like tenure. These included land reforms in Finland, Poland, Yugoslavia, Mexico, Bolivia, Japan, Taiwan, and South Korea, as well as the populous Indian states of West Bengal and Kerala.

In other land reforms, the great majority of land reform beneficiaries ended up in large state or collective farms that nearly always proved to be inefficient and authoritarian. These included land reforms in the countries of the former Soviet Union, Eastern Europe (other than Poland and the former Yugoslavia), Cuba, Nicaragua, Ethiopia, China, Vietnam and North Korea.

Roughly 130 million rural households (about 580 million people) who reside in these communist or formerly communist countries comprise the second category of families that could benefit from land reform measures. Over the last 20 years most of these countries have transitioned away from command economies towards market-oriented systems (the most prominent exceptions being Cuba and North Korea). In these transitional economies, the "second generation" land-reform challenge is to give a remaining roughly 130 million households who had been on state or collective farms the right and the wherewithal to farm as individual families with stable, long-term and transferable rights to their land. As we shall describe further below, these "second generation" land reforms have actually progressed much more rapidly in the past two decades than the land reform measures still needed for tenant and labourer families in traditional less developed countries.

There is also a third, much smaller, group of agricultural families who have insecure tenure, often on public lands. Some of these families are economically desperate "squatters" who have moved onto the land to cultivate it. Others are traditional landholders, often indigenous groups, who have occupied the land for decades or more and hold rights recognized by customary law, but not adequately recognized by the state's formal laws. We also include in this category some traditional pastoralists who use land in extensive or migratory patterns, but without sufficiently recognized group tenure rights to the land. In some cases, these holders are now in conflict with settlers, ranchers, loggers or miners who more recently came upon the land to extract resources or who otherwise disrupt existing use patterns (for example, in parts of Indonesia, Brazil and the Philippines). This third category of families who could benefit from land reform measures includes roughly 25 million households, or about 125 million people.

Table 1.1 broadly summarizes both those who have ownership or owner-like rights and those who are in the categories of potential land reform beneficiaries.

Table 1.1: Approximate Number of Rural Households Worldwide Subject to Land Tenure Problems (millions of families)

	Owners or owners-like rights	*Insecure use rights*		*Squatters and insecure traditional holders*	*Total*
		Insecure tenants, agricultural labourers	*Insecure individual use rights or still collective farmers*		
(a)	(b)	(c)	(d)	(e)	(f)
Traditional less-developed countries	180	100	n/a	20	300
Former command economics	95	n/a	130	5?	230
Total:	**275**	**230**		**25 ?**	**530**

Our principal focus in this report is upon: (1) households of agricultural labourers and insecure tenant farmers; and (2) households of current or former collective-farm workers who lack secure land rights.

These two categories—columns (c) and (d) of Table 1.1—comprise the great bulk of households lacking a secure and predictable relationship to the land. This report deals only briefly with some aspects of the problems "squatters" and insecure traditional holders face.

Nearly all of the foregoing tenure insecurity issues are essentially situations where the rights of cultivators, considered as unitary households, require definition and protection as against the rights or powers exercised by private landlords, plantation owners, collective farm managers, local cadres, the state, or other external actors. In addition, there exists an important intra-household land tenure issue—the issue of gender, which concerns the relative land rights and powers exercised by wives, daughters, widows, and female heads-of-household. We shall also address this often-neglected and important aspect of the land-tenure issue.

II. Potential Benefits of Land Reform

Agricultural land still constitutes the primary source of income, status, and security for over half a billion families outside of the developed world. As Table 1.1 reflects, nearly half of those families exist in highly unstable and unsatisfactory tenure relationships to the land on which they depend for their livelihood. It should not be surprising to find that in improving the stability of tenure relationships, land reform can improve lives in a number of ways. Decisively improving the relationship of cultivators to land through land reform has generally led not only to increased crop production and improved nutrition for poor households, but can produce other crucial benefits:

- "ladders out of poverty" and a foundation for sustained and inclusive economic growth;
- grass roots empowerment and movement towards more democratic societies;
- reduced social unrest and instability;
- reduced urban migration;
- better environmental stewardship; and
- creation of wealth.

In sum, secure and stable access to land provides a foundation for better livelihoods and living conditions. It also helps the rural poor to benefit from broad-based economic growth.

A. Increased Crop Production and Nutritional Welfare

With regard to crop productivity gains, there is cogent international evidence on four points. First, smaller holdings generally produce more than larger ones, whether measured hectare for hectare or according to total factor productivity. Second, and related, family-operated farms generally produce more than collective farms and farms largely dependent on wage labour. Third, on any given holding, a cultivator with ownership or secure, long-term owner-like tenure is far more likely to make long-term capital and "sweat-equity" investments that improve and conserve the land than is a cultivator with insecure tenure. Finally, a cultivator with ownership or owner-like tenure is more likely to use improved (and more expensive) seeds, fertilizer, and other inputs than is a tenant in the typical tenancy arrangement where the tenant pays for all inputs and receives only a portion of the output.

In sum, total production per hectare and total factor productivity are highest on farms owned and operated by a single family. In developing countries where the ratio of agricultural population to arable land is typically high, these farms are also small. When policy makers in developing or transitional countries argue (as they sometimes do) that there are inherent "economies of scale" in large farms, or envision (as they also sometimes do) big, mechanized farms as the "ideal" to be achieved, they are demonstrably wrong. Such views are contradicted by the vast weight of evidence. Such views disregard the comparative strengths of societies that lack substantial amounts of land and capital, but have abundant and underemployed labour, especially rural labour. This is the case in many if not most developing countries. In such settings, the highly motivated application of intensive family labour to a small piece of land owned or held securely by that family is precisely what makes economic sense. What typically makes little or no sense in such a society is large-scale mechanized farming, which requires scarce capital and displaces abundant labour that has no alternative employment.

Additionally, even apart from likely increases in crop productivity, research shows that access to land through ownership or owner-like tenure has a substantial impact on family nutrition. Land reform in traditional settings can improve the lives of beneficiary families, typically among the poorest of the poor, by letting them keep a significant portion of the crop that would otherwise go to the landlord or plantation owner or by enabling them to produce food on land (including house-and-garden plots) to which they previously did not have access.

B. Foundation for Economic Growth

Cogent comparative evidence also demonstrates that broadening access to land and strengthening cultivators' land rights can generate increases in overall economic activity. As a broad base of agricultural families benefiting from land reform receive higher incomes, they enter the marketplace to purchase goods and services, ranging from improved housing to schoolbooks, from bicycles to sewing machines. This increased demand stimulates the creation of non-farm employment. Thus, a dynamic family farming sector has significant forward and backward linkages to broader societal development. Research thus confirms that a broad-based distribution of land assets not only benefits the poor, but becomes a solid basis for sustained and inclusive economic growth.

C. Facilitating Democracy

Land reform, when it is implemented effectively, removes its beneficiaries from the "power domain" of the landlord, plantation owner, local cadre, or collective-farm manager. Further, as land-reform beneficiaries increase their incomes and become more economically secure and confident, their ability to participate in the political process is strengthened. Initially, land reform beneficiaries may be empowered to make demands for a fairer share of government-administered programmes and services. Land reform creates more secure and self-confident producers who are willing to challenge the inertia, elitism and neglect that frequently characterize the politics of underdevelopment. The contribution made by land reform to eventual democratization of previously authoritarian societies has perhaps been seen most dramatically in the cases of Taiwan and South Korea.

D. Reducing Instability and Conflict

In traditional developing countries, land reform has reduced political instability by eliminating basic grievances arising from the relationship between tenants or agricultural labourers and erstwhile landowners. Many of the past century's most violent civil conflicts ensued when land issues were ignored. Land reform can address the most basic rural grievances and increase citizen commitment to a system in which economic and social demands are negotiated peacefully.

Classic civil wars catalyzed and fed largely by land-based grievances such as those that occurred in Mexico, Spain, China and Vietnam may be less likely with the fading of revolutionary Marxism as a mobilizing ideology. Nonetheless, such land-based grievances still lend themselves to manipulation by ideologues or demagogues who seek domestic power in places like Nepal and Zimbabwe. More frequently, unsolved land grievances now lead to various forms of "low intensity violence", such as through spontaneous land invasions in Brazil, peasant alliances with drug lords in Colombia, or desperate migrants seizing indigenous lands in the Philippines or Indonesia. Research also confirms that overall inequality in assets is associated with higher crime rates.

More ominously, in this new century, such land-based grievances may now be increasingly mobilized by religious extremism and turned either inwards against domestic governments (such as Pakistan) or outwards against the United States and other democracies. The recruitment we are speaking of here relates not to the veneer of leadership at the top, often drawn from the wealthy or the middle class, but to the establishment of a mass base or rank-and-file, which gives terrorist movements much greater resilience and staying power. Often this recruitment process may be mediated through the kind of flight to the cities that results from the "push" of desperate poverty in the rural areas rather than the "pull" of urban job opportunities, and which is likely to affect the landless (and their young sons) disproportionately. For example, landlessness is the best predictor of rural poverty in India—better than illiteracy or caste—and it is likely the best predictor in other settings such as Pakistan, Egypt, and Indonesia.

E. Other Benefits

Land reform can provide at least three additional benefits. First, and closely related to the point just made, many landless families are driven by their poverty into the cities. Effective land-reform measures give landless agricultural families a stake in their village society, reducing pressures that lead to premature and excessive urbanization.

Second, long-term, secure rights to land set the stage for environmental stewardship and sustainable farming practices. Moreover, in specific settings such as Brazil, Indonesia or the Philippines, reallocation of secure rights to existing cultivated land may also have an important environmental impact through forestalling landless peasants from descending on, cutting down, and burning the forest in the desperate search for a piece of land to farm. The latter is a form of escape parallel to the desperate flight to cities.

Third, secure land rights that are transferable also acquire a predictable market value, and can be used as collateral, "cashed out" for non-agricultural investment or retirement, or passed on as wealth to the next generation. The wealth thereby created in the hands of the poor, and to the general benefit of society, can indeed be massive, as persuasively argued by Hernando de Soto in his oft-cited book, *The Mystery of Capital,* and as further confirmed by evidence from India on capital formation by the poor, discussed below.

The benefits of land reform described here apply, in general, not only to land reforms carried out to benefit tenant farmers and agricultural labourers in traditional less developed countries, but also to land reforms that permit former collective farm workers in transitional economies to obtain secure rights to land of their own. They too invest, increase production, gain income, consume more, become empowered and less aggrieved, increase their stewardship, and strengthen their rural attachments.

Secure long-term land rights are not, by themselves, sufficient to achieve all these multiple benefits; but in most settings—both traditional and transitional—they are a necessary precondition for the achievement of all or most of them. Perhaps no single, identifiable measure comes as close as land reform to cutting the Gordian knot of underdevelopment in so many societies.

III. Why then is Land Reform So Often Neglected?

Despite the importance of the rural poor's land grievances in so many diverse settings, and the multiple development goals served by land reform, the topic receives little notice in industrialized democracies. Foreign policy experts, the media and even foreign aid programmes pay relatively little attention to land reform issues. The rare exception arises in cases where the issue has become a focus of violence or demagoguery, as in Zimbabwe recently. Massive programmes of reform—such as the current programme to give long-term, secure land rights to China's 210 million farm families, now nearly halfway to completion—receive hardly a single line of print or frame of television coverage in the media of the developed countries.

Why has there been such a studied lack of attention from developed countries? First, substantial land grievances, unlike ongoing development issues such as eradication of disease, family planning and international debt, tend to arise in their conspicuous form only at discrete moments in time. Thus, the media, the public and foreign aid agencies do not consistently focus on the need for programmatic responses. The media and public, in particular, are sorely lacking in education on the continuing importance of meaningful land reform in developing and transitional economies.

Second, where the land issue does finally receive some attention, it is often because it has been seized upon and dramatized by demagogues or violent elements. This may be, in part, a result of ignoring an ongoing, systemic problem until the frustrations of the aggrieved finally "boil over." In these cases, the land reform issue, instead of being the subject of thoughtful study, planning, and response via a non-violent and democratic alternative, is likely to generate a negative reaction as being "what the bad elements want."

Third, developed countries often ignore the land reform issue because it is politically controversial among developing country elites. Local leadership and elites in developing countries are often opposed to any serious land reform measures because the reform may threaten, or is perceived to threaten, their individual or family interests. Such opposition and the related political controversy and sensitivity it engenders have tended to make both bilateral and multilateral foreign-aid agencies shy away from public focus on the land reform issue as too controversial, even where they privately recognize its importance. This is especially true in traditional less-developed countries where land reform options centre on redistributing private land. Second-generation reforms in the former command economies, because they typically involve state-owned land, involve relatively less political controversy and thus have received somewhat more attention from foreign-aid agencies.

Fourth, past flawed and unsuccessful land reform efforts sometimes taint "land reform" and cause policy makers to ignore it as a development option. Numerous countries have had less than successful experiences with land reform. The lack of success is typically due to design flaws, insufficient political will, or ineffective implementation, factors which are often interrelated. In some cases, these ineffective land reforms have led to real or perceived negative consequences, such as violence or highly regulated land markets. Based on failed and flawed earlier attempts, policy makers sometimes (and unfortunately) choose to ignore land reform as an option altogether rather than address the flaws in the earlier efforts.

Like many other needed structural or redistributive reforms, experience indicates that the long-term societal gains from effective redistributive land reform far outweigh the short-term pains relating to political sensitivity or controversy. Political realities, however, do not always allow for much-needed reforms that make long-term economic and social sense. In many cases, the political obstacles can be overcome, as discussed in Section V. Policy makers, activists, and potential beneficiaries should not despair, however, if the political obstacles to achieving land reforms that are sweeping and comprehensive cannot be overcome. Instead, it is crucial to recognize a variety of less ambitious land reform measures that may still be capable of providing widespread and effective benefits, as discussed in Section VII.

IV. "Second Generation" Land Reforms

The comprehensive 20th Century land reforms that were carried out without collectivization played a major role in fostering development and stability. Land reforms that led to collectivization proved almost universally to be failures. Many countries that previously conducted collectivized land reforms are now undertaking "second generation" reforms aimed at reorganizing state and collective farms into family-size units and introducing market-oriented land systems. These second generation land reforms now appear to be playing a major beneficial role in many formerly collectivized countries, most notably for the approximately 210 million farm households of China.

China's farm families worked on large communes or collective farms from about 1955 to 1980. During 1980-1983, China pursued a decollectivization process through which the state distributed virtually all land of the collectives to each farm family in individual landholdings. Unfortunately, the families received insecure rights to the land. Local officials could relocate them from plot to plot through periodic "readjustments" in the name of maintaining absolute equality of distribution as household size changed. Despite these shortcomings, the change from collective farming to individual (even though insecure) tenure created the conditions for increasing crop yields by more than 80 per cent in less than a decade. China is now nearly halfway through completing a major new land reform that is giving these families, totalling about 850 million persons, individual land contracts to secure and transferable 30-year use rights.

Extensive decollectivization, with the granting of individual ownership or long-term use rights to farmers, has also occurred in Vietnam, Lithuania, Estonia, Latvia, Romania, Albania, Bulgaria, Hungary, Kyrgyzstan, Georgia, and Armenia. Land reform has made much less progress in Russia and Ukraine, and still less in Kazakhstan, Uzbekistan, and Belarus.

V. Factors that Make Land Reform Possible

While the need for land reform remains great in many traditional less-developed countries and transitional economies, the political obstacles are often substantial. However, lack of political will or the existence of substantial political controversy were also present shortly before many non-violent, democratic land reforms in the latter half of the 20th Century. Factors that played important roles in overcoming political obstacles to land reform in recent decades may provide lessons as we assess prospects for 21st Century land reform efforts. Over the last half of the 20th Century, there have been at least seven such factors.

1. *Grass roots Support for Land Reform and Linkages with Political Parties*: Building or harnessing widespread grass roots support for land-reform measures (for example, through local civil society organizations) and forming linkages to political parties have been important factors in determining whether land reform measures will be adopted and whether land reform, once legislated, will actually be implemented. Grassroots support from civil society organizations and political linkages were significant in the success of reforms in El Salvador, and the Indian states of Kerala and West Bengal, as well as in many of the transitional economies of Eastern Europe where decollectivization and privatization of land have been carried out. Grass roots support is likely to be even more important in the future for two reasons. First, many countries where land reform is still needed are political democracies, at least *de jure* if not *de facto*. Second, advances in communications technology now allow NGOs to organize and publicize the conditions of the rural poor more effectively than they did in the past.
2. *Political will for Land Reform Vested in a Strong Central Authority*: Political will for land reform vested in a strong central authority, such as General MacArthur in Japan or Chiang Kai-Shek in Taiwan, has been important in the past. Authoritative action from the Centre may still play a key role in settings such as China, but in the future this variable seems likely to be of diminishing importance given to the international trend towards more democratic forms of governance.
3. *The Threat of Strong Revolutionary Movements Drawing Support from Grievances of the Landless*: The threat of

revolutionary uprising has sometimes played an important role in making land reform politically possible. Willingness to seriously consider land reform in the Philippines over past decades seems to have risen along with the fortunes of rural revolutionaries. The guerrilla threat was also a factor in making land reform politically feasible in both El Salvador in the 1980s, and South Vietnam in the early 1970s. With the general demise of revolutionary Marxism, this variable of impending revolution appears to have receded in significance and, in any case, has little relevance to "second generation" land reforms. If religious extremism comes to take the place of revolutionary Marxism in mobilizing the grievances of the landless, however, this variable could well resurface with considerable importance in countries such as Pakistan, Indonesia, Bangladesh, Egypt, and Afghanistan.

4. *Adequate Compensation Paid for Land Expropriated*: The amount of compensation paid to landowners from whom land is taken is a critical variable. Over time, evidence has accumulated that the "carrot" of paying adequate compensation to those whose land is taken can be more influential than the "stick" of a perceived revolutionary threat, or land occupations. Lack of adequate financing for land reform has often limited otherwise possible land reform efforts. In particular, providing adequate compensation for the taking of privately owned land may reduce landowner opposition and help legitimize reform for significant sectors of public opinion, thereby substantially reducing the amount of grass roots pressure or central-authority determination otherwise necessary to accomplish a given degree of reform. The feasibility of this "carrot" has been reduced in recent years, however, by substantial increases in the value of privately owned land in many countries, especially in settings where there is high population pressure on land. This has reduced the quantity of land that can be acquired with a given amount of financing. As a result, new and imaginative approaches to land reform design that take into account how much land must be distributed to provide essential benefits to most of the needy group must be considered. These implications are discussed in Section VI.
5. *Ideology*: Ideology has played an important role in democratic land reform in much of Eastern Europe and parts of the former Soviet Union. There, some post-communist governments have espoused an active policy of decollectivization, sometimes linked

to restitution of land rights to former owners of small and medium farms (or their heirs) who lost land during collectivization. However, non-restitutive approaches that allocate land to whomever presently works on the collective farm (whether or not they are descended from the former owners) are likely to be far easier to administer. They are also more likely to lead to small farms that are owner-cultivated rather than being leased-in from heirs (beneficiaries of restitution) who reside in urban areas.

6. *Public Education and Information*: Public education can be an important factor in making land reform possible, especially in democratic settings. Educating the public and policy makers that land reform can bring extensive economic and social benefits may help crystallize the political will to carry out land reform. Public education will be a crucial factor in the 21st century.
7. *Availability of Technical Assistance*: Likewise, the availability of technical assistance can be an important factor in making land reform possible. Fact-grounded advisory work that makes policy makers, as well as the public, aware of specific solutions to technical issues involved in particular land reform settings may not only help to crystallize the necessary political will, but help to ensure that a land reform programme is designed in such a way that it can actually be implemented. Such technical assistance can also be crucial to the development of training, monitoring, dispute-resolution, and other measures that help facilitate effective implementation of land reform programmes.

VI. Land Reform Design: Lessons from the 20th Century

Once the political will to carry out land reform has crystallized, what have been the key programme design lessons learned? We first highlight a dozen basic points relating to land reforms carried out in traditional less developed countries during the 20th century. Then, we turn to similarities and differences of "second generation" land reforms now underway in transitional economies.

A. Lessons from Traditional Less-Developed Countries

1. *If land reform is to provide benefits to a substantial portion of the rural landless, there has rarely been an effective substitute*

for the mandatory (but compensated) taking of private land as the basis for redistribution. Countries sometimes pursue other alternatives, including distribution of or "resettlement" on uncultivated public lands, strict regulation of the landlord-tenant relationship, or voluntary sale and purchase of land through "land bank" or "market-assisted" mechanisms. Such alternatives have sometimes provided marginal benefits, but they have rarely provided widespread beneficial impacts. Resettlement programmes have generally been costly, disruptive, and environmentally damaging. Ongoing regulation of the continuing landlord-tenant relationship has usually proven impractical and even counter-productive. (Egypt, the longest-running success story, reversed its protective regulations dating from the 1960s under landlord pressure in the mid-1990s. West Bengal has been a more recent success story, but even there large numbers of tenants were evicted before protections became effective for the rest.) "Market-assisted", "negotiated" or" community-led" land reform initiatives (in which poor households are provided subsidies to make private purchases of land) are relatively new and innovative efforts. They have not yet proven effective for redistributing substantial amounts of land, although they are worthy of further experimentation. Where political, social or economic realities do not allow for sweeping land reforms based on mandatory, compensated takings of private land, "second-best" or less comprehensive measures should be pursued.

2. *The state should provide reasonable compensation to private landowners whose land is taken for redistribution.* Failure to pay reasonable compensation virtually guarantees that landlords will evade the law, cause the law to be rescinded, or violently resist enforcement of the law. But there are also good legal and ethical arguments for reasonable compensation, regardless of the *de facto* coercive power of the government. This compensation need not always equal full market value, but the programme design should carefully consider both market value and the flow of income from the land. Payment of substantially less than "market value" may often be sufficient to produce the same ongoing flow of income that had been produced by the land taken, but this may be inadequate in those settings where there is a large gap between market value and the lesser capital fund needed to reproduce the annual income flow.

In many settings, particularly Asian settings, where population pressure on the land is fierce, market values may considerably exceed the capitalized flow of income from the land. It may then be essential to pay full market value, and to design the land reform in ways that make it nonetheless affordable, as discussed in the fifth point below. On the other hand, there may be other settings (particularly in Latin America or Africa) where much privately owned land is not intensively used, and where market values for such land are quite low and affordable. Four further points should be noted regarding compensation. First, if a portion of compensation to owners is deferred and paid in government bonds, those bonds should be inflation-adjusting, and former landowners should probably be able to use a portion of the bond as preferred collateral in obtaining loans for various productive investments.

Second, in most settings beneficiaries should also be required to repay the government something for the land, with the possible exception of cases where the very poor receive very small plots, discussed also in the fifth point below.

Third, land reform programmes that provide market or near-market value for land are more likely to be able to be implemented over varying and longer periods of time, if necessary, than land reforms that are significantly confiscatory. This is because the latter may be possible only in narrow "windows of time" when special factors exist, such as a very strong central authority with limited ties to rural landowners, or a threat of revolutionary upheaval drawing upon the grievances of the landless.

Fourth, democratic land reforms should be carried out in the spirit of an eminent domain proceeding in which it is understood that certain lands need to be acquired for application to an important social purpose. Democratic land reform should not be conducted in the spirit of judging the present large landowners to be "class enemies" or deserving of financial penalization.

3. *If possible, land reform should be of sufficient scope and universality to reach most potential beneficiaries.* This means that agricultural labourers, tenants, and non-permanent labourers as well as permanent ones, should benefit from the process. Benefits, however, need not be uniform. For example, in the Keralan land reform in India, tenant farmers received full-sized

parcels, while some agricultural labourers received small house-and-garden plots.

4. *Providing even very small plots can produce substantial benefits.* Agricultural labourers in the Indian state of Kerala who received tiny house-and-garden plots of 1/10 acre (.04 ha or about 4350 square feet) found themselves considerably better off in terms of income, family nutrition, and status. Similar findings have come from recent research in the Indian states of Karnataka and West Bengal. In Karnataka, agricultural labourer families who received government-granted house-and-garden plots of only 1/25 acre (.016 ha or about 1730 square feet) were able to produce most of the family's nutritional needs for vegetable, fruits, and dairy products and obtain cash income equivalent to one full-time adult wage from plant and animal products on the tiny plot. Strong supporting evidence on the benefits of owning small house-and-garden plots exists from such diverse settings as Russia, Indonesia, and the Caribbean.
5. *The maximum per-family quantity of land allocated to beneficiaries should be the amount a family can intensively farm with its own labour and modest capital.* Some land reforms have provided per-family holdings far larger than necessary. The distribution of large holdings can severely limit the number of potential beneficiaries. Rather than providing an idealized holding size, it is generally better to calculate the area likely to be available under the particular land reform's acquisition formula and divide that amount by the number of families needing land. The resulting quotient should be the maximum for per-family allocations. (One private report criticized the 1980-1981 land reform in El Salvador for not giving a "vital minimum" of 9 hectares to each beneficiary family, without realizing that such a pattern of distribution would have required distributing twice the land area of the entire country.) Governments should recognize that every recipient family will be made substantially better off than they presently are, even if the size of their new holding is not "ideal" or "viable" as a sole source of income. Land reform neither creates nor destroys land: what it does is put an existing agricultural population into a relationship with an existing land base that is fairer and more productive than it was before.
6. *Minimize costs per beneficiary family in order to benefit the*

maximum number of landless rural families. Available resources should be focused initially on the *sine qua non* of acquiring land for redistribution. Land reform beneficiaries often require some ancillary support for interventions such as credit and technical assistance. However, programme designers must avoid allocating excessive resources per-family to such ancillary programmes and may need to omit entirely some costly benefits such as new housing. (A universal experience in successful land reforms is that families that gain ownership, even of a very small plot, build their own house, starting with something very simple and progressing to a more substantial dwelling over time. Faced with a choice between spending "x" amount of resources to give one family land and a house, or four families land but no house, it seems clear that programme designers should opt for land-only benefits to the maximum number of families.)

7. *Focus on intra-household allocations of land rights*. Most past land reform programmes, even those cited as successes, have ignored gender issues. Women possess limited rights to land throughout much of the world, even though they are often responsible for producing much of the food. It is important to understand how reform or reallocation of land rights may impact women differently than men, and to ensure that the reform avoids discriminating against women. Providing women with strong legal rights to land can have important security, efficiency, welfare, and employment effects. One basic step toward promoting equitable intra-household allocation of resources is to ensure that any land distributed through land reform programmes is jointly titled in the names of both husband and wife or even independently titled in the name of a woman, rather than in the single name of the "head of household", who is usually male.
8. *Understand, consider, and—to the extent possible—incorporate customary law and land tenure regimes when undertaking land reform initiatives*. Some land reforms have involved attempts to replace customary land-tenure regimes that still exist in some parts of the world—especially areas of less population density—with more "modernized" land rights. While some changes over time may be appropriate, it is vital to recognize that most customary land-tenure regimes have evolved over long periods, and that such regimes can function with a high degree of equity

and environmental awareness. Rather than wholesale replacement of customary regimes in the name of "modernization", new written law should generally attempt to incorporate or defer to customary law and legal institutions, at the same time recognizing that "custom" itself has the characteristic of adaptability and often changes over time.

9. *Programme administration should involve beneficiaries, serve beneficiaries, and be well-publicized to beneficiaries.* Farmer-dominated, local-level committees with strong representation of the beneficiaries are in the best position to determine who owns the land, who has been cultivating it, and what it produces. In some settings, grassroots farmer unions or organizations of agricultural labourers may play this role. Beneficiaries should not be required to leave the locality to fill out applications for benefits, nor should they be required to fill out applications under the watchful (and often intimidating) eyes of local elites. The applications should be simple and the programme should include repeated publicization of the land-reform programme and beneficiaries' rights at the local level using local media and other means that will effectively communicate such information.
10. *Avoid excessive restrictions on land sale and lease markets.* In a post-reform environment, such restrictions can have perverse results even if they are intended to "protect" the beneficiaries. Land reform programmes have often imposed restrictions on selling and leasing to prevent improvident sales by the new, small landowners or to thwart the growth of new "exploitative" tenancy arrangements. While some limited and especially time-bound restrictions on land markets can be warranted under special circumstances, excessive restrictions can deprive farmers of access to credit (by preventing mortgage of the land rights) and can reduce the ability of farmers to adjust the size of their farm in response to economic and agronomic conditions. Paralleling the right to sell, mortgage should be allowed, for both productive investments and the purchase of additional land.
11. *Do not force or pressure land reform beneficiaries to adopt a particular mode of organizing their farms.* Most beneficiaries are likely to choose to create individual family farms (90% of Mexican land reform beneficiaries chose family farming and 10% chose collective farming). Some, however, may choose extended family operations or collective modes of production,

at least initially. Farmers are best suited to make organizational decisions regarding their farms. The legal framework should support a variety of alternatives for farm organization.

12. *Collect, monitor and analyze data during all phases of the programme.* Before completing design of the programme, designers should collect overall data regarding the number and composition of landless and land-poor families, distribution and concentration of land ownership, as well as production, rents and income from land. In addition, after programme implementation begins, it is important for programme administrators to conduct ongoing monitoring to evaluate both the progress and the impact of the reforms. Such systematic monitoring and evaluation is crucial for identifying problems and providing information necessary for programme refinement.

B. Lessons from Transitional Economies

"Second generation" land reforms in transitional economies involve many, but not all, of the same programme design issues. The land to be redistributed is normally owned by the state or by the dissolving collective, so it is usually unnecessary for the state to purchase the land. Some programme design issues, however, deserve emphasis:

1. *Aim for universal or near-universal coverage of the potential beneficiaries.* Universality can be achieved in two ways. First, all members of the state farm or collective farm, including pensioners, should be given a right to receive land individually. Some Eastern European countries have complicated this step by giving preference to pre-collectivization landowners and their heirs. The second factor to consider in achieving universality is to include all or nearly all state farm and collective farm land in the redistribution. A few countries, such as Russia, have exempted large portions of this land from the redistribution programme and thus limited the programme scope.
2. *Land market development issues are more important in "second generation" reforms.* In particular, it is important to build the necessary legal and policy framework and institutional mechanisms to enable farmers to transfer land rights by lease, sale, inheritance, and mortgage (while including, where needed, safeguards to prevent improvident transfers). The creation of

such frameworks and institutions can be difficult in settings where land markets have not existed for decades.

3. *Allow state or collective farm members to exercise various options on a continuing basis, including options to farm collectively, to withdraw their land as part of a small group, or to withdraw their land to create a family farm.* This is particularly important in settings where the state farms or collective farms have not broken up quickly or decisively. At a minimum, state or collective farm members (or heirs of former small owners under a restitution programme) should not have to make an irrevocable decision at the beginning of the land reform process regarding how he or she will exercise the new land rights.
4. *Permit and foster efforts by state or collective farm members to conclude interim arrangements.* These transitional arrangements could include short-term leasing to the state or collective farm (or its corporate successor) while the member decides what to do with the land in the longer term. State or collective farm members must be adequately informed of their alternatives and not forced by the farm management into a particular choice.
5. *Permit collective farm members to withdraw a fair share of the collective farm's animals, machines, and other non-land assets together with the member's share of the land.* Collective farm members are unlikely to be able to pursue other options without such non-land assets. Moreover, the concept of collective ownership should provide the basis for their right to claim a share of these collectively owned assets.
6. *Important principles of gender equality, participatory administration, programme monitoring, and sensitive treatment of customary law also apply to "second generation" reforms in transitional economies.* These important programme design issues, highlighted in Section VI.A in the context of more traditional land reform settings, are equally important for "second generation" reforms. Women should receive at least equal rights to distributed land, programmes should be administered by local, beneficiary-dominated committees, programme progress and impacts should be carefully and regularly monitored, and customary law regimes should be treated sensitively and not summarily replaced.

VII. New Land Reform Options in the 21st Century

For most of the 20th century, the dominant methodology for conducting "land reform" in non-Marxist settings was compulsory acquisition of defined lands from private landlords or plantation owners, often at a sharp discount from market value, and redistribution of such lands to tenant farmers or agricultural labourers in farms whose size approximated that of small-to-medium sized farms in that country. Sometimes "land reform" also meant compulsory regulation of the landlord-tenant relationship in an effort to elevate and protect tenants' rights, or resettlement of tenants or labourers onto (supposedly) uncultivated public lands. The latter approaches, in particular, were nearly always unsuccessful.

Although land reform faded as a development priority in many parts of the globe during the 1970s and 1980s, the world witnessed a renewed interest in land reform near the end of the 20th century. It seems likely that this interest in land reform will be sustained through the first part of the 21st century.

The land reform approaches in the new century, however, involve both new challenges (such as those in the decollectivizing and market-creating transitional economies) and new responses. The approaches are likely to be substantially revised from those that dominated the past century. Such a new set of approaches should begin by reaffirming that the provision of secure land rights (ownership or owner-like tenure) to the roughly 230 million households in the developing and transitional economies is one of the highest economic development priorities. Policy makers should also recognize that the provision of secure land rights may be necessary for the maintenance of political stability in many countries.

In pursuing the new land reform approaches, policy makers should learn from cumulative past experience and be advertent to current needs, in a number of ways:

- In traditional developing countries, large-scale land reform programmes that redistribute a large proportion of a country's privately owned agricultural land to tenant farmers or landless labourers are unlikely to be politically feasible in most (but not all) settings.
- "Second generation" land reform programmes that undo the damage caused by involuntary collectivization in the transitional economies will continue to be a central challenge. Such

programmes are likely to include: (1) provision of longer-term, more secure and transferable land rights in countries that have already decollectivized, such as China and Vietnam; (2) measures to facilitate voluntary decollectivization in countries where that has not yet happened, such as Russia and Ukraine; and (3) measures to introduce alternatives to collectivization for the first time in the few settings (such as North Korea) where such alternatives do not yet exist.

- In many settings, especially in transitional economies but also in some traditional developing countries, it will be necessary to establish or strengthen land market prerequisites, including secondary institutions such as land registration and mortgage that will help support, protect and add value to land rights.
- In many traditional less developed countries, political and financial factors will likely make distribution of micro-plots—including house-and-garden plots—more feasible and desirable than continuing efforts to distribute "full sized" small farms. Even tiny plots can, in many settings, provide a substantial increment to family nutrition and income, increased status and wage bargaining power, freedom from moneylenders, and a place for one's own house. Such measures can be affordable even if the state must pay relatively high market values for land to be redistributed. Although a redistribution of very small plots cannot carry the full range of benefits described in Section II, the existing evidence strongly suggests that there is a rather steep "benefits curve" in which even a modest initial land endowment produces a high proportion of many of these benefits.
- Corresponding with the point just made, developing country policy makers should consider making substantial reallocations of resources from less-crucial uses, such as rural housing schemes, to the highly affordable and replicable programmes to provide small house-and-garden plots of a sufficient size, averaging around 0.06 acre up to 0.25 acre (0.03 to 0.10 hectare).
- The failed approach of resettling land-poor farmers and labourers onto distant uncultivated public land should be largely abandoned except in very special and compelling circumstances.
- Failed attempts to regulate the landlord-tenant relationship should also be abandoned, except in special and compelling circumstances. For example, in West Bengal, the state should either retain tenancy regulation for remaining tenants, since the

costs of evictions have already been borne, or, even better, provide the protected tenants with realistic options for becoming full owners. In settings where past tenancy reform approaches have unintentionally led to negative consequences, the state may need to undo aspects of the failed reforms. For example, Indian states should consider removing the most onerous legislative restrictions (and prohibitions) on tenancy in order to provide greater opportunity and freedom for land-poor farmers to access land through rental markets.

- States should pay more attention to safeguarding and assuring customary land rights of the poor, including rights to forestland and grazing land. Thus, in addition to land reform that helps the poor to gain access to land, efforts are needed to help the poor formalize or otherwise protect their existing customary rights to land. States are increasingly recognizing that group or communal rights may be the best option for traditional pastoralists who use grazing land in extensive or migratory patterns, accompanied with appropriate licensing or access restrictions.
- Policy makers have recently and belatedly begun to pay attention to intra-household allocation of land rights. This should continue and increase, with efforts to ensure greater rights for women in the enjoyment, disposition, partitioning, and inheritance of land.
- Policy makers in many countries will need to undertake more systematic attempts to find timely compromise solutions to forestall violent, confiscatory and anti-democratic efforts to secure land for the poor (including efforts that may lead to backlashes that further undermine democracy, the rule of law, and the position of the poor).
- Rare cases may still exist where land is cheap and plentiful enough (such as unused or grossly underutilized land in parts of South America or Africa) to allow the state to employ the old approach of redistributing "full-sized" farms to a large number of beneficiaries. In such cases, however, the state should compensate existing owners whose land is mandatorily acquired by paying the owners either the market value or very close to the market value of their land.
- Policy makers should also recognize that there could be a potentially important, supplementary role for "market assisted land reform" or "negotiated land reform." In these approaches, the state removes policy distortions that inflate land prices and

provides funding to needy families to acquire private land being voluntarily offered for sale. Such schemes require further testing in pilot projects before being widely pursued.

- The state should assist land reform beneficiaries with simple complementary measures (other than housing), but not attempt to be the provider of all services to land reform beneficiaries. For example, the state may need to provide some support for credit, extension, or basic infrastructure (such as extending simple, local roads and electrification lines from existing residential areas to newly distributed house-and-garden plots). However, all such interventions should be conducted with an eye to keeping costs-per-beneficiary at a level that allows for universal or near-universal coverage to those requiring benefits. Private providers and the beneficiaries themselves have important roles to play.

VIII. Conclusion

The case for land reform is compelling. The experience of many countries shows the crucial role of land reform in providing not only a source of income, security, and status for the non-landowning rural poor, but also as a foundation for broader rural development and political stability. Accumulated experience also provides responses to the current challenges of land reform that are highly likely to be both affordable and politically feasible over the coming years.

Land reform in many settings thus appears both necessary and achievable. The stakes are considerable, both for the traditional developing countries, the transitional economies, and, not least, the industrialized democracies. Successful efforts to assure secure rights to land, even in small quantities, to the roughly 230 million farm households on our planet who stand in need of such land rights could make a mid-to long-term difference of almost incalculable magnitude to the kind of planet we will live on and that our children will inherit.

2

Land Policies for Social Inclusion in India

T. Haque

Introduction

Land is an important source of livelihood as well as social recognition and status for the majority of the Indian population. Over 50 per cent of the country's total workforce and 70 per cent of the total rural workers depend on agriculture for livelihood. The manner in which land is owned and utilized in a secure and sustainable manner lies at the core of the livelihood and food security of people, particularly in rural areas. Ensuring access to land for landless agricultural workers and land security for marginal and small farmers, women, tribals and other vulnerable groups is crucial for their food security and sustainable livelihood, especially as they cannot access adequate non-farm employment opportunities for lack of appropriate education and skills. The rural poor's access to land is important as there is high incidence of poverty among landless and semi-landless households. Food insecurity, hunger and poverty often result from the imbalances in the present process of development which hinder access to land and other natural resources and livelihood assets in a sustainable manner. (FAO: 2006). It is now widely recognized that traditional land reform policy in India has failed to deal with the issue of high incidence of landlessness and rural poverty, while growing land acquisition for various non-agricultural purposes has tended to alienate many marginal and small farmers, tribals and other vulnerable groups from land, thereby threatening their livelihood security. Besides, in the absence of appropriate land-use policy, there is no protection against conversion of fertile agricultural land for non-agricultural purposes, which is likely to have an adverse impact on the agricultural growth and food security of the country. While the demand for land is rising for diverse economic activities, social exclusion is also emerging from that process. In view of these facts, the present paper analyses various positive and negative aspects of important land policies adopted in India in the past few decades, especially in the context of inclusive rural growth.

Evolution of Land Policy

There are three distinct phases in the evolution of land policy in India. First, the land policy followed by the British Government before independence, aimed mainly at securing maximum land revenue, although in the later part of the British rule some measures of land reform were undertaken to provide relief to the tenants. Second, some land policies took concrete shape after independence and were aimed at abolition of intermediaries, abolition or regulation of tenancy, imposition of ceilings on land holdings, redistribution of ceiling-surplus land, consolidation of holdings, prevention of alienation of tribal land, etc. Third, in the wake of economic liberalization in the past few years, several state governments have adopted land acquisition policies which are not consistent with the philosophy of inclusive growth; while taking no interest in traditional land reform measures.

Land Policy before Independence

Prior to independence, the colonial rulers in India followed a land policy which aimed mainly at securing maximum revenue for the British government by means of land tax (Haque and Sirohi: 1986). The East India Company which gained political and economic power from the Nawab of Bengal in 1757 abolished the system of fixed rent and created a system in which zamindari was auctioned temporarily to highest bidders who, in turn, had the freedom to collect as much rent as they saw fit. In fact, the peasantry was subjected to such tyranny and exploitation that 10 million out of a total population of 30 million in Bengal died in the great famine of 1769-70 (Government of India: 1945). This was followed by other devastating famines in the years 1784, 1787 and 1790, respectively. The frequent occurrence of famines also adversely affected the internal market for British goods in rural India which was one of the motivating factors for Lord Cornwallis to introduce permanent settlement in 1793, the main objectives of which were to induct moneylenders and traders as zamindars who could invest capital for agricultural development and also act as reliable agents of revenue generation for the British rule. In reality, however, none of these objectives was fulfilled. By early nineteenth century, therefore, Regulation VII of 1822 provided for periodic resettlement in most parts of United Provinces and other areas, while in parts of United Provinces and later in Punjab, Regulation VII of 1822 and Regulation IX of 1833 provided for mahalwari settlements with the entire village, which required each peasant of the village to contribute to total revenue demand of the village on the basis of size of holding. Besides, the Bengal Tenancy Act, 1885, Bihar Tenancy Act, 1885 and Orissa Tenancy Act, 1914 made legal provision for occupancy rights to tenants who were in possession of land continuously for twelve years. Also, the Oudh Rent (Amendment) Act of 1921, Agra Tenancy Act of 1926 and Madras Estate Land Act of 1908 provided for protection of ryots from eviction. In 1894, the British government passed a land acquisition bill in order to facilitate the acquisition of agricultural land for infrastructure and other public purposes. Another important aspect of land policy of the British government was in respect of consolidation of holdings. The settlement Department of Bombay decided as early as in 1847 not to recognize or enter plots below a certain area in the revenue records in the hope that this would discourage fragmentation and ensure continuance of compact and sizeable plots. This view was further endorsed by the Royal Commission on Agriculture in 1928. In reality, however, the land policy of the British government up to 1931 could not ensure efficient cultivation of land or harmony in land relations, as the policies were largely biased in favour of the intermediaries, while the actual cultivators lacked any incentive or encouragement to cultivate land efficiently.

During 1931 to 1947, the freedom movement in India was at its peak when the demand for pro-poor land reform by various peasant organizations as well as the Indian National Congress also gained momentum. The Indian National Congress at its Karachi session in 1931 and Faizpur session in December 1936 drew up a land reform policy which focused on reduction of rent and exemption of uneconomic holdings from payment of rent. The Faizpur session also stressed the need for fixity of tenure with heritable rights for tenants and abolition of all feudal levies and forced labour. Further, the National Planning Committee which was set up in 1938, under the chairmanship of Jawaharlal Nehru proposed that cultivation of land should be organized in complete collectives on Government land and in co-operatives elsewhere (Government of India: 1948).

Land Policy since Independence

The main features of the post-independence land policy are as follows:

1. Abolition of Intermediaries

During the 1950s almost all state governments in India enacted legislation for the abolition of intermediaries (zamindari), as a result of which about 20 million cultivators were brought into direct contact with the state. However, a number of factors led to only partial success of the measures undertaken to abolish zamindari in India. First, barring West Bengal and Jammu and Kashmir, the legislation on abolition of intermediary tenure was not accompanied by simultaneous legislation on land ceiling. As a result, there was enough time left for manipulations by ex-intermediaries to evade the law through: (i) fictitious partition of properties, (ii) influence of the village officials for changes in land records, and (iii) registration of sharecroppers or tenants as farm labourers. In Tamil Nadu, the laws were made as late as 1964, 1969 and 1972. In Gujarat, the law to abolish *devasthan inams* was passed only in 1969. In Kerala, *sreepandaravalla* lands could be abolished under an enactment made in 1971, while intermediary rights on *kanom* lands, *oodupally* lands and service inam lands have not been abolished so far. In Uttar Pradesh, the Zamindari Abolition Act, 1950 had allowed ex-zamindars to continue to possess their *Sir* and *Khudkasht* (self-cultivated) lands with the status of *bhumidars* having absolute security of tenure. Besides, the term personal cultivation was defined vaguely to allow even absentee landlordism. For instance, in Uttar Pradesh, *bhumidars* had neither legal obligation to supervise or undertake cultivation of land in person, nor any requirement to reside within reasonable distance from the place where the land was located. Nevertheless, it is generally recognized that as a result of the laws passed by various state governments on abolition of intermediaries, large intermediaries have been mostly abolished and the socio-economic power of the landlords and exploitation of the tenants have greatly diminished overtime.

2. Abolition or Regulation of Tenancy

The three-fold objectives of tenancy reform laws adopted by various state governments since independence were to (i) confer ownership or occupancy rights on tenants with respect to non-resumable tenanted land, (ii) prohibit or restrict leasing and sub-leasing of land in future, and (iii) fix rent. It has been estimated that about 12.6 million tenants were conferred either ownership or occupancy rights over 6.7 million hectares of land up to September 2006. In several states, tenants living on the land of the landlords were given permanent homestead right. However, it was also observed that restrictive tenancy laws provoked the landlords to evict tenants, sub-tenants and sharecroppers in most places. Tenants were evicted from about 30 per cent of the total land, while they could be given either ownership or occupancy right over only 4 per cent of the land. (Appu: 1996). At the same time, informal, albeit insecure tenancies continued to exist in most places with all their growth-retarding features. The fair rents fixed by various state governments ranged between 1/2 to 1/6 of gross produce under different situations, though actual market rents were higher in most places. (Haque and Sirohi: 1986).

3. Ceilings on Landholdings

As already stated, the states of West Bengal and Jammu and Kashmir imposed ceilings on landholdings along with abolition of intermediaries in the 1950s. All other states enacted ceiling laws during 1960 to 1972 before any national guidelines were laid down and subsequently revised them in 1972 after the adoption of national guidelines. The present ceiling level ranges from 3.68 hectares to 28.33 hectares for various categories of land (Table 2.1). The main objectives of ceiling on landholdings were to reduce the unequal and unproductive character of the agrarian economy and to address the issue of landlessness and unemployment in rural India.

Till September 2006, about 4.9 million acres of ceiling-surplus land was distributed among 5.4 million beneficiaries of which 2.1 million belonged to Scheduled Castes, 0.9 million were Scheduled Tribes and 2.4 million were of other castes. The state of West Bengal alone distributed 1.02 million acres among 2.85 million beneficiaries. The main reasons for poor implementation of ceiling laws in most states were lack of political will, litigation, lack of proper land record and administrative inefficiency. In the meantime, the number and area of both large and medium holdings which could yield some ceiling-surplus land for redistribution with improvement in implementation, have substantially declined and the average size of land in all categories has reduced due to population pressure on land. Also, in view of the total absence of political will-power on the part of almost all states as well as lack of proper land records, one does not know whether the actual ceiling-surplus land can be properly identified for their effective acquisition and redistribution. In fact, redistribution of about 0.81 million hectares of land which has been declared surplus, but has not yet been distributed due to litigation or other reasons, itself poses a challenge.

Table 2.1: Ceiling on Landholdings by State

State	*Level of Ceiling (hectare)*
Andhra Pradesh	4.05 to 21.85
Assam	6.74
Bihar & Jharkhand	6.07 to 18.21
Gujarat	4.05 to 21.85
Haryana	7.25 to 21.85
Himachal Pradesh	4.05 to 12.14 (a)
Jammu & Kashmir	3.68 to 7.77
Karnataka	4.05 to 21.85
Kerala	4.86 to 6.07
Madhya Pradesh/Chhattisgarh	4.05 to 21.85
Orissa	4.05 to 18.21
Punjab	7.00 to 21.80
Rajasthan	7.25 to 21.85 (b)
Tamil Nadu	4.86 to 24.28
Tripura	4.00 to 12.00
Uttar Pradesh	7.30 to 18.25
West Bengal	5.00 to 7.00

(a) = In certain specified areas of HP up to 28.33 hectares.
(b) = In certain areas of Rajasthan up to 70.82 hectares.
Source: T. Haque and A.S. Sirohi, *Agrarian Reforms and Institutional Changes in India*, Concept Publishing Company, New Delhi, 1986, p. 84.

Distribution of Government Land

Several state governments, not being able to acquire and redistribute ceiling-surplus land, opted for a softer version of land reform by either allocating available government land or regularizing encroachments on such land. Till September 2006, various state governments in India distributed nearly 14.7 million acres of government land to rural landless poor families (Govt. of India: 2007). The state of Andhra Pradesh distributed about 4.2 million acres of government wasteland, followed by Uttar Pradesh which distributed about 2.5 million acres of government land to the landless poor. However, these numbers have to be read carefully as field investigations in Andhra Pradesh had shown that about 30 per cent of the reported beneficiaries did not have legal as well as physical possession of the allocated land (Tim *et al.*: 2008). Also much of the land distributed was of inferior quality which needed substantial investment for improvement to be used productively. It has been estimated that the total area of government wasteland at present is about 157.7 million acres, 30 per cent of which is land with or without scrub, 22 per cent is underutilized/degraded notified forest land and 7.8 per cent lies in sandy and coastal areas. The degraded pastures and grazing lands constitute nearly 4.08 per cent of the total government wastelands (Govt. of India: 2009). Although such lands are not fit for cultivation, they provide livelihood support to rural and forest population in a number of ways. Currently there is a ban on the encroachment of government forest lands imposed by the Supreme Court with effect from 23rd November, 2001. Besides, in the past few years, some state governments, especially Tamil Nadu, has deviated from the practice of allocating such land to the landless poor. Instead, the Government of Tamil Nadu launched a Comprehensive Wasteland Development Programme in 2003, the major component of which was to develop five million acres of government wasteland by involving the corporate sector for commercial agriculture. Such a step is not only in conflict with national land reform policy, but also prejudicial to the employment and livelihood interests of the poor.

Distribution of Bhoodan and Gramdan Lands

Bhoodan and Gramdan movements were launched by Acharya Vinoba Bhabe in 1951 and 1957, respectively to persuade land owners and leaseholders in villages to renounce their land rights after which all the lands would become the property of a village association for egalitarian redistribution for purposes of joint cultivation. According to the latest available data, about 4.59 million acres of land was donated all-over India, out of which 2.32 million acres was distributed to about 0.315 million families. Of the total area donated about 1.10 million acres was found not fit for cultivation. (Govt. of India: 2009). However, these figures need to be properly verified before any effort can be made to redistribute bhoodan and gramdan lands to the landless poor, as there are discrepancies between the records maintained by Revenue Departments and the Bhoodan Yagna Committee.

Consolidation of Landholdings

Consolidation of landholdings refers to amalgamation of fragmented land parcels so that all plots of a landowner can be brought together in a compact area. This is intended to promote agricultural efficiency and productivity. Prior to independence, about 0.45 million acres of land was consolidated between 1940 and 1947, in the central provinces, united provinces, Punjab and Jammu and Kashmir. After independence, almost all states except Tamil Nadu, Kerala, Manipur, Nagaland, Tripura and parts of Andhra Pradesh enacted legislations for consolidation of holdings. While it was compulsory in the states of Punjab, Haryana, Uttar Pradesh and Maharashtra, it was voluntary in most other places. So far, about 1.72 million acres of land has been consolidated in the country, mostly in Maharashtra, Uttar Pradesh, Haryana and Punjab. Meanwhile, the scheme has been discontinued in several states, including Maharashtra (since 1993), Gujarat (since 1992), Rajasthan (since 1965) and Andhra Pradesh (since 1966) because their implementation affected marginal and small farmers and tenants adversely in several cases. In fact, the main reasons for slow progress in the consolidation of holdings in most places, include: (i) lack of legal provision for compulsory consolidation in most places, (ii) problem of proper valuation of land and conflicts arising thereof, (iii) cases of eviction of tenants and alienation of small and marginal farmers, (iv) low size of holding, (iv) lack of up to date land records, (vii) lack of trained staff, and (viii) financial difficulties faced by various state governments.

Protection against Alienation of Tribal Land

According to the 2001 census, tribal population constitutes nearly 8.1 per cent of India's total population. They are indigenous groups, largely concentrated in the states of Madhya Pradesh, Maharashtra, Orissa, Gujarat, Rajasthan, Jharkhand, Chattisgarh, Andhra Pradesh, West Bengal and Karnataka. Land and forests are the main sources of livelihood in tribal areas. Yet in the pre-independence period, colonial government as well as non-tribal intermediaries have historically encroached upon the land and forest rights of the tribals. Although the land tenure systems in tribal areas varied from region to region, alienation of tribals from land under various market and non-market forces seemed common in most places. This resulted in tribal uprisings in several places. As a response, some state governments enacted laws for prevention of alienation of tribal lands to non-tribals even prior to independence. For example, the Bengal Eastern Frontier Regulation 1873, Santhal Parganas Tenancy Act, 1854, Agency Tracts Interest and Land Transfer Act, 1917 and Chhota Nagpur Tenancy Act, 1924 provided some measure of protection, though inadequately. After independence, all the concerned states passed comprehensive laws for the prevention of alienation of tribal land, the summary of which is given in Table 2.2.

Despite such protective legal measures, however, the problem of land alienation in tribal areas continues unchecked. Till recently, about

Table 2.2: Legislations for Protection of Tribal Land

S. No.	*State*	*Legalization in force*	*Main features*
1	Andhra Pradesh	(a) The Andhra Pradesh (Scheduled Areas) Land Transfer Regulation, 1959, amended by The Andhra Pradesh (Scheduled Areas) Land Transfer (Amendment) Regulation, 1970, 1971, and 1978	Prohibits all transfer of land to non-tribals in Scheduled Areas. Authorizes government to acquire land in case a tribal purchaser is not available. There is, however, no legal protection to ST land outside the scheduled areas.
2	Assam	The Assam Land and Revenue Regulations 1886, amended in 1981	Chapter X of regulation prohibits alienation of land in tribal belts and blocks.
3	Arunachal Pradesh	Bengal Eastern Frontier Regulation, 1873, as amended	Prohibit transfer of tribal land
4	Andamans and Nicobar Islands	Andaman and Nicobar Islands (Protection of Aboriginals' Tribes) Regulation, 1956	Protects tribal interest in lands
5	Bihar/Jharkhand	(a) Chhota Nagpur Tenancy Act, 1908. (b) Santhal Parganas Tenancy Act (Supplementary provision) 1940. (c) Bihar Scheduled Areas Regulation, 1969.	Prohibits alienation of tribal land and provide for restoration of alienated land.
6	Chhattisgarh	Sections 165 and 170 of Madhya Pradesh Land Revenue Code, 1959. (b) Madhya Pradesh Land Distribution Regulation Act, 1964	Sections 165 and 170B of the code protect STs against land alienation. The 1964 Act is in force in the scheduled areas.
7	Dadra & Nagar Haveli	Dadra & Nagar Haveli Land Reform Regulation, 1971.	Protects tribal interest inlands
8	Gujarat	Bombay Land Revenue (Gujarat Second Amendment) Act, 1980.	Prohibits transfer of tribal land and provides for restoration of alienated land.
9	Himachal Pradesh	The Himachal Pradesh Transfer of Land (Regulation) Act, 1968.	Act prohibits transfer of land from tribals to non-tribals.
10	Karnataka	The Karnataka Scheduled Caste and Scheduled Tribes (Prohibition of Transfer of Certain Lands) Act, 1975.	Act prohibits transfer of land assigned to SCs and STs by government. No provision to safeguard SC/ST interest in other lands.
11	Kerala	The Kerala Scheduled Tribes (Regulation of Transfer of Land and Restoration of Alienated Land) Act, 1975.	Act of 1975 made applicable with effect from 1st June, 1982 by notification of January, 1986 prohibits transfer of land of tribals and provides for its restoration.
12	Lakshadweep	Lakshadweep (Protection of Scheduled Tribes) Regulation, 1964	Prohibits transfer of tribal land.

(*Contd.*)

(*Contd.*)

S.No.	*State*	*Legalization in force*	*Main features*
13	Madhya Pradesh	(a) Sections 165 and 170 of Madhya Pradesh Land Revenue Code, 1959. (b) Madhya Pradesh Land Distribution Regulation Act, 1964.	Sections 165 and 170B of the code protect STs against land alienation. In the scheduled area of Madhya Pradesh and Chattisgarh, the 1964 act is in force.
14	Maharashtra	(a) The Maharashtra Land Revenue Code, 1966, as amended in 1974. (b) The Maharashtra (Restoration of Lands to Schedules Tribes) Act, 1974.	Prohibits alienation of tribal land and provides for restoration of both illegally and legally transferred lands of a ST.
15	Manipur	The Manipur Land Revenue and Land Reforms Act, 1960.	Section 153 forbids transfer of land of STs to non-STs without permission of DC. Act not been extended to hill areas and hill area tribals not covered.
16	Meghalaya	Meghalaya Transfer of Land (Regulation) Act, 1971.	Prohibits alienation of tribal land.
17	Nagaland	Bengal Eastern Frontier Regulation, 1873 and Assam Land and Revenue Regulation, 1866, as amended vide Nagaland Land and Revenue Regulation (Amendment) Act 1978.	Prohibition of land transfer of tribals.
18	Orissa	The Orissa Scheduled Areas Transfer of Immovable Property (STs) Regulation, 1956. The Orissa Land Reforms Act, 1960.	Prohibits transfer of ST land and provides for its restoration.
19	Rajasthan	The Rajasthan Tenancy Act, 1955, The Rajasthan Land Revenue Act, 1956.	Section 175 and 183B specifically protects tribal interest in land and provides for restoration of alienated land to them.
20	Sikkim	Revenue Order no. 1 of 1917. The Sikkim Agricultural Land Ceiling and Reform Act, 1977.	Order of 1917 still in force. Chapter 7 of 1977 restricts on alienation of lands by STs but is not in force.
21	Tamil Nadu	Standing Orders of the Revenue Board BSO 15-40. Law against land alienation not enacted.	BSO 15-40 applies only to Malayali and Soliga tribes. Prohibits transfer of assigned land without approval of DC.
22	Tripura	Tripura Land Revenue and Land Reform Act, 1960, as amended in 1974.	Act prohibits transfer of ST land to others without permission of DC the collector. Only land transfer after 1.1.1969 are covered under restoration provision.
23	Uttar Pradesh/Uttarakhand	U.P. Land Laws (Amendment) Act, 1981, amending Uttar Pradesh Zamindari Abolition and Land Reforms Act, 1950.	Provide protection of tribal land. But amending Act is not applied and stayed by Allahabad High Court in Swaran Singh Vs State Govt. 1981.
24	West Bengal	West Bengal Land Reforms Act, 1955, as amended.	Chapter II-A prohibits alienation of tribal land and provides for restoration.

Source: Govt. of India, 'Report of the Committee on State Agrarian Relations and the Unfinished Tasks in Land Reforms', Ministry of Rural Development, New Delhi, pp. 137-138 (mimeo).

3,75,164 cases of land alienation were filed, involving 8,55,282 acres of land, out of which 1,62,650 cases only were decided in favour of the tribals, involving 4,47,314 acres. Still about 57,521 cases were pending in court, involving 43,961 acres of land, while 1,54,993 cases were rejected, involving 3,63,493 acres (Govt. of India: 2004). The main sources of land alienation include (i) land acquisition by the government for highways, mining, urbanization, special economic zones etc. for which there is lack of a strong 'land for land' rehabilitation, (ii) manipulation of record because of defective surveys and settlements and non-recording of possession, (iii) permission easily granted for purchase of tribal land as well as conversion of agricultural land to non-agricultural land by local level competent authorities, (iv) benami purchases, (v) gift by tribals, and (vi) long-term leases etc. Unless the legal loopholes are plugged and the Gram Sabha is made responsible for dealing with all kinds of tribal land cases, there is little hope for the situation to improve, especially since the Government officials and local politicians are insensitive to the problem. Besides, there should be proper survey, settlement and updation of land record and also complete ban on land acquisition for private companies in the scheduled areas, except for genuine public purposes with concurrence of Gram Sabha.

Land Acquisition and Rehabilitation and Resettlement

Both Central and State governments acquire land for various development projects, invoking mostly the Land Acquisition Act, 1894 which gave absolute power to the state without leaving any choice for the landowner. In 1984, the scope of the Act was extended to include land needed for private companies. The Special Zone Act of 2005 further empowered the state to acquire land in any area of any size for private companies. In fact, in the past few years, several State governments have ventured to acquire land for so-called public purpose, but actually private interests, including real estate, mining, etc. unsettling millions of farmers and farm workers that met with resistance by the people. In response, the Union Government has now introduced a new Bill in the Parliament, called the National Land Acquisition and Rehabilitation and Resettlement Bill, 2011. As per its preamble, the main objective is to balance the need for facilitating land acquisition for industrialization, development of essential infrastructure facilities and urbanization, while at the same time also meaningfully address the concerns of farmers and those whose livelihoods are dependent on the land being acquired. The proposed Bill, if passed, will be applicable to the whole of India except the state of Jammu and Kashmir.

Box 2.1: Main Features of the Draft Land Acquisition, Rehabilitation and Resettlement Bill, 2011

i	The Act will apply to the whole of India, except the State of Jammu & Kashmir.
ii	Government can acquire land (a) for its own use, to hold and to control, (b) to transfer it for the use of private companies for stated public purpose and (c) for immediate and declared use by private companies for public purpose.
iii	Land acquisition for sub-section (b) and (c) can take place where at least 80 persons of the affected families give their consent to the proposed acquisition.
iv	The Bill rules out the acquisition of multi-cropped irrigated land, unless unavoidable.
v	The Bill shall be compliant with (a) PESA, 1996, (b) Forest Rights Act, 2006 and other regulations to Schedule V areas.
vi	The compensation for the land will be calculated based on the minimum land value as specified in the Indian Stamp Act, 1899 for the registration of sale deeds in the area where the land is situated or the average of the sale price for similar type of land situated in the village or vicinity, as calculated from 50 per cent of the sale deeds registered during the proceeding three years or whichever is higher.
vii	The building, trees, wells, crops etc. will be valued at market prices and compensated at not less than six times the original market value in rural area and twice that of the market values in urban areas.
viii	The R & R entitlements for landowners are (a) subsistence allowance of Rs. 3,000 per month per family for 12 months, (b) Rs. 200 per month per family as annuity for 20 years, with appropriate index for inflation, (c) if house is lost, a constructed house of plinth area of 150 sq mts of house site in rural areas or 50 sq mts plinth area in urban areas, (d) one acre of land to each family in the command area, if land is acquired for an irrigation project, (e) Rs. 50,000 for transportation, (f) whose land is acquired for urbanization, 20 per cent of the developed land will be reserved and offered to landowners in proportion to their land acquired, (g) upon every transfer of land appreciation shall be shared with the original owner whose land has been acquired, and (h) mandatory employment for one member per affected family or Rs. 2 lakh if employment is not offered and offer of shares up to Rs.25 per cent of the compensation amount. The R & R package extends to other livelihood losers, including tenants and landless, excepting the provision under (e) and (f) above. Besides, they will be entitled to one time resettlement allowance of Rs.50,000.

However, the proposed LAR&R bill is undergoing changes due to pressure from various concerned groups. It also suffers from several inadequacies and weaknesses. First, it provides for land acquisition for declared use by private companies for public purpose, without defining the term 'public purpose' properly. Given the system of poor governance that India has, one is not sure whether the land security and livelihood issues of small and marginal farmers and tenants, tribals and other affected persons will actually be addressed by the new law. Second, the definition of public purpose is quite wide; this leaves scope for manipulation, e.g. defining infrastructural projects to include mining in tribal areas which witnessed agitation on this issue recently. Besides, provision of land acquisition for village or urban site, may be desirable, but is likely to be misused as a loophole to acquire land for builders and real estate agents and therefore, necessary safeguards need to be provided. Third, while multi-crop irrigated land may be largely excluded from the scope of land acquisition, other categories of irrigated and fertile land and forest land should also be excluded in order to maintain national food security, ecology as well as livelihood security of the farmers. Further, the compensation offered is not 'just' as the price of the land is determined by the local officials based on 50 per cent or more of the sale deeds registered during the proceeding three years. Such prices are always undervalued. Price for land should reflect the potential rather than historic value. Fifth, clause 5 of the Fifth Schedule of the constitution, requires the Governor of each state to review all laws and adopt them to suit the interests of the Scheduled Tribes before extending their application to scheduled areas. Also, the decision to set up any development project in scheduled areas should be with the tribal Gram Sabha/Tribal Advisory Council and ownership of any such projects should be vested in them. Finally, although the R&R component of the Bill looks people friendly, its effective implementation may pose a problem in the absence of good governance.

Land Rights for Women

According to the National Sample Survey Organization (Govt. of India: 2006), there are about 69.4 million women farmers in India, majority of whom do not have any secure land rights. Nearly 11.6 per cent of the total agricultural holdings are operated by women covering 9.1 per cent of the total land. But there is no data to suggest that the land operated by women are also owned by them. In fact, women farmers' lack of secure land right stands in the way of their cultivating land efficiently as they cannot access institutional credit facilities, extension, marketing and other support services. Also, in the absence of land right they remain more vulnerable.

As a matter of fact, women have suffered due to both gender-biased inheritance laws and land reform policies of the government. The Hindu Succession Act, 1956, as amended in 2005, provides equal right to men and women in matters of inheritance. But the law is hardly implemented anywhere mainly due to various socio-cultural barriers. Also, the Muslim personal law and tribal customary laws are not free from bias against women. Unfortunately, even the redistributive land reform policies followed in India since independence have not addressed the issue of landlessness among women. The revenue minister's conference held in New Delhi on 14 March, 1992 recommended that at least 40 per cent of distribution of ceiling-surplus land should be made in the name of female members and the remaining jointly in the name of husband and wife. The ceiling-surplus land already distributed should be converted into joint *pattas* in the name of husband and wife (Govt. of India: 1992). But almost all the states failed to follow such guideline. The Government of West Bengal issued a circular dated 17/20 August, 1992, which suggested that if a female member of the beneficiary household is eligible for distribution of agricultural land, it should be distributed in favour of the said female member. In other cases, the land should be jointly distributed in favour of husband and wife of the beneficiary family to the extent possible. Cases where distribution has already been made, shall not be reopened for this purpose. The wording of the circular itself was wrong, as it used the words 'if' and 'to the extent' possible which showed lack of seriousness on the part of the government. In practice, no such joint *pattas* were issued. In most other states, even an executive order was missing. No doubt, in recent years, the policy statements made in the Ninth Five Year Plan declare that government would endeavour to provide joint title to both husband and wife, when distributing agricultural land and house sites. The Eleventh Five Year Plan (Govt. of India: 2007) mentioned that all landless families with no homestead land as well as those without regularized homestead should be allotted 10 to 15 cents land each, with households headed by females being given priority. In reality, however, no such effort has been made by any of the states, other than Karnataka (Namma Bhumi–Namma Thota), West Bengal (CDPA/NGNB) and Odisha (Vasundhara) to some extent, with support from Landesa.

Updation and Digitisation of Land Records

Before independence, there was a reasonably good system of preparation and maintenance of land records, showing ownership of different plots, area and boundaries of each plot, crops grown, amount of revenue payable, etc. This was mainly for the purpose of assessment and collection of land revenue. After independence and more so, in the past few decades, the importance of land revenue as a source of income for the state governments has declined and consequently, there is no incentive to keep records up to date. Second, the revenue officials have been burdened with multiple duties, including preparation of electoral rolls, conduct of elections, implementation and monitoring of development schemes, implementation of land reforms, management of natural disasters, law and order, etc. Third, positions of revenue officials at the lower and middle levels have not been filled up for several years in many states. Finally, there has been overall deterioration in governance in most places. As a result of all these inadequacies, there is total neglect of land administration and land management, including survey and settlement and updation of land record in the majority of the states.

Since independence, very few revisional surveys have been concluded. The Government of West Bengal started revision survey operations in 1974; they have not been completed yet. In Bihar, the process was started in the 1950s and it took more than 30 years to complete in just 12 districts (Saxena: 2005). In the districts of Darbhanga and Bhagalpur, the survey operations started in 1964 and continues till date without any hope of conclusion (Govt. of India: 2009). Table 2.3 shows the years of original and revision settlement.

Table 2.3: Years of Original and Revision Settlement

State	*Year of Original Settlement*	*Year of Last Revision Settlement*
Andhra Pradesh	1861-1888	1892-1911
Assam	1900-1927	1991, still continuing
Karnataka	1806-1904	1966-1978
Kerala	1886-1934	Not attempted
Madhya Pradesh	1901-1950	1990-2000, but stopped without completion
Maharashtra	1840-1880	1917-1920
Orissa	1922-1932	Done in a few districts in the last 3 decades
Punjab/Haryana/ Uttar Pradesh	1889-1912	Survey is merged with consolidated
West Bengal	1889-1942	1956-1961, 2nd revision began in 1974, still continuing

Source: N.C. Saxena 'Updating Land Records: Is computerization sufficient?' *EPW*, Jan. 22, 2005, p. 316.

In fact, due to the unsatisfactory nature of the output of survey operations and the exploitative nature of the whole process, most state governments have not shown much interest in these exercises. In view of various deficiencies in the traditional survey method, the Committee on State Agrarian Relation and the Unfinished Tasks in Land Reforms (Govt. of India: 2009) pointed out that individual states should review the position of the survey operations within their states and where it is found that the records so created have outlived their utility, the operations may be concluded as they would serve little purpose. Besides, advanced technological methods like satellite imagery coupled with verification of ground reality may be tried because it is less expensive and less time consuming. Moreover, the states have to devise a method by which record of rights may take place without there being requirement of survey operations in the present mode. One of the methods can be combining sub-registrar's office with that of the tehsildar so that all the transfers of land through registered deeds are immediately mutated and corrected accordingly. However, the final data should be approved by the Gram Sabha having a minimum attendance of 80 per cent.

The Ministry of Rural Development, Government of India initiated a scheme on computerization of land records initially on pilot basis in 1988-89 and as a full-fledged programme in 1992-93. Since then, various states are being supported by the Central Government for computerization of land records, although the progress achieved in different states varies widely. Table 2.4 shows the states of computerization of land records and digitization of village maps/field measurement books in selected states.

The Eleventh Five Year Plan (Govt. of India: 2007b) points out that digitization of land records after undertaking comprehensive land survey would be completed by the Twelfth Plan Period and then move towards the torrens system which is 'title' based rather than deed based in the Thirteenth Five Year Plan. However, the Govt. of India has in the meantime already prepared a proposal on the Land Titling Bill, 2011, the objective of which is to provide for the establishment, administration and management of a system of conclusive property titles through registration of immovable properties. The preparation of register of titles will be taken up based on the available land records data. It is generally recognized that the scheme of computerization of land record has been successful to some extent in states like Karnataka, Andhra Pradesh, Gujarat, Rajasthan, Maharashtra and Tamil Nadu where there is a culture for information technology uses. In other states, the progress is quite slow. At the same

Table 2.4: Summary on Status of Modernizing Land Administration in Selected States

State	*Registry computerized*	*Computerization of land records*	*Digitization of village maps/Field Measurement Book*
Andhra Pradesh	Completed	Not operational	<10% of VM vectorized; FMBs scanned
Bihar	Piloting	Data entry	No activity
Gujarat	Roll-out	Fully operational, manual records banned	<10% of VMs vectorized
Himachal Pradesh	Piloting	Roll-out; manual records not banned anywhere	No activity
Karnataka	Completed	Fully operational, manual records banned	50% village maps vectorized, FMB scanning in pilot phase
Kerala	Only indexes operational	Data entry stage	No activity for village maps; Piloting digitization of FMBs
Maharashtra	Completed	Fully operational, manual records not banned	All VMs scanned, vectorized and geo-referenced; Scanning of FMBs ongoing
Madhya Pradesh	Only indexes operational	Operational, manual records not banned	VM digitization at pilot stage, problems with area
Orissa	Piloting	Data entry and piloting	Piloting of digitization of VMs and scanning of FMBs
Punjab	Piloting	Data entry	Limited piloting
Rajasthan	Near completion	Fully operational, manual records banned	No activity
Tamil Nadu	Roll-out	Fully operational in all taluks, manual records banned	VMs scanned; not vectorized; Piloting digitization of FMBs
Uttar Pradesh	Piloting	Fully operational in all taluks, manual records banned	No activity
West Bengal	Piloting	Limited pilots	Limited piloting of both

Note: Pilot stage implies that less than 10 per cent has been completed; roll-out means that 10 per cent-75 per cent has been completed; and near complete means that more than 75 per cent has been completed.

Source: *The World Bank: Land Policies for Growth and Poverty Reduction*, Oxford University Press, 2007, p. 53.

time, it is also true that e-governance is not a substitute for good governance. Hard core jobs such as correcting land records, recording encroachments, verifying mutations in the field and observing changes in land use requires dedicated field effort with proper supervision (Saxena, 2005), which is presently missing in the land-governance structure in most states.

Some Recent Innovative Land Reform Initiatives

In recent years, the state governments of West Bengal, Karnataka, Andhra Pradesh, Odissa and Kerala have launched some special schemes for non-ceiling based land redistribution, the key features of which are summarized in Box 2.2.

Key Challenges and Conclusions

It is evident from the foregoing discussion that there are some land policy challenges for inclusive growth in India. These include (i) improvement in the rural Poor's land access, (ii) protection of land and forest rights of tribals, (iii) provision of legal and institutional framework for adequate and secure land rights for women, and (iv) formulation and implementation of farmer friendly land acquisition policy and preparation of accurate records of land rights and their updation and digitization on regular basis. It should indeed be a cause for concern that: despite various measures of land policy reforms undertaken since independence, the agrarian structure in India continues to be unequal, though it is not as exploitative and unproductive, as before. About 10 per cent of rural households in the country are landless, while nearly 70 per cent farm holdings are marginal, having an average size of less than 0.40 hectare (Govt. of India: 2006). These landless and semi-landless households are also the poorest of the poor in rural India. In the absence of proper education and skill they also do not have much scope for upward mobility through participation in off-farm and non-farm sources of employment and income in the organized sector. Therefore, improving their access to land, along with provisions of irrigation water, credit, etc. would be crucial for their food and livelihood security, at least in the short to medium run. This poses a challenge because most state governments do not seem to have the necessary political will or administrative infrastructure to enforce radical redistribution of land through effective implementation of ceiling laws. Under the present circumstances, we have following options:

Box 2.2: Key Features of Selected New Land Redistribution Initiatives

(i) Karnataka	The state of Karnataka launched a scheme, called Namma Bhoomi – Namma Thota (My land – My Garden) in 2005-06, which provided for purchase of land and allocation of 4500 sq feet of dry land or 2250 sq ft. of wet land for a family which is landless for two generations. The main objective was to improve the livelihoods of landless poor agricultural labourers through the allocation of small plots. Title to the land was to be granted jointly in the name of husband and wife or jointly in the names of the members of an eligible self help group. About 50,000 landless families obtained plots under the scheme. Although the scheme has been discontinued for no valid reasons, there is now effort to ensure convergence between Indira Awas Yojana (IAY) and Namma Bhoomi – Namma Thota (NBNT) so that beneficiaries of NBNT could obtain house construction benefit under IAY. The field observations by RDI indicate that allocation of small plots under NBNT has helped the beneficiaries to improve their food security and social status, the state government should consider reviving the scheme.
(ii) West Bengal	In 2006, the Government of West Bengal initiated a scheme, called Chas o Basobaser Bhumidan Pariklpa (Cultivation and Dwelling Plot Allocation Scheme) with the aim of allotting small plots of land to rural landless and houseless families by purchasing land from open market. The scheme provides for 0.04 acre of homestead plot to all rural BPL households without houses and 0.16 acre of cultivable land to landless agricultural labourers. So far, about 60,000 beneficiaries have been identified under the scheme. The new Trinomul Congress led government has partially modified the scheme and given it a new name 'Nijo Griho – Nijo Bhumi.' Because of its popularity among the landless poor in places where it has been implemented, the Government should scale up such scheme which has tremendous potential for improving food and nutritional security of the landless households as well as poverty reduction rate in West Bengal.
(iii) Andhra Pradesh	In 2002, the state of Andhra Pradesh launched Indira Karanthi Patham Project (IKP) which subsidized the purchase of about 4539.24 acres of land by 5303 women. However, due to increase in land prices, the land purchase component of IKP has been discontinued with effect from 2010. Further, the Rural Development Institute, Bangalore has been assisting the Govt. of Andhra Pradesh to secure land rights to the poor, using the services of paralegals and community resource/persons in a low cost and effective manner. It has a target of ensuring secured land rights to 60,000 families by March 2012.
(iv) Odisha	In 2004, the Govt. of Odisha launched Vasundhara scheme, the objective of which was to provide homestead plot of at least 4 cents in size to all the homesteadless rural families in Odisha, in about 3 years period. So far, the state has identified 2,50,000 beneficiaries under the scheme. Recently the Govt, has partially modified the scheme to opt for land purchase where government land is not available for allocation and also increase the size of plot from 4 cents to 10 cents, where feasible. Rural Development Institute, Bangalore is assisting the Government to improve homestead land allocation by providing additional capacity to revenue inspectors, using Community Resource Persons (CRPs). The CRP model was piloted in 36 villages and now scaled in 2500 villages.

Source: Rural Development Institute, Bangalore and Respective State governments.

First, government should consider purchasing land from the open market and allocate at least 10 to 15 cents of land to all the 15 million or so landless families in the name of women or jointly in the name of wife and husband, with no expected resistance from anyone. This will enable a beneficiary family to construct a house if she/he does not have one and also use the additional space for taking up some economic activities such as cultivation of vegetables, fruit trees, goat rearing, poultry, etc, thereby enhancing her/his family's food and nutritional security as well as social status.

Second, the new guidelines of the Indira Awas Yojana which now includes the concept of land purchase for house construction for eligible rural landless families should be modified to provide at least 10 cents of land and not just 3 cents or so for this purpose. A guaranteed access to even homestead plots of 10 cents in size would create a sense of livelihood security for the landless poor.

Third, land leasing should be legalized within ceiling limits in all states to improve the rural poors' access to land through growth of land lease market. Currently, land leasing is either banned or restricted in most states, even though informal, albeit insecure tenancies exist to some extent. Currently, about 11.5 per cent rural households in India lease in land which accounts for 7.1 per cent of the total area. (Govt. of India: 2006). The majority of the informal tenants comprise landless labourers (35.8 per cent), marginal farmers (55.7 per cent) and small farmers (4.7 per cent). The existing tenancy laws discourage landowners to lease out land and undertake diversified economic activities for upward mobility and also increase the rural poor's accessibility to land through leasing. Another important reason why land leasing would be a better option is because rising land prices tend to curtail the small farmers and landless workers' ability to purchase land.

Fourth, there should be special land tribunal to expeditiously dispose of all ceiling surplus land related cases and distribute nearly 2.0 million acres of land which is presently under litigation. Similarly, in some states, there is some identified cultivable *bhoodan* land available which can be distributed on priority basis.

Fifth, there is a huge amount of government land which is encroached by both poor and rich people. While the encroachment by the landless and semi-landless poor households can be regularized at least for the purposes of homestead plot, encroachment by the rich as well as the land mafia should be freed for redistribution among the eligible landless and semi-land poor families.

Regarding protection of land and forest rights of tribals, the measures that may help improve the situation would include:

(i) Special drive by all the concerned state governments to complete records of rights in all the scheduled and other areas in a time-bound manner with active participation of Gram Panchayat and Gram Sabha;

(ii) Ban on any transfer of tribal land to non-tribals, through sale, lease or mortgage etc., and not even with permission by any so-called competent authority on any ground whatsoever. In case it becomes necessary for any tribal land owners to sell land due to unforeseen adverse circumstances, the government should purchase it at the existing market price, and allocate it to a landless tribal family.

(iii) Complete ban on land acquisition in scheduled areas by government for private entities for industry, mining, real estate, etc. Land acquisition may be permitted with concurrence of Gram Sabha only for genuine public purpose such as roads, public hospitals, educational institutions, etc. along with provisions for alternative homestead plot and fair compensation.

(iv) Special drive for the detection and restoration of land belonging to Scheduled Tribes in a time-bound manner.

(v) Effective implementation of PESA, 1996 and Indian Forest Rights Act, 2006 for conferring real ownership and utilization of rights over forest resources by the tribals and other forest dwellers, for which massive legal aid and awareness-building campaign among tribals would have to be initiated.

There are several legal, administrative and socio-cultural barriers to women's effective access to equal and secure land rights. The first and foremost challenge in this respect is to create awareness among women about their rights as well as overall social awareness and mobilization across the country on the need for women's equal land rights. The Hindu Succession Act (Amendment), 2005 provides for equal rights to men and women in the matter of inheritance of all immovable properties. But this remains ineffective because of lack of awareness and social mobilization. Secondly, there should be a national law, empowering women of all religions, castes and regions which provides equal ownership and inheritance rights to women over agricultural and non-agricultural land, in respect of both parental and husband's land. Third, there could be adequate incentive for any family to register land in the name of women if stamp duty in such cases is either substantially reduced or abolished.

While the proposed Land Acquisition Bill, 2011 rightly provides that irrigated land capable of producing multiple crops will not be acquired under most circumstances, this provision should be extended to all other categories of fertile land as in coastal or Indo-gangetic plain regions. Second, the term 'public purpose' should be properly defined to include only genuine public utilities. Third, land for land should be given as compensation, where feasible, as most farmers may not have the skills to take up non-farm enterprises. At least allocation of homestead plots of 10 to 15 cents to each unsettled family would have to be ensured.

Finally, it is important to understand and recognize that the preparation of accurate records of land rights and their updation on regular basis is necessary not only for land security of farmers, but also for effective implementation of land reforms and overall agricultural development. It particularly helps the small landowners to access credit and other support services for agriculture. It also facilitates the development of land market as well as private capital investment in agriculture. However, the methods of traditional surveys, resurveys and records may not be of much help, as these are time consuming and also not very cost-effective. Therefore, digitization of land record and village maps coupled with verification of ground truth in a time-bound manner with participation of Gram Sabha would hold the key for the future. Ensuing transparency in verification and accuracy of the data are the real challenges. Second, any transfer of land through registered deeds should be simultaneously mutated on time. Third, the establishment of a system of conclusive property titles in place of mere registration of deeds poses a real challenge, for which the revenue department of all states needs to be strengthened and modernized.

References

Food and Agriculture Organisation (2006): Final Declaration at International Conference on Agrarian Reform and Rural Development, Proto Alegre, Brazil, March.

Government of India (2009): Report of the Committee on State Agrarian Relations and the Unfinished Task in Land Reforms, Ministry of Rural Development, New Delhi.

Government of India (2007): Annual Report, Ministry of Rural Development, New Delhi.

Government of India (2007): *Eleventh Five Year Plan*, Vol. III, Planning Commission, New Delhi.

Government of India (2006): Reports on Household Ownership Holdings and Operational Land Holdings in India, NSSO, 59th Round, New Delhi.

Government of India (2004): Reports of the Expert Group on Prevention of Alienation of Tribal Land and Its Restoration, Ministry of Rural Development, New Delhi.

Government of India (1992): Report of the Revenue Ministers Conference, Ministry of Rural Development, New Delhi.

Government of India (1948): Report of the Sub-committee on Land Policy, Agricultural Labour and Insurance, National Planning Committee, New Delhi.

Government of India (1945): Famine Inquiry Commission, Report on Bengal.

Haque, T. and A.S. Sirohi (1986): *Agrarian Reforms and Institutional Changes in India*, Concept Publishing Company, New Delhi.

Saxena, N.C. (2005): "*Updating Land Records: Is Computerisation Sufficient*?" *Economic and Political Weekly*, January 22, Mumbai.

The World Bank (2007): *Land Policies for Growth and Poverty Reduction*, Oxford University Press, New Delhi.

Tim, Hanstad, T. Haque and Robin Nielson (2008): Improving Land Access for India's Rural Poor, *Economic and Political Weekly*, Mumbai, Vol. XLII, No. 10.

3

Small Homegarden Plots and Sustainable Livelihoods for the Poor

Robert Mitchell and *Tim Hanstad*

1. Introduction

This chapter examines ways in which the poor can use small amounts of land to establish homegardens to advance important livelihood objectives. The chapter considers the potential benefits of homegardens in light of policy, financial and cultural constraints, and provides a framework for planners to consider whether (and which) homegarden interventions are appropriate for improving livelihoods of the poor.

In many settings in the developing world, lack of access to land may be the most critical obstacle to creating homegardens useful to the poor. Land is a critical resource for most families in the world. Land has inherent value, and land can be a critical component in the creation of value. Even relatively small plots of land substantially supplement the physical, financial and nutritional security of poor households. Land is an important basis for identity and status of individuals within a family and of families within a community. Land can also be the foundation for political power (Agarwal: 1994, Deere and Leon: 2001, Mearns: 1999). As such, programmes that allocate small plots of land for homegardens can provide benefits far beyond those derived directly from the homegardens.[1]

The chapter is divided into three main sections. Within the context of the sustainable livelihoods analysis, the second section of the chapter examines the benefits that homegardening may bring to the poor. The third section discusses various factors bearing upon the establishment of new homegardens and the improvement or expansion of existing homegardens. The fourth section of the chapter discusses policy and programme considerations from the perspective of planners evaluating various strategies for using homegardens to address sustainable livelihoods objectives.

Sustainable Livelihoods

Sustainable livelihoods analysis provides a framework for examining the significant role homegardens and homegarden plots play in the livelihoods of the poor. The sustainable livelihoods approach focuses on the capabilities of people, and highlights interrelationships among people and the assets they develop and on which they rely. The analytical process focuses on the strengths of people and their aspirations as they pursue their livelihood objectives.

> A livelihood comprises the capabilities, assets, and activities required for a means of living. A livelihood is sustainable when it can cope with and recover from stresses and shocks and maintain or enhance its capabilities and assets both now and in the future, while not undermining the natural resource base (Carney: 1998: 4).

The sustainable livelihoods approach seeks to increase the sustainability of the lives of the poor by promoting six core objectives: (1) more secure access to, and better management of, natural resources; (2) more secure access to financial resources; (3) a policy and institutional environment that supports multiple livelihood strategies and promotes equitable access to competitive markets; (4) better nutrition and health; improved access to high quality education, information, technologies, and training; (5) a more supportive and cohesive social environment; and (6) better access to basic and facilitating infrastructure (DFID: 2001)[2]

Within this framework, homegardens, and particularly homegarden plots, can play many roles. Secure access to a homegarden plot can be a livelihoods objective. Homegardens are also a natural asset through which other livelihood objectives, such as gender equality and sustainable use of resources, may be achieved. In addition, land can be a route or opportunity through which a multitude of other assets become accessible to the household (Baurnan: 2002). Access to land sufficient to establish a homegarden can enable a household to produce foods for consumption or trade. Skills learned in production increase the family's human assets. Consumed foods improve the family's nutritional status and food security. Sale of foods improve the family's financial status. Trade, exchange of information and cooperation with other villagers strengthens the family's relationships with others.

The sustainable livelihoods approach also recognizes that policies, institutions, and processes influence access to and use of assets, which ultimately impacts livelihoods. For example, land law, government land policies, government and customary institutions, as well as factor and product markets will impact whether a family can access and productively use a plot of land that is large enough and otherwise suitable to create a homegarden.

Homegardening in World Practice

Homegardens have usefully been defined as "a small scale, supplementary food production system by and for household members that mimics the natural, multi-layered ecosystem." (Hoogerbrugge and Fresco: 1993). Homegardens appear to have developed independently in the Indian subcontinent, Indonesia and other parts of Southeast Asia, the tropical Pacific islands, the Caribbean, and various parts of tropical Latin America and Africa (Brownrigg: 1985, Landauer and Brazil: 1990), and "can be found in almost all tropical and subtropical ecozones where subsistence land-use systems predominate" (Nair 1993: 86). Temperate climate homegardens were important for many years in the USSR and continue to provide an important safety net for families in the successor nations of the region (e.g., tho Seeth and others: 1998).

Published analyses of homegardens generally refer to four identifying characteristics (Brownrigg: 1985). First, the garden is located near the residence. Second, the garden contains a high diversity of plants. To this criterion some add that the garden recycles nutrients in a sustainable manner, that plants are planted densely, and that plants are layered to mimic natural forest. Third, garden production is a supplemental rather than a main source of family consumption or income. Fourth, the garden occupies a "small" area.

The issue of homegarden size is particularly important for our analysis. Although most commentators identify homegardens as occupying "small" plots, this criterion is applied to a wide range of plot sizes, varying from a few square metres to more than one hectare.[3] Because the purpose of this chapter is to examine the ways in which poor households can obtain and beneficially use homegarden plots, we focus here on homegarden plots falling on the smaller end of the range. If limited public resources are to be used to provide the poor with land for gardening, the size of parcels distributed will determine the number of households benefited.

A fifth distinguishing characteristic of homegardens that is important to our analysis is offered by Marsh (1998), who states that homegardens are a production system that the poor can easily enter at some level since it may be done with virtually no economic resources, using locally available planting materials, natural manures and indigenous methods of pest control. To the extent a poor family can afford to make beneficial use of homegarden plots, the plots are more likely to make a sustainable contribution to the family's livelihood objectives.

2. Benefits of Homegardens

Homegardens are believed to provide a number of benefits to families, ranging from improving nutrition and providing a source for additional household income, to improving the status of women in the household. Potential environmental benefits of homegardens may be important not only for homegardening households, but for the broader society as well. Where the creation of homegardens includes distribution of land to otherwise landless and land poor families, families are believed to benefit in several additional important ways, including improved family status and improved bargaining power over wage income (see Box 3.1). This section surveys the various ways in which homegardens may contribute to improve the livelihood of poor families.

(a) Plantings and Family Health

It has been established that even moderate and mild energy malnutrition contributes to child mortality, and micronutrient deficiencies are associated with increased risk of child and maternal mortality (Kiess: 2001). It is estimated that, worldwide, 53 per cent of the approximately 10 million child deaths every year can be attributed to malnutrition (Black and others: 2003). Vitamin A deficiency is the leading cause of preventable childhood blindness, and is associated with an increased risk of mortality and an increased severity of infectious diseases livelihoods framework.

Homegardens are one strategy for addressing malnutrition and micronutrient deficiencies. Even though animal products are the best sources of micronutrients, vegetables and fruits may be the only source of micronutrients that are reliably available to poor households (Talukder and others: 2000, Bloem and others: 1998, Reddy: 1995). A number of studies have reported that homegardens produce a high percentage of fruits and vegetables consumed by homegardening families.[4] Although it is relatively straightforward to determine whether a homegardening programme has increased production and consumption of fruits and vegetables, it is not a simple matter to determine the impact of homegardening on nutritional status (HKI/AP: 2003).[5] Nevertheless, a number of studies have concluded that homegardens are associated with better household nutrition.

A large-scale homegardening project implemented by Heller Keller International in Bangladesh found that families who grew more fruits and vegetables and families who grew a larger variety of fruits and vegetables were likely have a higher intake of vitamin A (HKI/AP: 2001). A study of homegarden consumption in rural Bangladesh found that fruits and vegetables were the most important factor associated with higher intake of vitamin A by women of reproductive age, that consumption of fruits and vegetables contributed more to vitamin A intake than consumption of animal products, and that the number of varieties of fruits and vegetables produced in the homegardens was significantly associated with a higher vitamin A intake (Bloem: 1996). The fact that the highest intake of vitamin A was associated with homegardens that were divided into several scattered plots, even though such scattered plots were cultivated by the poorest families, may suggest a tendency of the poorest households to make the most efficient use of the limited resources available to them (Bloem: 1996).

Food-based approaches to combating vitamin A deficiency disorders have several advantages over other strategies, such as distribution of vitamin tablets: (1) homegardening programmes allow benefits to reach everyone in the family, not just young children or some other particular group; (2) homegardening programmes can be sustained by households and communities, reducing realiance on outside agents; and (3) homegarden fruits and vegetables can provide other nutritional benefits, helping to prevent degenerative diseases and mortality (HKI/AP: 2003). A successful homegardening strategy places the poor at the Centre of decision making over how best to satisfy their livelihood objectives, and makes them less reliant on outside forces.

In addition to their usefulness in combating vitamin A deficiency, homegardens are associated with a number of other nutritional benefits, some of which have tended to be overlooked (Marsh: 1998). Homegardens have been reported to provide 18-40 per cent of household calories on Java (Christanty: 1981 and Stoler: 1978, cited in Hoogerbrugge and

Fresco: 1993) and 50-58 per cent of the recommended daily allowance for calories in the Philippines (Sommers: 1978, cited in Christanty: 1990).[6] In Lima, women who must feed a large household are reported to value production of starchy crops such as sweet potato, taro and maize, which have a high energy value and high "filling" quality (Ninez: 1985). This comports with the findings of Hoogerbrugge and Fresco (1993), who stated that homegardens primarily produce energy foods rather than vitamins, and that vegetables and fruits play a relatively minor role in classic homegardens; they suggested that it is a mistake to promote homegardening as vegetable and fruit gardens alone.

Homegardens can provide important protection against family food insecurity. On Java, climatic conditions mean that owners of homegardens have something available to harvest throughout the year, either for consumption, for home industry or for sale, and this availability is especially important to the economic stability of poor households, particularly during the period between rice harvests (Soemarwoto: 1985). A detailed study of Russian households concludes that for the very poorest households, which account for approximately 20 per cent of all Russian households, "gardening is absolutely necessary, serving as insurance against food insecurity" (tho Seeth and others 1998: 1621). Homegardens may become the principal source of household food and income during periods of stress, as in Kampala, Uganda after the civil war, where urban agriculture is reported to have substantially fed the city (Marsh: 1998). In the context of a study of land-poor households in Kerala, India, homegardening production has been observed to have a "buffering effect" on household consumption when there are shortfalls in wage income (Kumar: 1978).

(b) Animals and Family Health

With the increasing awareness among nutrition experts that fruits and vegetables contribute less to improved vitamin A status than previously assumed, the focus of homegarden programme has begun shifting to include analysis of the benefits of animal husbandry, poultry and fishponds (HKI/AP: 2001).[7] Animal husbandry extension projects underway in Nepal, Cambodia (poultry and eggs) and Bangladesh (poultry, eggs and fish) promote household egg production by introducing improved breeds of birds that produce more eggs, along with vaccinations and assistance with proper housing and feed, promote improved grass fodder and deworming tablets for milk cows, and are introducing fast growing fish cultivars and plant sources of fish feed (HKI/AP: 2003).

Homegardens on Java have been reported to provide 14 per cent of household protein requirements (Hoogerbrugge and Fresco: 1993, citing Christanty: 1981). A study of homegardens in Ghana found that households could potentially produce substantial amounts of meat and related income per year (Asare and others: 1990).[8] Certainly in India it is very common to combine gardening with poultry and livestock on the homegarden plot. Typically the household ties up, fences in or keeps the animals in a shed located on the plot. The household uses manure as fertilizer for the garden and as fuel source. In Javanese homegardens, animals are not confined and receive only minimal feeding—chicken range freely and eat leftovers from the kitchen and "whatever they can find in the garden", while buffalo, cows, goats and sheep graze on village common lands and are fed additional food at night from grasses cut from dykes of rice fields and other areas (Soemarwoto: 1985).

To date there is very little written regarding dietary intake and nutritional status of children in households that integrate production of fish, small animals and vegetables (Schipani: 2002). However, a survey in Bangladesh found that, even after accounting for household socio-economic status, young children in households that raised chickens in the homegarden had the lowest incidence of nightblindness (as compared to households not raising poultry, or raising poultry without a homegarden) (HKI/AP: 2001, citing Kiess: 1998).[9]

Families who raise animals use the homegarden plot as a place to keep the animals, either throughout the day, or only at night after the animals return or are brought back from foraging. The homegarden thus serves not primarily as a source of fodder for animals (and may not be the main source of fodder for most households), but is a place for keeping animals. A study of homegardening households in Karnataka, India found that 93 per cent of households who had livestock kept their livestock on their homegarden plot exclusively, while a similar study of homegardening households in West Bengal found that 90 per cent of respondents with livestock kept them on the homegarden plot at least part of the time (Hanstad: 2004). These percentages would be even higher if poultry were included.

(c) Household Income

The livelihood benefits of homegardens go well beyond those related to nutrition and subsistence. In many cases, the sale of products produced on homegardens significantly improves the family's financial status.

It is a common misconception that homegardens are exclusively subsistence oriented, whereas in fact homegardens provide households with cash crops as well as food crops (Hoogerbrugge and Fresco: 1993). In fact, returns to land and labour are often higher for homegardens than for field agriculture (Marsh: 1998). Homegardens can contribute to household income in several ways. The household may sell products produced in the homegarden, including fruits, vegetables, animal products and other valuable materials such as bamboo and wood for construction or fuel. The household may use the homegarden site to conduct cottage industries to produce crafts or small manufactures that can be sold (Marsh 1998).

The volume of homegarden production actually sold appears to be highly variable, with studies reporting that between 9 per cent and 51 per cent of production is sold (Hoogerbrugge and Fresco: 1993). Urban homegardens in Papua New Guinea are reported to have expanded over a period of several years (in one neighbourhood from a mean area of 125 m^2 in 1974 to a mean area of 817 m^2 in 1981), which the households attributed to opportunities to sell homegarden produce in local markets (Vasey: 1985). In the studied sample, 42.1 per cent of squatter households sold homegarden produce at the market, as compared to a district average of 25.8 per cent, indicating that the more impoverished households depended on homegardens more than the average (Vasey 1985).

Livestock and tree crops produced on homegardens in south-eastern Nigeria accounted for over 60 per cent of family cash income in one study (Okigbo: 1990). A study of urban and rural households in three Russian provinces found that two-thirds of all households obtained some income from agricultural home production, and in rural areas the market value of home production (computed using average local prices reduced by calculated market transaction costs) exceeds household labour income (tho Seeth and others 1998).[10] In the Helen Keller International (HKI) pilot homegarden project in Bangladesh, 54 per cent of households reported selling homegarden products and earning the cash equivalent of 14.8 per cent of total average monthly income (HKI/AP: 2003).

In addition to direct earnings from sale of homegarden production, production consumed by the household free up household earnings for other purchases. In the Bangladesh HKI homegarden project, the income value of homegarden production increased from 14 per cent of average monthly income to 25 per cent after taking into account purchased fruits and vegetables (Marsh: 1998).

In some cases, a portion of the cash income from homegardens is used to purchase additional food for household consumption. A study of urban homegardens in the Philippines revealed that homegardening families spend less on food than non-gardening families, while homegardening families who plant a larger number of varieties of fruits and vegetables spend even less (Miura: 2003).

In Cambodia and Nepal, 31-65 per cent of income (31 per cent in the case of Nepal and 65 per cent in the case of Cambodia) derived from sale of poultry raised on homegardens was used to purchase other foods, while other proceeds were used to invest in production, education, savings and other purposes (HKI: 2003). Urban homegardeners in Papua New Guinea sell various fruits at local markets and obtain cash that allows them to purchase rice that produces several times the food energy of the sold fruits (Vasey: 1985). Thus, homegardens provide households with a number of options by which they can satisfy their livelihood objectives, and each household can determine for itself what combination of consumption, trade and sale of homegarden production best fits its livelihood strategy.

(d) Wage Security and Household Status

Ownership of the homegarden plot can make important contributions to improved and sustainable livelihoods in ways that often overlooked, including improved leverage in labour markets, enhanced social status and greater political participation.

One objective to providing ownership to house and garden plots in some settings is to free agricultural labourers (or other wage labourers) from exploitive land-labour linkages. In the 1940s, the Puerto Rican government distributed small homestead plots of between half an acre and one acre to the families of agricultural labourers. The idea behind the law was, in the words of Section 241 of the law, that it is a

> "fundamental human right of all the human beings who live exclusively by the tilling of the soil, to be the owners of at least a piece of land which they may use to erect thereon… their own homes, thereby delivering them from coercion and leaving them free to sell their labour through fair and equitable bargaining." [28 L.P.R.A. (1955), quoted in Rosenn 1963: 344].[11]

Roughly 50,000 families are reported to have received homestead plots under the law, which is further reported to have provided a degree of "peace and spiritual satisfaction" to the families (Pico: 1964). Access to small plots of land also allowed the agricultural labourers to participate in elections without selling their votes to the landlord.[12] Several states in India have provided ownership of (typically) small house plots to agricultural labourer families in order to remove them from feudal-type dependence on employers on whose land they had been living.

Households who own the plot on which their home is constructed also enjoy an immediate increase in status within the village. A survey of rural households in Karnataka, India revealed that among households that had received small plots to construct homes and gardens, increased status within the village was the most cited benefit of ownership, surpassing even income and nutrition benefits; poorer households cited increased status even more often than other households (Hanstad and others: 2002). In West Java, homegardens are an important symbol of social status, and households who are forced to build their house on homegardens owned by others are considered to be of low status (Ahmad and others 1980, cited in Soemarwoto (1987). Increases in household status not only provide psychological benefits to household members, but are believed to provide households with better access to trade relations within the village, as well as better access to government programmes serving village households.

Describing the role of homegardens in the Saraguro community of Ecuador, homegardens were observed to "make a contribution far greater than that to diet, ritual life and remedy; the gardens are themselves a manifest representation of the community's most deeply held values: autonomy, status, religious piety, and personal investment in family" (Finerman and Sackett 2003: 477). In Java, urban homegardens function as a status symbol and expression of the owner's self-image and aesthetic sense, while unfenced rural homegardens can provide a social place for neighbours to socialize and children to play (Christanty: 1990). Homegardens are thus seen to contribute to a cohesive social environment, another core objective of the livelihoods approach.

(e) Benefits to Women

Although women's labour constitutes an important input, it is a mistake to conclude that homegardening is primarily a female activity (Hoogerbrugge and Fresco: 1993). In fact, the role of women varies widely among cultures. In Africa most homegardening tasks seem to be performed by women, in Sri Lanka women provide labour only at peak times, in Indonesia men prepare the land, cultivate tree crops and market

Box 3.1: Benefits of owning a house plot.

For otherwise landless families, ownership of a plot used and establishment of a homegarden can provide beyond those derived directly from the homegarden itself: for construction of a house numerous livelihood benefits

1. *Place for residence.* Although perhaps the most obvious, this benefit should not be overlooked when millions of households lack secure rights to land for a house. Secure legal rights to the plot also provide the family with proper incentives to construct a quality house and make other long-term improvements to the plot.
2. *Status.* Studies in India indicate that recipients of government-allocated house plots cite increased status as the most important benefit from the plot (more important even than increased income and food consumption) (Hanstad and others: 2002).
3. *Wealth generation.* House plots and occupying structures are typically the most important source of wealth of poor households. As these poor households build and improve their house, build other structures (cattle sheds, fences, wells, etc.), plant trees, and make other labour-intensive improvements to their plots, they create wealth for themselves. Households who lack secure rights to sufficiently-sized house plots are constrained in developing their asset portfolio.
4. *Bargaining leverage in labour markets.* Agricultural labourers who do not own their own house site often rely upon their employers for a place to live. This often creates a dependency relationship that severely limits the labourers' bargaining leverage for wages. We interviewed a group of landless women of Madhya Pradesh state in India who had been living on their landlord's land for decades. Although they did not pay rent, the landlord paid them only 50% of market wage rates, did not allow them to work for other farmers, and at times even prevented them from leaving or entering their homes.
5. *Post-harvest activities and storage.* In many settings, the homegarden plot is the site for important post-harvest activities such as drying and threshing. The plots also typically provide space for storing food, tools and other capital assets.
6. *Non-agricultural income generation activities.* Owning a homegarden plot with some extra space can enable poor households to pursue other non-agricultural production, service or retailing activities such as handicraft production, blacksmithing or petty shops.
7. *Access to credit.* In a study of government-allocated house-and-garden plots in Karnataka, India, more than one-third of respondents reported that obtaining the plot had increased their access to credit and nearly one quarter reported actually receiving credit as a result of owning the plot (Hanstad and others: 2002).

homegarden production, while women and children cultivate annual crops (Hoogerbrugge and Fresco: 1993). Women do most of the work in Bangladeshi homegardens (Talukder and others: 2000), as well as on Russian home production plots, including working the land, planting, weeding and carrying irrigation water, which requires considerable effort and time (tho Seeth and others: 1998). However, in traditional settings, any member of the household can be found in the garden, and adult men and women, children and the elderly often have specialized roles in homegardening (Brownrigg: 1985).

The role of women in homegardening may be socially determined, as is apparently the case among the Saraguro people of Ecuador, where homegardens are the exclusive domain of women and women decide what to plant and when to harvest, decide how to exploit homegarden resources, make decisions on sales of production and perform daily maintenance (Finerman and Sackett: 2003). Or the role of women may be a function of other factors, as in Papua New Guinea, where although women were identified as 61.9 per cent of gardeners and 66.7 per cent of principal gardeners in urban areas (Vasey: 1985), this fact was attributed to household composition and the secondary status of women in the workforce rather than gender (Vasey 1985: 1990).

Even where women play a primary role in homegardening, it can be important to involve the entire family in projects to promote homegardening, especially in cultures where women have little contact with outsiders and may hesitate to become involved in projects without the approval of their husbands (Marsh: 1998).

Sales of homegarden produce may be one of the only sources of independent income for women, and such sales may be an important source of income for women (Marsh: 1998). Finerman and Sackett (2003) report— that having an abundant homegarden is an important source of status for Saraguro women, and conclude that homegardens demonstrate the woman's freedom from dependence on vendors and neighbours, her ability to expend resources on developing the garden demonstrate her fiscal standing, her production of flowers to adorn the church demonstrate her piety, and her investment in cultivation demonstrates her devotion to family.

Where women control homegarden resources, this may improve household nutrition, especially nutrition of children (Kumar: 1978, Talukder and others: 2000). In the urban homegardens of Lima, whereas men gardeners typically are interested in producing crops that have a high market value, women gardeners tend to want to produce food for family consumption (Ninez: 1985).

In at least some cultural settings, education and information may play an important role in determining the degree to which women control homegarden production. In a study of an HKI pilot homegarden project in Bangladesh researchers found in participating households, 65 per cent of the time women made decisions about distribution of garden produce for consumption or sale, as compared to 25 per cent of the time in non-participating households, and women in participating households received and controlled the income from sales of garden production 67 per cent of the time, as compared to 31 per cent of the time in non-participating households (Marsh: 1998).

(f) Environmental Benefits

Diversity of plant species and the layered canopy of species are the most striking features of homegardens, with all homegardens generally consisting of "a herbaceous layer near the ground, a tree layer at upper levels, and intermediate layers in between" (Nair: 1993: 91). Plant diversity seems to decrease with altitude, length of dry season, share of cash crops, population density, labour shortage within the household and distance to urban areas (Hoogerbrugge and Fresco: 1993). Traditional Thai homegardens are reported to contain multiple and sometimes rare varieties of each planted species and represented "insitu reservoirs for biodiversity at all levels: genetic, species, and ecological", all of which helps to prevent pest and weed outbreaks (Gajaseni and Gajaseni: 1999: 19). The high density of homegarden plants also provides habitat for wild animals such as insects, reptiles, birds and small mammals (Christanty: 1980). Describing the diversity of Indonesian homegardens (here called "mixed gardens") in the Nineteenth Century, Sollewijn Gelpke observed:

> "He who enters a mixed garden... with a botanical eye, sees before him a diversity of plants of which the uninitiated can form no idea. From the greatest inhabited heights to the shores of the sea, on clay and sand, in marshes and on dry land, the whole wealth of the tropics is laid open. That wealth of vegetation is all the more striking when the observer regards it from an economic point of view. He sees palms, bamboos, bananas and a number of fruit trees, all seemingly much alike and with various winding plants clinging to them." (J.H.F. Sollewijn Gelpke: 1901, quoted in *G.J.A.* Terra: 1954: 33).

Nutrient recycling is the principal determinant for ecological rationality of homegardens (Gajaseni and Gajaseni: 1999). A detailed study of four traditional Thai homegardens found that the households refrained from harvesting everything that could be harvested, and that this ensured minimal nutrient export from the system. In addition, the Thai homegardens contained more plant litter than a typical tropical forest, which should contribute to a highly efficient nutrient recycling (Gajaseni and Gajaseni: 1999).

> "[A] village with its home gardens is not merely a dwelling-place but also an important agro-ecosystem. It is an integrated unit in which the solar energy is channelled through the plants to animals and man, and matter is cycled and recycled. This cycling and recycling process, together with the layered plant cover, protects the soil of the home garden from exhaustion, leaching, and soil erosion." (Soemarwoto: 1985: 2).

Not much is known regarding how homegardens cycle nutrients and conventional analytical and research procedures cannot adequately describe the functioning of homegardens; this leads many researchers to conclude that homegarden systems are unsustainable despite that fact that such systems appear to have persisted for long periods without apparent symptoms of soil nutrient depletion (Nair: 2001). More needs to be known regarding the contribution that homegardens make to improved management of natural resources.

Homegardens may be considered to improve or exacerbate public sanitation, depending upon the care with which household wastes are handled. In West Java, it is common for homegardens to contain fishponds. Fish are fed kitchen waste and the pond is fertilized by animal and human waste, including waste from toilets built above the fishpond (Soemarwoto: 1985). These households do not use the fishpond water for any household needs. Livestock waste is also used to manure the garden and other fields. Urban homegardens may improve public sanitation at virtually no cost to the larger community by using organic wastes from slaughterhouse manure, treated sewage sludge and wastes from fisheries and breweries (Vasey: 1990). However, mishandling of wastes, particularly human feces, may compromise sanitation (Vasey: 1990, Soemarwoto: 1987). The potential public sanitation benefits should be considered in any assessment of the benefits and costs of expanding the public water delivery system to accommodate homegardening.

Another potential impact of homegardening is land conservation. Terraced homegardens have been recommended to preserve soils on sloping areas (Terra: 1954). Fruit trees, bamboo and other trees can be used to rcjuvenate infertile soils. Tree roots that penetrate as far as 10 meters can bring mineral constituents into the topsoil, while fallen leaves can provide a natural protective mulching cover and bring more humus into the soil, helping to prevent exhaustion of soils (Terra: 1954). However, it is important for homegardening families not to remove ground litter or engage in excessive weeding of the homegarden, which can increase the risk of soil erosion (Soemarwoto: 1987).

Distribution of homegarden plots may also have beneficial off-site environmental effects. For example, where population pressures and lack of arable land threaten to push families to resettle in forests and wetlands, distribution of homegarden plots to landless and land poor families can reduce pressures to migrate. This not helps to reduce conversion of lands better left as forests and wetland, but also allows families to remain in areas with established social services and markets for surpluses produced on the homegarden. In addition, distribution of homegardens may reduce the need for land-poor families to gather fodder and fuelwood from marginal lands, contributing to the sustainability of such lands. See Mitchell (forthcoming).

3. Factors Related to Developing Productive Homegardens for the Poor

A number of factors combine to determine whether homegardens are an appropriate strategy for improving the livelihood of poor families. Among these, access to suitable land—i.e., a land plot that is large enough and sufficiently fertile for a family to establish a homegarden, and to which the family has ownership or ownership-like rights—is perhaps the most fundamental factor. The importance of access to land is occasionally mentioned in the homegarden literature, but does not appear to have been addressed in any detail. Where access to suitable land is not a constraint to establishing homegardens (or once planners have arranged to overcome the issue by providing secure access to land), other important factors become relevant, including access to water, access to know-how, and access to stocks of appropriate plants and animals. Cultural acceptance of homegardening and access to sufficient capital and labour are equally important. This section reviews these factors.

(a) Lack of Land

We are interested in how homegardens can benefit the poorest of the poor, which usually includes families without adequate access to land. A number of commentators have noted that lack of access to land is a serious constraint to homegardening (Hoogerbrugge and Fresco: 1993, Marsh: 1998, Vasey: 1985). Across cultures, ownership of land appears to be a significant inducement to gardening and landlessness and tenant status have been identified as constraints that effectively eliminate homegardening as a viable development strategy (Brownrigg: 1985). Indeed, for families without adequate and secure access to land, lack of land is the single most important barrier to homegardening.

> "Arrangement of adequate access to land and security of tenure should be a primary consideration in the design of projects which promote home or community gardens, as is true for other kinds of agricultural production development." (Brownrigg: 1985: 111).

The poor sometimes have adequate access to land, as for example in Russia and most other countries that emerged from the Soviet Union. In these countries access to land for homegardening has generally not been an issue, and all village families have access to plots used for homegardening. Such land is usually located immediately adjacent to the village house, and vegetable plots arranged on arable fields near the village. During the harvest season, owners take turn guarding the vegetable plots against theft.

However, inadequate access to land remains a significant problem for poor families worldwide, even in areas that commentators commonly associate with homegardens, such as Java. On crowded Java, the great majority of homegarden plots (known locally as "pekarangan") are smaller than 200 m^2 (Prosterman and Mitchell: 2002).[13]

Even where a family has nominal access to land, the insecurity of their rights to such land may dissuade the family from making any long-term investments in improving the land, such as by planting trees, improving drainage, installing fencing or building a fishpond. Squatters and families who occupy land merely with the permission of a landlord (often the head of household's employer) may even worry that their improvement of the land they occupy may lead to eviction as others seek to cash in on the investment. Thus, it is not only the quality and sufficiency of the land itself, but the quality of the family's right to control the land which are critical.

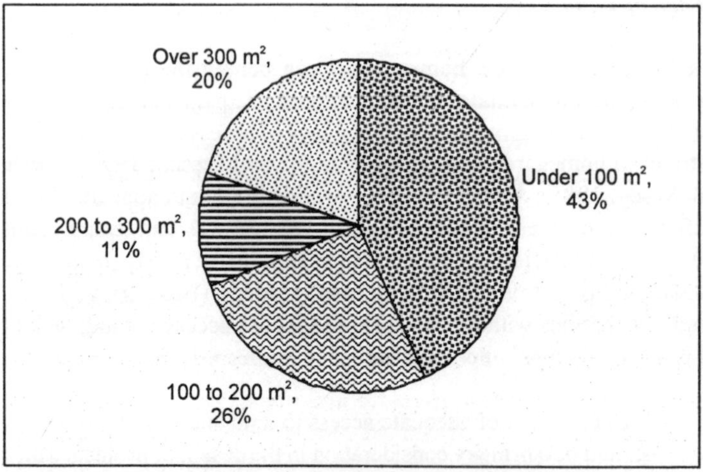

Fig. 3.1: Size of Household Pekarangan Plots on Java

Source: Computed using 1995 Housing and Settlement Statistics (Indonesia Statistic Centre Bureau, 1995), cited in Arifin (2002), App. 2.

If lack of access to land is a constraint in establishing a homegarden, how much land do families need to establish a homegarden? The size of homegardens varies considerably across cultures, and even within the same community. For example, Hoogerbrugge and Fresco (1993) survey a number of studies that report variations in size from 10 to 120 m^2 in one Zambian study to between 5,000 and 20,000 m^2 in another Zambian study and variations of 172 to 500 m^2 in one Java study to between 200 and 1700 m^2 in another Java study. In Papua New Guinea, houseplots of 300-400 m^2 are often too small to meet the ambitions of household gardeners, who extend their gardens beyond the allotment or establish second gardens away from the house (Vasey, 1985).

While the average size of existing homegarden plots is instructive regarding social and economic norms in a given community, there is evidence that distribution of even relatively small amounts of land to land poor families can provide a base for important improvements to the household nutrition. In a study of Kerala wage earning families who cultivate homegarden plots occupying a fraction of an acre, Kumar (1978) found that the value of homegarden production was the most consistent positive predictor of child nutrition, and was an especially strong predictor during the slack employment season, as well as in households in which the mother is not employed outside the home.[14] Kumar goes on to report:

> Results of the study indicate that the proceeds of the produce from even small family plots of land, if intensively cultivated, lead to large increments in child nutrition. This is especially true when maternal labour force participation is absent. Additionally, during slack employment seasons, availability of garden produce, or income thereof, seems to provide a buffer against reductions in child nutrition during this period. Even though the exact mechanisms by which this occurs are not clear, the conclusion is that provided intensive cultivation of distributed land is possible, even small family plots can have an important nutritional impact for young children. (Kumar: 1978: 60-61).[15]

The larger role of women in Kerala homegarden production, as well as the fact that women can more easily and better care for children at home, may contribute to the fact that better child nutrition is strongly associated with more productive homegardening (Kumar: 1978). Given the strong correlation between homegardening and the presence of mothers not engaged in wage labour, distribution of land to landless and land-poor families may be a particularly fruitful strategy where employment opportunities for women are limited for societal or economic reasons (Kumar: 1978).

In a study of 62 rural homegardening households in Karnataka, India, researchers found that intensity of tree growing increased markedly once homegarden plot size reached 1800 sq. ft. (See Fig. 3.2). Of 14 households that grew 5 or more trees, only one had a homegarden plot smaller than 1800 sq. ft., while of the 16 households with plots 1800 sq. ft. or larger, 13 households (72%) grew 5 or more trees. This suggests that there is likely to be a critical minimum plot size above which households will begin planting more trees. The footprint of the house itself, which commonly occupies 500 sq. ft. or more of the land plot, is presumed to reduce greatly the number of trees that can be planted on the smallest parcels.

In a separate study, researchers identified Karnataka households that appeared to making very intensive use of land. The researchers found that, after taking into account the size of the parcels, families who receive land from the government appear to be as likely to plant trees or raise animals as are families who inherit or purchase the homegarden plot. None of the interviewed families had received government extension advice or assistance with planting trees or raising livestock (Hanstad and others, forthcoming). This data suggests that at least for some poor families

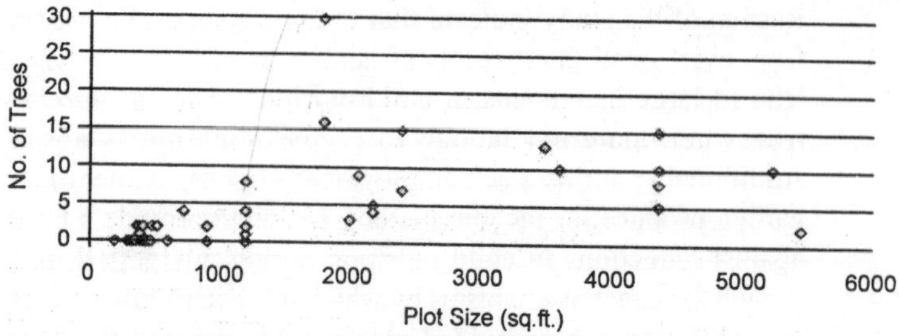

Fig. 3.2: Trees Planted by Karnataka Homegardening Households

Source: Hanstad and Lokesh 2003 (unpublished data).

in Karnataka, access to land is a primary barrier to tree planting and animal raising.

In a sample survey of 97 land poor households that had received houseplots under government housing schemes in West Bengal, researchers found that the productive value of plots increased significantly with plot size until plot size reached approximately 3000 sq. ft. (270 m^2) (Hanstad and Lokesh: 2002). More detailed interviews of 45 similarly situated West Bengal households revealed similar results. In the smaller sample, homegardens smaller than 1000 sq. ft. (90 m^2) were found to provide the fewest benefits. Productive value and reported benefits increased significantly for homegardens of 1000-1999 sq. ft. (90-180 m^2), and increased further for homegardens of 2000-2999 sq. ft. (180-270 m^2), but then plateaued or even decreased for plots larger than 3000 sq. ft. (270 m^2) (Hanstad and Lokesh: 2002).

In Papua New Guinea, one study suggests that an area of only 150 m^2 is sufficient to supply adequate fresh vegetables for a family of four, although another researcher argues that it would require 3,100 m^2 (Thaman 1990). In a study of Philippine urban homegardens, Miura (2003) found that size of homegarden plot did not affect the number of varieties of fruits and vegetables grown in the garden. Out of 103 households, 54 (52%) with plots larger than 100 m^2 had planted more than five varieties of fruits and vegetables, while 26 of 49 households (53%) with plots smaller than 100 m^2 had planted more than five varieties. Of course, the quantity of production of the larger and smaller plots is likely to differ.

Another important consideration is the cost of distributing small plots of land to land poor households. Assuming the government acquires land at market prices; can governments afford to distribute small parcels? Again, affordability cannot be considered in the abstract, but must be evaluated by looking at the cost and effectiveness of other programmes designed to address the needs of land poor families.

Hoogerbrugge and Fresco (1993) suggest that homegardens generally occupy "marginal" plots; that is, plots that are too small to be used for field cropping or grazing, or are on land that is too steeply sloped to be used for field cropping. Of course, to the extent that this is so, it is likely a consequence of locating rural housing, and the housing of the urban poor, on hillsides and other lands that are less suitable for field crops and urban development. To the extent that homegardens are cultivated by poor households, it is natural to suppose that the gardens and housing will be located on economically marginal land.

Box 3.2: Cost of Obtaining Land: Evidence from India

The cost of obtaining suitable land on which to allocate homegarden plots is likely to be affordable, especially in rural areas. In Karnataka, India, a November 2001 sample of 400 rural households in four districts estimated the value of unimproved and non-irrigated agricultural land to be between 21,000 and 44,000 Rupees per acre, with an average of 33,250 Rupees per acre (Hanstad and others 2002), which equates to approximately US$ 694 per acre (US$ 1,714 per hectare) at November 2001 exchange rates. This represents an average cost of approximately $86 per family benefited if each family receives 500 m^2 of land (20 families per hectare).[16] These estimates reflect the likely purchase price of acquiring agricultural land at market prices, but do not include administrative costs of acquisition and allocation, or costs of constructing simple roads and drainage. Although such costs are likely to be low in comparison to land acquisition costs, they are not negligible.

Another consideration is whether land is available near the families who need the land. Vasey (1985) describes experience in Papua New Guinea in which a government programme to allocate gardening land away from the house failed since it was too difficult for the cultivators to guard against theft and vandalism. In Indonesia, government planners express concern that in some cases, land located near rural villages on Java is prime rice paddy land, the terracing of which has been undertaken at great social cost. Strict policies are in place to prevent conversion of such land to other uses, though the conversion of rice paddy land to residential land continues to occur.

At least one state in India (Karnataka) is initiating a programme that will establish several-acre colonies of house-and-garden plots with 10 plots per acre, and allocate the plots to landless and land poor families.

The land will be located within one kilometer or less of the village in which the recipients currently reside, and such land will either be existing government land or land that the government purchases.

(b) Lack of Water

Although homegardens are primarily rainfed, it is common for homegardeners to irrigate during the dry season. Watering depends on the type of crop and can vary from twice daily to twice annually (Hoogerbrugge and Fresco: 1993). Several studies have found that drawing, transporting and hand irrigating the homegarden are the most onerous and time consuming gardening tasks. For example, Russian women expend considerable effort and time carrying water to irrigate home production gardens (tho Seeth and others: 1998), although Russian gardens are much larger on average than tropical home gardens and are thus likely to require much more water. Homegardens that require even a few gallons of water per day during the dry season may require too much labour to be worthwhile (Brownrigg: 1985).

In some areas, lack of water may be the major factor limiting homegardening (assuming the household has access to land). In Papua New Guinea, potable, piped water for irrigation is the most expensive input for urban homegardeners, and water is especially expensive during the dry season (Vasey 1985). In the United States, subsidies for water delivered to agriculture, including homegardening, is the most important government subsidy that agriculture receives (Brownrigg: 1985, citing Cleveland: 1982). Ninez (1985) reports that in Lima, Peru, urban homegardeners used waste water from the kitchen to irrigate their garden, but sometimes found they had to purchase small amounts of additional water. In 1983, the cost of water to irrigate a garden of 200 m^2 was approximately US$3.50 per month, but in the poorer neighbourhoods water was sometimes not available during the warmer months, and gardens suffered greatly.

It can be prohibitively expensive for the household to install a system for bringing water to the house, and there is a large social cost for providing irrigation water to households, especially in urban areas. Homegarden irrigation is rarely a consideration in the design of a reticulated water system, and it may be too expensive to make capital improvements to an existing system necessary to accommodate demand for irrigation water (Vasey: 1990). For urban homegardens in Papua New Guinea, use of unmetered water for irrigation during the dry season exceeded the value of the crops produced, while metered households found it uneconomic to use water for irrigation (Vasey: 1990).

However, even if water costs exceed the value of production, low cost irrigation water might be justified as an appropriate subsidy for poor neighbourhoods since it contributes to social safety net while reducing moral risks associated with government handouts. The cost-benefit analysis of subsidizing irrigation water for homegardens should not consider purely the economic value of homegarden production, but should take into account the broader social benefits of helping poor families to gain a degree of economic independence and the accompanying enhancement of self-image and social status. Development projects that bring potable water to the village could free up labour spent hauling water, allowing the household to devote saved labour to homegardening (Brownrigg: 1985).

Households may address the need for irrigation water using various strategies. In drought prone areas of the African Sahel, simple wells and artisanal irrigation are used. In West Bengal, rural communities are sometimes sited near a rainfed natural or artificial community pond that families can draw from for irrigation and other household water needs. Rainwater harvesting may be an affordable means of capturing, storing and applying water for homegardens (Agarwal and Narain: 1999). International Development Enterprises—India (2004) has developed and successfully marketed several low-technology, low-cost micro-irrigation systems in India that are appropriate for homegardens ranging in size from 20 to 1000 m^2 and cost from US$5 to US$90. Households may also use household wastewater (Marsh: 1998).

Water conservation strategies can reduce homegarden demand for water. Such strategies include terracing, trenching, deep mulch and surface mulch (including living mulch and ground cover creepers). Plant spacing and mulch may be used to conserve moisture (Vasey: 1985), and drought-tolerant plants can reduce the impact of water shortfalls (Marsh: 1998).

In areas of high rainfall, canopy layers, raised beds and drainage canals may help to prevent flooding (Brownrigg: 1985), while use of water-loving plants and plastic coverings can help to reduce the effects of water abundance (Marsh: 1998).

(c) Lack of Capital

Most homegardeners are unwilling or unable to invest much capital in their homegardens (Vasey: 1985, Mendez: 2001). Where households have access to capital, it is not surprising to find that homegardens are more productive. For example, one reason that homegardens cultivated by rural households in Russia are more productive than those cultivated by urban households is that the rural households have easier access to inputs and implements that originated from the former collective farms, some of which they may receive as a part of their wage for work on agricultural enterprises that replaced the collective farms, and some of which they may "divert" from such enterprises (tho Seeth and others 1998).

Homegardening materials need not be prohibitively expensive. In India, a "kitchen garden kit" developed for a pilot activity containing seven varieties of tree seedlings, high-quality vegetable seeds and five hybrid chicks currently costs 600 Rupees, or about US$12 at current exchange rates.

An important limiting factor discussed in relation to capital is the use of inputs to improve soil fertility. Lack of capital is likely to be more of a barrier to homegardening where soils do not contain sufficient nutrients to support gardening and where households are not familiar with composting and other methods for improving soil fertility. In such cases extension advice should include instruction on accelerated composting and the possibility of gardening in containers while establishing better soils (Brownrigg: 1985).

In addition to composting, households may construct terraces and plant leguminous trees to improve soil fertility (Marsh: 1998). These methods may be labour intensive, but do not require large amounts of capital. Where gardening tools and other inputs are absolutely necessary, the government may find it useful to establish small local stores to sell simple tools and supplies at affordable prices.

Where fencing is required to reduce foraging by animals or theft of homegarden production, live fencing can be used to reduce costs. Plants used in fencing can also provide additional products for household use or sale (Marsh: 1998).

(d) Cultural Barriers

Distribution of land for homegardens is most likely to lead to successful establishment of homegardens if homegardening is already successfully practiced in the vicinity where the new homegardens are to be established. The presence of viable homegardens in the immediate vicinity is the best predictor of success since such homegardens demonstrate that homegardening is socially and culturally acceptable, and is valued by households (Marsh: 1998). Although projects to promote gardening often prefer to work with communal organizations, household-level food production must be a family undertaking since labour, space and time are valuable resources to poor households and "cannot be risked on the uncertain participation of a number of individuals", even on a small scale. (Ninez: 1985).

Cultural preferences may inhibit households from taking up homegardening. Households may associate homegardening with poverty and therefore decline to establish gardens (Miura: 2003). In Nepal, dark green leafy vegetables are often considered low-status foods, which might help to explain why Shankar (1978) found that consumption of vitamin A-rich foods did not increase along with homegarden size in the studied groups. In Bangladesh, dark green leafy vegetables are widely believed to be bad for young children, and this belief does not depend upon the household's income or access to land. (Cohen: 1985).

Noting that the smaller size of homegardens in some western and eastern areas of Java, as compared to larger homegardens in the central part of the island, cannot be explained ecologically, Terra (1954) concluded that the differences must therefore have an ethnographic explanation. He opined that the difference has to do with the tendency of lands in matriarchal societies of the region to evolve into family-owned plots, while lands in patriarchal societies of the region tend to remain controlled by the head of the dominant clan. (Terra: 1954).

Lifestyle preferences can limit the ways in which households use their house plot. For example, although it is common to combine gardening and small livestock production in Indonesia and other South East Asian cultures, small livestock are said to be the main ravagers of gardens in other areas (Brownrigg: 1985). Households may thus feel compelled to choose between gardening and raising small livestock. Where local culture places a high value on free access to house sites by neighbours, households may be unwilling to construct fences necessary to keep out animals (or keep in animals) (Brownrigg: 1985).

Cultural values may evolve over time to cause changes in the way homegardens function. Soemarwoto (1985) reports a trend of moving the household toilet from above the fishpond to inside the house in order to improve aesthetics and comply with ideas of modernization, even though this means that human waste is no longer be recycled in the fishpond where it had been converted into protein, but is instead be flushed into streams where it will contribute to harmful algae blooms.

(e) Lack of Information on Nutritional Benefits of Homegardening

Public information campaigns can play an important role in encouraging families to plant and consume more vegetables. Many families are not aware that vegetables and fruits are nutritional. In a study of Philippine urban homegardens Miura (2003) found that mothers generally had no knowledge of vitamins and iron in foods until informed by community health workers. Before they learned about the nutritional value of vegetables, some Filipino families mistakenly believed that micronutrient tablets distributed by the government were more desirable than consumption of vegetables, and that vegetables were the poor man's substitute for tablets. Brun (1989) recounts the case of a village studied in West Senegal, where mothers cultivating homegardens did not seem to understand that vegetables were good for their children and most mothers stopped growing carrots when their children snuck through homegarden fences to eat them raw.

Miura (2003) found that urban Filipino families who obtained health information from NGO-funded local health improvement programmes and community health workers had planted more fruits and vegetables, including dark green leafy vegetables, than families who had not received such information, regardless of socio-economic status, size of the homegarden plot and general knowledge of nutrition. Miura concluded that the fact that socio-economic status did not affect the degree to which families adopted advice received from NGO-funded local health improvement programmes and community health workers indicates that home gardening is a suitable strategy for poor communities. Even where families are already consuming vegetables and animal proteins, educational campaigns can help to increase their consumption by mothers and children. One such campaign in Central Java, conducted through radio, banners, billboards, posters and face-to-face communication, resulted in an increase in consumption of eggs and vegetables by mothers and young children across all socio-economic groups, leading to an improvement in serum retinal levels (de Pee and others: 1998).

(f) Lack of Agricultural Extension Advice

It is common to find that government agricultural extension programmes ignore homegardens in favour of working with commercial field crop producers. This should not be surprising given that most agriculture ministries are focused on production of staple foodstuffs and export commodities rather than subsistence agriculture. On the other hand, it is rather strange that governments that spend large sums on "household food security" programmes—in India these take the form of redistributing food resources through public distribution systems, fair price shops, food for work programmes, etc.—should so neglect improvement of homegardening plots that would not only add additional food to the system, but would deliver that food to households that most need it.

Agricultural extension can contribute significantly to homegarden production. At the beginning of the Helen Keller International pilot homegardening project in Bangladesh, 50 per cent of households reported having a garden with mean size of 61 m^2 and growing an average of 3.1 varieties of vegetables, whereas after two years with the project 100 per cent of households reported having a garden with mean size of 138 m^2 and growing an average of 17 varieties (Marsh 1998). This demonstrates that families can learn to cultivate homegardens more intensively.

The background of extension agents can play an important role. Brownrigg (1985) recounts the experience of a project in Ghana in which senior extension agents were young women with graduate degrees in agriculture and training in nutrition, while junior agents were local young women and men from the target communities. She reports that the local extension agents were accepted much more easily than either foreigners or non-local national staff.

One common approach to providing extension advice for homegardening is for a project to establish a "demonstration garden." Brownrigg believes this can be a poor strategy, especially if the gardener in charge is either from the village elite (which causes people to view homegardening as an elite activity) or is a hired labourer from the lower rungs of the village social ladder (which causes people to view homegardening as a low status activity).[17] If a demonstration garden is used, implementers may want to convince at least one of the better focal home gardeners to work with the project to improve an established homegarden (Brownrigg: 1985).

Gardening promotion projects are likely to be most successful where homegardens already exist. Homegardens must be introduced gradually in areas where they do not exist; an overnight "revolutionary" approach has proved problematic in many homegarden campaigns (Ninez: 1985).

(g) Lack of Appropriate Plants and Livestock

Interventions that seek to advise families on appropriate techniques for improving homegarden production may find that homegardening families do not have adequate access to seedlings and other necessary materials. This is more than a question of access to information or even access to capital. The materials necessary for homegardening may not be locally available.

The introduction of improved inputs can make a significant contribution to homegardening productivity. A useful example is a project undertaken by Heller Keller International to improve animal husbandry in rural Nepal, Cambodia (poultry and eggs) and Bangladesh (poultry, eggs and fish). The projects targeted household egg production by introducing improved breeds of birds that produce more eggs, along with vaccinations and assistance with proper housing and feed, targeted livestock production by promoting improved grass fodder and deworming tablets for milk cows, and targeted fish household fish production by introducing fast growing fish cultivars and plant sources of fish feed. In all three countries, ten to twelve months after the improved breeds were introduced, households in which chicken liver had been consumed within the past 7 days rose from 21 per cent of households to 35 per cent of households. In addition, the number of eggs consumed in the household rose from a weekly average of 5 eggs to a weekly average of 12 eggs, while the number of eggs consumed by all household children rose from a weekly average of 2 eggs to a weekly average of 3 eggs (HKI/AP: 2003).

It is important for planners to avoid the unsustainable practice of importing planting materials and breeding stock and instead establish "sustainable local mechanisms for the procurement, production, and distribution of seeds and other planting materials" (Brownrigg: 1985: 104). Dozens of successful gardening projects have independently concluded that the solution lies in establishing nurseries, seed multiplication units and seed beds under local control (Brownrigg: 1985). Perhaps the largest homegardening improvement project ever undertaken is the NGO Gardening and Nutrition Education Surveillance Project, in which Helen Keller International works with local NGOs to establish village nurseries to provide seedlings and seeds to rural homegardeners in Bangladesh. The programme was scaled up over ten years and was reaching 8,00,000 households as of 2001 (HKI/AP: 2001).

Before a homegardening project identifies which plants and livestock to promote, planners should study the preferences and capacities of local homegardening families. The choice of appropriate plants and livestock must depend upon what local people prefer to eat, as well as what plants and livestock are well suited to the local environment and resource profile of households (land, labour, capital), especially during the "hungry season" when stores of foods are scarce (Brownrigg: 1985).

(h) Lack of Available Labour

Homegardens typically are cultivated using "marginal" labour, which is marginal in the sense that it is flexible and its use reflects low opportunities for alternative employment (Hoogerbrugge and Fresco: 1993). Although there is not much data on volume and timing of labour inputs (Hoogerbrugge and Fresco: 1993), commentators tend to agree that most traditional gardening practices involve only a few days of preparation and less than an hour per day for maintenance and harvesting. During the five-month growing season, it was estimated that homegardens in urban Lima required an average of 50 minutes per day to prepare soil, plant, cultivate, water and harvest (Ninez: 1985).

Poorer families may garden more than better off families. In the urban capital district of Papua New Guinea, squatter settlements and government housing estates, which have a high proportion of unemployed and underemployed workers, contain homegardens averaging 469-523 m^2, which is 29-35 per cent larger than the average for the district as a whole (Vasey: 1985).

However, even poor families may rationally decide that the opportunity cost of homegardening is too steep. For example, it is reported that many poor families in Java must engage all household members in wage labour and therefore have no time to spare for gardening, even though the family is aware that they could obtain additional income or nutritious food from the homegarden (Soemarwoto: 1987).

In this respect it is important for planners to appreciate that the opportunity cost of spending time on gardening is not zero, and labour intensive technologies may not be appropriate in many contexts since household members do not have unlimited time available for gardening (Brownrigg: 1985). Homegardens are more sustainable if labour requirements are low and somewhat flexible (Marsh: 1998). Planners contemplating homegardening projects should consider the availability of marginal labour among households targeted for assistance.

4. Policy and Programme Considerations

In determining whether to adopt homegardening as a strategy for improving the livelihood of poor families, planners must begin by defining the class of prospective beneficiaries.[18] Although homegardening may be a beneficial intervention for both urban and rural poor populations, the opportunities are likely to differ based on a number of considerations.

Suitable land is much more likely to be available in sufficient quantity and quality in rural areas, either because poor families already possess suitable land or because planners can obtain suitable land and distribute it to target families. Once the general class of beneficiaries is defined, we propose that planners undertake the following analysis, summarized in Figure 3.3.

(a) Preliminary Assessment of General Suitability of Homegardens

After planners define the general class of beneficiaries, a threshold question is whether homegardens are likely to provide benefits to the target population. To answer this question, planners should ask whether homegardens presently provide benefits to any families (including the non-poor) in the vicinity of the area where the target population resides (the target area). Areas in which households have historically been unable to overcome climatic, economic or cultural constraints to operating homegardens may not be appropriate areas for homegardening interventions (Marsh: 1998, Ninez: 1985).

If homegardens do not appear to exist or do not appear to be providing substantial benefits to any families in vicinity of the target area, the question then becomes whether homegardens are providing benefits in settings that planners deem to be analogous to target setting, including settings in other communities with similar climates and similar resource constraints. The existence of homegardens in the vicinity of the target setting (or in analogous settings) would suggest that water, soils, climate, cultural preferences and other factors are not a barrier to establishment of productive homegardens in the target area, and further suggests that homegardens may be a viable strategy for addressing livelihood needs of the target population.

In making the preliminary assessment on whether homegardens might provide benefits to the target population, planners will likely benefit from field inquiries or studies, using rapid or participatory appraisal methods. Such field inquiries or studies might apply one or more of three approaches:

(1) *What are existing typical uses and benefits of homegardens by target population*? Focus should be on a representative sample of the target population to determine how home plot is used, what benefits accrue, and what constraints are faced.

(2) *What are potential uses and benefits do homegardens offer target*

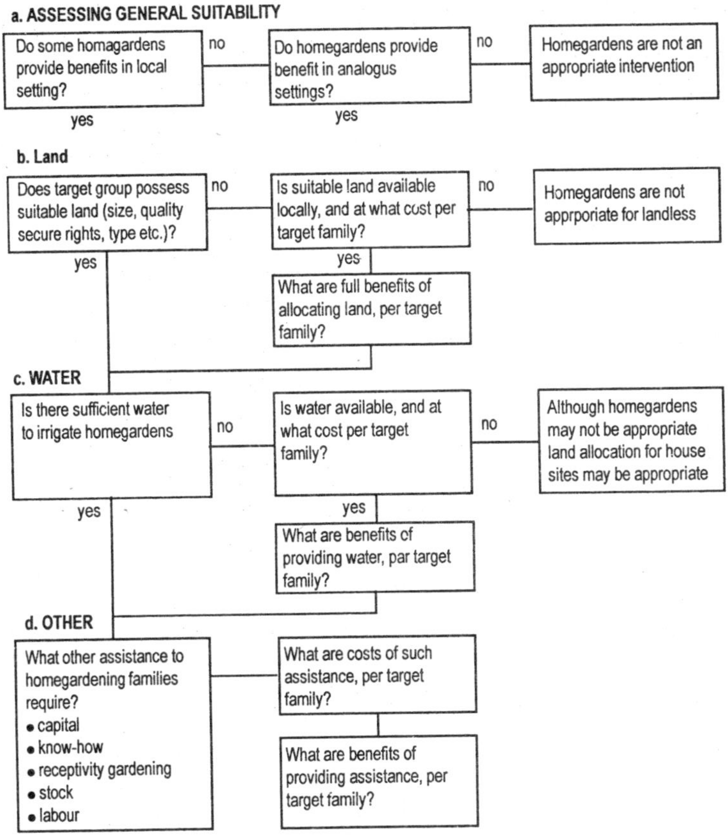

Fig. 3.3: Analyzing Appropriateness of Homegardens

population? Focus should be on a purposively selected sample of the target population that have well-developed homegardens. Investigators should determine the benefits and how and why this portion of the target population has been able to use homegardens more productively.

(3) *What best practices have emerged from NGOs or other interventions related to homegardens in the target area*? It is possible or even likely that NGOs or other government departments have made or are making small-scale homegarden interventions in part of the target area. These deserve study so lessons can be learned, successes replicated and mistakes avoided.

(b) Access to Land

Once planners make a preliminary determination that homegardens might provide benefits to the target population, planners should determine what proportion of families in that population have access to suitable land on which to construct a house and establish a homegarden. It is important to address the issue of access to land separately from access to other inputs for several reasons. Where families do not possess suitable land or their rights to possessed land are not sufficiently secure, this is the single largest barrier to establishing homegardens. While the cost of obtaining and allocating land to target families may be quite affordable, the nature of land—the fact that it is immovable, and the fact that access to land can have important social and political implications—makes its allocation more complicated than allocation of other homegardening inputs such as water, know-how, appropriate plant and animal stocks, etc. For these reasons, it is appropriate for planners to evaluate the costs of identifying, obtaining and allocating land independent of other costs.

At the same time, planners should attempt to assess the special benefits associated with allocating homegarden plots to landless and functionally landless beneficiaries. Provision of land to a landless or functionally landless family can have benefits far beyond the benefits ordinarily associated with homegardening, including improved status in the village and the ability to demand reasonable wages in the job market without risk of eviction. (See Box 3.1) Thus, planners should assess the value of special benefits associated with allocating land to landless families separate from their evaluation of other homegardening benefits.

In determining what proportion of families in that population have access to suitable land on which to construct a house and establish a homegarden, planners should assess suitability of land in relation to two important factors. First, the parcel on which the family resides must be large enough to support a homegarden that can provide benefits to the family.[19] Second, the family must have ownership-like rights to the land rather than use the land as a squatter or with the permission of a landlord. The existence of ownership-like rights determines whether the family has the right to improve the land and whether the family has a reasonable expectation of receiving the benefits of any long-term improvements, such as planting trees.

For that portion of the target group already possessing suitable land for homegardening, planners should proceed to consider the sufficiency of water and other inputs such as family capital, family know-how, etc., as described in following sections. For that portion of the target group that does not possess suitable land (as defined above), planners must undertake a critical analysis to determine the availability of appropriate land in the vicinity of the target families. Planners should consider the quantity of land available from various sources, including government land and land that the government might purchase from private owners.

In the case of assessing the availability of public lands for reallocation, planners should take special care to verify that the land is not functioning as a common property resource providing benefits essential to the local population. This is a significant risk in many settings where the government claims ownership of common property resources such as community forests or wastelands located near population centres. Planners frequently underestimate or are unaware of the extent to which poor families use such resources to graze livestock, hunt forest animals, and gather plants for consumption, fodder, medicine and fuel (Jodha: 1990, Blair: 1996, Meinzen-Dick and others: 2001). Even where planners recognize that such activities occur, they may underestimate the importance of these activities in the household economy of the poorest families. It would be counterproductive to eliminate a common property resource that provides a low level of support to a large number of poor families in order to distribute homegarden plots to a smaller number of families, even if the total economic use of the land would be enhanced by such a use conversion. The detrimental impact on some poor families would not justify the benefit to other poor families. Planners should also satisfy themselves that available land is not located in a place where residential uses would create sanitation or other environmental problems.

If planners determine that appropriate land is available in the vicinity of the target population, they should then calculate the cost of purchasing such land (if it is necessary to purchase it), as well as administrative costs of obtaining and allocating the land, calculated in terms of costs per family benefited. (See Box 3.2) Planners must also consider whether the beneficiaries should share in the costs of land purchase (as well as other costs), and, if so, to what extent. Important factors to consider include: affordability for what are likely to be among the society's poorest households, administrative costs of collection (relative to benefits of such collection), the desirability of cost-sharing by beneficiaries to promote their "ownership" of the programme activities, and related moral risk issues.

Because homegardens must be established near where target beneficiaries live and because appropriate land is likely to be more available for acquisition and allocation in rural areas, homegardening—and allocating land for homegardens—may be a particularly viable strategy in rural areas. Programmes that enhance the livelihoods of poor rural families may have the added advantage of reducing the incentives that members of poor families have to migrate to cities in search of better livelihoods.

Another consideration is whether creation of homegarden plots is likely to reduce the amount of arable land, causing a fall in overall agricultural productivity or exacerbating food security concerns. Even assuming that arable land is used to create homegardens, these risks are likely to be insubstantial for several reasons. First, because even very small plots can provide important benefits to families, the total amount of land needed for homegardens is likely to be modest. Assuming that 5 per cent of the acquired land is used to construct roads, drainage and other infrastructure, a one hectare plot could provide 380 m^2 plots to 25 families. To place this in perspective, in India, distribution of 380 m^2 plots to each of the nation's estimated 15 million completely landless families would require only 6,00,000 hectares of land, which is approximately 4/10 of 1 per cent of the nation's 161.8 million hectares of arable land (FAO: 2002).

Second, if used with even modest intensity to produce vegetables, fruits and animal products, homegarden plots are likely to produce at least as much agricultural value per unit area as had been produced on the arable land. A study of Javanese home gardens found that net income per square meter was higher for homegardens than for rice fields. The same study found that relative costs of production on home gardens were much lower than for rice fields (15.1 % of gross income versus 55.9% of gross income for rice fields) (Danoesastro 1980, cited in Christanty 1990).[20] The latter is particularly important for poor households, who typically have less access to credit and are less able to insure against risk. A study of well-developed homegardens in Karnataka indicates that the income per square meter is several multiples higher than arable land in the same area used for grain crops. (Hanstad and others, in preparation). On a macro-scale, homegardens in Russia comprised about 4.5 per cent of total agricultural land in 2000, and yet were reported to have produced 64 per cent of all produce in the country, including 90 per cent of all potatoes and 75 per cent of fresh vegetables (Borisova: 2000).

Third, even if homegardens do not produce as much per hectare as arable land, the effect of any reduction in the total amount of food produced is likely to be offset by the fact that homegardens efficiently provide foods to one of the most food-insecure segments of the population. Homegardens are likely to make such foods available to the poor more efficiently than other government food programmes since the poor themselves control distribution of homegarden products. Whatever production poor families do not consume they will trade or sell to others.

In evaluating social value of distributing homegarden plots, it is useful to evaluate the efficacy of alternative programmes targeted to assist the same populations. In India, for example, we have found that government resources devoted to constructing housing for landless and land poor families might be better spent, and would reach more beneficiaries, if some portion of programme resources were diverted from housing construction and instead used to obtain larger house sites that provide space for homegardening. Our own field research in India indicates that even the poorest rural households are able to accumulate the resources to construct a house (in stages) if they have secure rights to an adequate house site. For example in one study of 45 households who had received house sites from the government, 32 had constructed housing without government assistance, indicating that these families had the personal incentive, as well as access to sufficient materials or sufficient savings or access to credit to construct the house once they received land (Hanstad and Lokesh: 2002).

Several other considerations may be considered in planning the acquisition and allocation of land for homegardens:

- Ownership of the land should be registered and certificated in the name of both spouses. This provides a level of additional protection to the wife in case of spousal separation or death of the husband. In addition to this important security protection, the literature indicates that legal land ownership by women facilitates increased respect within their families and communities, as well as a means of earning and controlling income, feeding their families and accessing government programmes (Agarwal: 1994, Deere and Leon: 2001, Deininger: 2003).
- Although it is important to register ownership of the land and issue appropriate ownership documents, the lack of a full-fledged land registry for conducting land transactions should not impede allocation of land to landless and land poor families. Families who receive small parcels for construction of a house and establishment of a homegarden are unlikely to want to transfer such land in the near term, which reduces the importance of the

transaction registry. However, such families must be able to prove their rights to the land to protect against uncompensated taking by the government.

- In order to reduce the threat that local elites, including large landowners, will seek to capture benefits of the land allocation programme, planners should keep the benefits relatively small for each beneficiary family. Most elites will not be interested in receiving small parcels measuring, for example, 380 m^2. In addition, if the parcels are sited in colonies, elites are likely to be even less interested in obtaining such land since its location within the colony reduces the alternative uses to which the elites could put the land.
- Before the land is allocated, planners must resolve any questions of prior ownership and address any land use restrictions to ensure that the land can be allocated for residential and homegardening uses. Where the land acquired is zoned as agricultural land, it may be necessary to redesignate the land for non-agricultural uses.
- Planners must consider typical habitation patterns and related customs when allocating land for homegardens. Sociological research may be required to inform the planning process. In deciding where to site new homegarden plots, planners should be sensitive to the issue of isolating groups according to caste and ethnicity. In deciding the shape of the allocated parcels, planners should determine how close to one another families in the region generally prefer to build their houses—longer parcels may allow for a more intimate siting of houses than square parcels. If beneficiary families are willing to relocate from their existing village, how close must the new colony of residential homegarden parcels be to that village? In India, for example, land poor families state that they would prefer to receive a new residential plot of land within one half to one kilometer of the existing village.

(c) Access to Water

Water deserves special consideration. In many environments, water for homegardening is likely to be the most important consideration after land. In some environments, water may be even more scarce than land and more expensive to supply during the driest months of the year. Where water is scarce throughout the year, its absence may preclude homegardening as a viable strategy.

On the other hand, the amount of water needed for homegardening is often not great, and carrying water to the garden may be a reasonable solution. In addition, water availability should be considered when identifying which trees and other plantings to promote in homegardening. Project planners may also be able to reduce water demand of homegardeners by providing extension advice on water conservation and subsidizing low-cost techniques homegardeners can use to collect, store and efficiently use rainwater and household wastewaters.

Even with conservation and such self-help approaches to irrigation, the increased need for irrigation water may stress public water delivery systems, particularly in urban areas and particularly during dry months. Clearly, if homegardening irrigation by some families causes other families to lose access to adequate amounts of water necessary for household needs, this will not be an acceptable trade off. If planners do not plan for the increased demand for water that comes with homegardening, water shortages could result.

Planners may therefore wish to explore the cost of upgrading systems for delivering water to households, or establishing new delivery systems where homegarden plots are being allocated to landless families. Once planners determine that it is technically feasible to supply adequate amounts of water to the planned homegardens, they must calculate the costs per family benefited to determine whether the cost is socially justified. It is quite possible that the public cost of supplying irrigation water to homegardens will exceed the economic value of products produced in the homegardens. Under such conditions homegardening will prosper only if the public subsidizes the cost of delivering irrigation water to homegardens. Such a public subsidy may be socially justified when viewed against other public subsidies targeted at providing social assistance to the poor.

(d) Access to Other Inputs

Once access to land and water are secured, planners must consider what other factors require attention to promote homegardening, and what issues are likely to arise with respect to each. Most literature dealing with homegardens applies to these questions.

(i) Capital

As was noted at the beginning of this paper, one important and distinguishing characteristics of homegardening is that it is the poor may enter without making great investments of capital (Marsh: 1998). Ultimately, homegardens are likely to be a useful strategy for improving the livelihoods of poor households only if such households value homegardens to the extent that they are willing to invest scarce savings in improving and maintaining the homegarden. To be sustainable, homegarden inputs must be affordable within the typical household budget of the poor. In India, we have met with very poor agricultural labourer families who invested scarce family capital to construct housing, plant trees and raise poultry and livestock once they obtained secure ownership of small plots of land. These families reported receiving no government assistance in purchasing inputs for homegardening.

As with many development interventions, programmes to distribute homegardens to the poor bring with them the risk that beneficiary families will not value what they are given for free. One solution is to design programmes such that recipients of homegarden plots are obligated to invest their own time and labour in making improvements to the plot, such as construction of a house. Families are likely to value such assets if they make active investments of this type.

In many situations planners may determine that some level of public subsidy of homegarden inputs is justified, at least with respect to homegarden start-up costs such as the purchase of seedlings and seeds. If the project subsidizes these or other homegardening inputs, it may be useful to use "introductory" schemes in which inputs are sold at a reduced price for a limited time, rather than outright gifts of inputs. This approach has been found to legitimize projects as something other than an attempt to "help the poor", which poor families often resent (Ninez: 1985). Planners should avoid any project design that calls for ongoing subsidy of inputs.

(ii) Know-How and Receptivity to Gardening

Although the poor may be able to find capital to purchase building materials, planting materials, poultry and livestock, in some cases it may be much more difficult for them to access reliable information on how to maximize the productivity of plants and animals. This is likely to be particularly true in urban areas where many families do not have direct experience in agriculture. Agricultural extension can help to fill information gaps.

It is important to integrate nutrition awareness and education into garden planning so that homegardening families have full information upon which to base decisions of whether and to what extent they will use the homegarden to produce foods to supplement and improve the family diet, as opposed to using the homegarden primarily as a source of additional household income. When designing an effective means for communicating an appropriate nutrition strategy, planners must understand the traditional diet and food taboos, seasonal food shortages, food storage practices, food cooking practices, and distribution of food within the household. It is most effective to integrate gardening technique education with nutrition education so that households can plant varieties that comply with taste preferences and that will supply nutrients year-round (Marsh: 1998).

Even where homegardening families have experience in producing various plants and animals, they may not fully understand the long-term consequences of various production techniques. For example, homegardeners who do not take proper steps to preserve soil fertility may eventually find that the soil is exhausted. Extension agents can explain the benefits of using animal manures and composting of kitchen wastes can help restore nutrients to soils. The fact that homegardens have persisted for generations in some societies without the addition of artificial fertilizers suggests that viable low-cost strategies do exist for preserving soil fertility. Programmes promoting homegardening should include agricultural extension component that helps families to appreciate the importance of soil fertility and affordable techniques that will preserve soil nutrients.

Agricultural extension and nutritional education can serve another important purpose: helping families to understand and value the economic practicalities of homegardening, as well as the ways in which homegardening can contribute to family food security, family nutrition, family income and family social status. Ideally, the local community will value homegardening as an appropriate strategy for all families rather than as a leisure activity of wealthier households, or a mark of household poverty. Although it may seem somewhat contradictory, homegardens may be a better strategy for improving the livelihoods of the poor if homegardens are presented as a universal strategy for improving household nutrition and household independence, rather than as a "poor man's" strategy of subsistence. To the extent non-poor families participate in homegardening, this removes the social stigma of gardening as an activity of the poor, and improves the relative status of poor families by providing them with another activity they have in common with non-poor families. For these reasons, planners should consider agricultural extension should target not only poor households, but all homegardening households. Brownrigg (1985) suggests that it may be better to avoid creating demonstration homegarden plots, and focus instead on working with a local family who is willing, in the spirit of experimentation and cooperation, to allow their homegarden to be used as a demonstration garden for the village.

Intervention projects should promote the economic benefits of homegardening rather than focus solely on benefits of home consumption. Increased income can also be used to purchase more nutritious foods. Other potential economic benefits include: (a) returns to land and labour are often higher for homegardens than for field agriculture, (b) homegardens can supply fodder for animals, fuelwood, supplies for handicrafts, (c) household processing of homegarden fruits and vegetables can increase their market value and preserve them for later consumption, and (d) sale of homegarden produce may be one of the only sources of independent income for women (Marsh: 1998).

(iii) Stocks

Homegardening planners should avoid introducing plant species that are locally unknown, no matter how nutritious and economical the plants are. Where it is uncertain whether particular plants can be grown easily in the local environment, or it is uncertain that they are accepted in the local diet, there may be high hurdles to their adoption (Brownrigg: 1985).

Homegardening projects are more likely to succeed by promoting existing local species and by making local planting stocks more readily available to the target population. The most sustainable approach is likely to be one in which local nurseries can be established to satisfy the needs of homegardeners while earning a profit for the nursery. Programmes to help establish local nurseries may help to demonstrate their profitability. In the Helen Keller International programme to promote homegarden development in Bangladesh, implementers found that the lack of a regular supply of quality seeds and other gardening inputs was a constraint to improving homegarden production. Their solution was to establish a series of privately owned village-level nurseries to provide information and plant stocks to homegardeners:

"Local partner NGOs work with their community groups to

establish village-level nurseries and homestead gardens. The village nurseries serve as a community support service network in such a way that they are the focal point for demonstration and training on low-cost, low-risk gardening practices for nursery holders, the leaders of the NGO women's groups, and household gardeners. In addition, they are the source and distribution centres for seeds, seedlings, and saplings, the sites for demonstration of new plant varieties, and the centres for community mobilization and organization. The majority of the village nurseries are operated as small enterprises and are a significant source of income for the household." (Talukder and others: 2000: 167)

5. Conclusion

Homegardens represent an especially useful strategy for promoting sustainable livelihood objectives of the poor, including secure access to land and water, improved financial security, improved leverage in wage bargaining, improved nutrition, improved social status and political status, and better access to basic infrastructure. Where poor families lack secure rights to homegarden plots of suitable size and quality, programmes to obtain and allocate land to such families will often be found to be a constructive and socially beneficial use of government resources. For such families, the benefits of obtaining ownership of land on which to construct a house and garden go well beyond other benefits normally associated with homegardening.

Where families already have secure rights to homegarden plots of suitable size and quality, governments should seriously consider investing in water infrastructure, agricultural extension and nutritional education, as well as programmes to ensure that appropriate stocks of plants and animals are available to homegardening families. Although some public funds will doubtless be necessary to establish or strengthen homegardening for landless, land poor and otherwise poor families, a successful homegardening intervention will be one in which the homegardens become a self-sustaining intervention, satisfying the particular livelihood objectives of the homegardening family, while reducing the family's continued dependence on public resources.

Government as well as International donors should consider advancing the issue of homegardening, including the issue of allocating land to establish homegardens, in three principal ways: (1) through supporting research; (2) by promoting consensus building among donors, government planners and project implementers; and (3) by directly supporting government or NGO homegarden intervention projects. Each approach may be undertaken simultaneously in any given development setting. Research and intervention projects can both drive the consensus building process and can benefit from the consensus reached at any given stage.

References

Agarwal, A. and Narain, S. (1999): "Making Water Management Everybody's Business: Water Harvesting and Rural Development in India." *International Institute for Environment and Development*, Gatekeeper Series No. 87.

Agarwal, B. (1994): "A Field of One's Own: Gender and Land Rights in South Asia." Cambridge University Press, Cambridge.

Arifin, H.S. (2002): "Ecological and Socio-Economic Benefits of Pekarangan Land."

Asare, E.O., Oppong, S.K. and Twum-Ampofo, K. (1990): "Home Gardens in the Humid Tropics of Ghana." In Landauer, K., and Brazil, M. (eds.) (1990). *Tropical Home Gardens*. United Nations University Press, Tokyo, Japan, pp. 80-93.

Baumann, P. (2002): "Improving Access to Natural Resources for the Rural Poor: A Critical Analysis of Central Concepts and Emerging Trends from a Sustainable Livelihoods Perspective." Food and Agricultural Organization of the United Nations.

Baumann, P. and Sinha, S. (2001): "Linking Development with Democratic Processes in India: Political Capital and Sustainable Livelihoods Analysis," 001 Natural Resources Perspectives Paper No. 68. 001, London.

Black, R., Morris, S. and Bryce, J. (2003): "Where and Why are 10 Million Children Dying Every Year?" 361 The Lancet 2226-2234.

Blair, H.W. (1996): "Democracy, Equity and Common Property Resource Management in the Indian Subcontinent", 27(3) *Development and Change*, 475-499.

Bloem, M., de Pee, S. and Darnton-Hill, I. (1998): "New issues in developing effective approaches for the prevention and control of vitamin A deficiency." 19(2) *Food and Nutrition Bulletin*, 137-148.

Bloem, M., Huq, N., Gorstein, J., Burger, S., Kahn, T., Islam, N., Baker, S. and Davidson, F. (1996): Production of fruits and vegetables at the homestead is an important source of vitamin A among women in rural Bangladesh. 50 (suppl. 3) *European Journal of Clinical Nutrition*, S62-S67.

Borisova, Y. (2000): "Land Ownership Remains Russian Privatisation's Final Frontier." *The St. Petersburg Times* (Russia). 11 April.

Brownrigg, L. (1985): "Home Gardening in International Development: What the Literature Shows." The League for International Food Education (LIFE), Washington, DC.

Brun, T., Reynaud, J. and Chevassus-Agnes, S. (1989): "Food and Nutritional Impact of One Home Garden Project in Senegal." 23 *Ecology of Food and Nutrition*, 91-108.

Carney, D. (1998): "Sustainable Rural Livelihoods." Russell Press Ltd., Nottingham, UK.

Christanty, L. (1990): "Home Gardens in Tropical Asia, with Special Reference to Indonesia." In Landauer, K., and Brazil, M. (eds.) (1990). *Tropical Home Gardens*. United Nations University Press, Tokyo, Japan, pp. 9-20.

Cohen, N., Jalil, M.A., Rahman, H. and others (1985): "Landholding, Wealth and Risk of Blinding Malnutrition in Rural Bangladeshi Households." 21 (11) *Social Science Medicine*, 1269-1272.

Danoesastro, H. (1980): "The role of the home garden as a source of additional family income." Seminar on the Ecology of Homegardens, III. Institute of Ecology, Bandung, Indonesia.

Deere, D. and Leon, M. (2001): "Empowering Women." University of Pittsburgh Press, Pittsburgh.

Deininger, K. (2003): "Land Policies for Growth and Poverty Reduction." World Bank, Washington, DC.

Department for International Development (DFID) (2001): "Sustainable Livelihoods Guidance Sheets." DFID, London.

de Pee, S., Bloem, M., Satoto, Yip, R., Sukaton, A., Tjiong, R., Muhilal and Kodyat, B. (1998): "Impact of a Social Marketing Campaign in Promoting Dark-green Leafy Vegetables and Eggs in Central Java, Indonesia." 68 *International Journal for Vitamin and Nutrition Research*, 389-398.

de Pee, S., West, C.E., Muhilal, Karyadi, D. and Huatvast, J.G.A. (1995): "Lack of improvement in vitamin A status with increased consumption of dark-green leafy vegetables." 436 *The Lancet*, 75-81.

Ensing, B., Freeks, G. and Sangers, S. (1985): "Homegardens and homegardening in the Matara district: the present situation and future prospects." MSc thesis, Social Science and Economics Department, University of Leiden, Netherlands.

Finerman, R. and Sackett, R. (2003): "Using Home Gardens to Decipher Health and Healing in the Andes." 17(4) *Medical Anthropology Quarterly*, 459-482.

Food and Agriculture Organization of the United Nations (2002): Production Yearbook.

Gajaseni, J. and Gajaseni, N. (1999): "Ecological rationalities of the traditional homegarden system in the Chao Phraya Basin, Thailand," 46(1) *Agroforestry Systems*, 3-23.

Government of India (2002): *Tenth Five-Year Plan*. New Delhi.

Hanstad, T. (2004) "Summary of Findings from Karnataka and West Bengal Rural Household Surveys" (Memorandum on file with Rural Development Institute, Seattle, Washington, USA).

Hanstad, T., Brown, J. and Prosterman, R. (2002): "Larger Homestead Plots as Land Reform: International Experience and Analysis from Karnataka." 37(29) *Economic and Political Weekly*, 3053-3062.

Hanstad, T. and Lokesh, S.B. (2002): "Allocating Homestead Plots as Land Reform: Analysis from West Bengal." RDI Reports on Foreign Aid and Development #115.

Hanstad, T., Lokesh, S.B. and Arun, M. (in preparation): Reaping the Rewards of Homestead Gardens: Nutritional and Income Benefits from Homestead Plots in Karnataka.

Helen Keller International/Asia-Pacific (2003): "Integration of Animal Husbandry into Home Gardening Programmes to Increase Vitamin A Intake from Foods: Bangladesh, Cambodia and Nepal." Available at http://hkiasiapacific.org/_downloads/APO%20Special%20issue%20Jan%202003.pdf.

Helen Keller International/Asia-Pacific (2001): "Homestead Food Production—A Strategy to Combat Malnutrition and Poverty." Helen Keller International: Jakarta, Indonesia.

Hoogerbrugge, I. and Fresco, L.O. (1993): "Homegarden Systems: Agricultural Characteristics and Challenges." *International Institute for Environment and Development*, Gatekeeper Series, No. 39.

Immink, M.D.C. (1990): "Measuring Food Production and Consumption, and the Nutritional Effects of Tropical Home Gardens." In Landauer, K., and Brazil, M. (eds.) (1990). *Tropical Home Gardens*. United Nations University Press, Tokyo, Japan, pp. 126-137.

International Development Enterprises (2004): "Micro Irrigation Programme." Available at (http://www.ide-india.org/micro_programme.htm). Site last visited March 19, 2004.

Jodha, N.S. (1990): "Rural Common Property Resources: Contributions and Crisis", *Economic and Political Weekly*, 25(26): 65-78.

Kiess, L., Bloem, M.W., de Pee, S., Hye, A., Khan, T.A., Huq, N., Talukder, Z., Haque, M. and Ali, M. (1998): "Bangladesh: Xeropthalmia Free: The Result of an Effective Vitamin A Programme and Homestead Gardening in Reducing Vitamin A Deficiency [abstract]." In American Public Health Association 126th Annual Meeting Report, Nov. 15-18. Washington, D.C. Reprinted in Appendix 13 of Helen Keller International/Asia-Pacific (2001). *Homestead Food Production—A Strategy to Combat Malnutrition and Poverty*. Helen Keller International: Jakarta, Indonesia.

Kiess, L., Moench-Pfanner, R. and Bloem, M.W. (2001): "Food-Based Strategies: Can they Play a Role in Poverty Alleviation?" 22(2) *Food and Nutrition Bulletin*.

Kumar, S.K. (1978): "Role of the Household Economy in Child Nutrition at Low Incomes: A Case Study in Kerala." Occasional Paper No. 95. Department of Agricultural Economics, Cornell University, Ithaca, New York.

Landauer, K. and Brazil, M. (1990): Introduction, in Landauer, K. and Brazil, M. (eds.), *Tropical Home Gardens: Selected papers from an international workshop held at the Institute of Ecology, Padjadjaran University, Bandung, Indonesia*, 2-9 December 1985). United Nations University Press, Tokyo.

Marsh, R. (1998): "Building on Traditional Gardening to Improve Household Food Security." *Food, Nutrition and Agriculture*, No. 22, Food and Agriculture Organization.

Marten, G.G. (1990): "A Nutritional Calculus for Home Garden Design: Case-Study from West Java." In Landauer, K., and Brazil, M. (eds.) (1990). *Tropical Home Gardens*. United Nations University Press, Tokyo, Japan, pp. 147-168.

Mearns, R. (1999): "Access to Land in Rural India: Policy Issues and Options," World Bank Policy Working Paper. World Bank, Washington DC.

Meinzen-Dick, R., Knox, A. and DiGregorio, M. (eds.) (2001): "Property Rights and Devolution of Natural Resource Management: Exchange of Knowledge and Implications for Policy." *DSE/ZEL*, Feldafing, Germany.

Mendez, V.E., Lok, R. and Somarriba, E. (2001): "Interdisciplinary Analysis of Homegardens in Nicaragua: Micro-Zonation, Plant Use and Socio-economic Importance." 51 (2) *Agroforestry Systems* 85-96.

Mitchell, R. (2004): "Legal Tools to Promote Environmentally Sound Management of Farmland and Forests." World Bank (forthcoming).

Miura, S., Osamu, K. and Susumu, W. (2003): "Home Gardening in Urban Poor Communities of the Philippines." 54(1*) International Journal of Food Sciences and Nutrition.*

Nair, P.K.R. (1993): *An Introduction to Agroforestry*. Kluwer Academic Publishers, London.

Nair, P.K.R. (2001): "Do Tropical Home Gardens Elude Science, or is it the Other Way Around?" 53(2) *Agroforestry Systems*, 239-245.

Ninez, V. (1985): "Working at Half-Potential: Constructive Analysis of Home Garden Programmes in the Lima Slums with Suggestions for an Alternative Approach." 7(3) *Food and Nutrition Bulletin*.

Okigbo, B. (1990): "Home Gardens in Tropical Africa", In Landauer, K., and Brazil, M. (eds.) (1990). *Tropical Home Gardens*. United Nations University Press, Tokyo, Japan, pp. 21-40.

Pico, R. (1964): "Problems of Tenure Reform in Latin America." 6(2) *Journal of Inter-American Studies*, 143-156.

Prosterman, R. and Mitchell, R. (2002): "Concept for Land Reform on Java." Paper presented at the Seminar "Rethinking Land Reform in Indonesia", Jakarta, 8 May 2002. On file with the Rural Development Institute, Seattle, Washington.

E. Rajagukguk (1988): "Agrarian Law, Land Tenure and Subsistence in Java: Case Study of the Villages of Sukoharjo and Medayu." Ph.D. dissertation, University of Washington.

Reddy, V. (1995): Letter to the editor. 346 The Lancet 1634-1636. Reprinted in Appendix 7b of Helen Keller International/Asia-Pacific (2001). *Homestead*

Food Production—A Strategy to Combat Malnutrition and Poverty. Helen Keller International: Jakarta, Indonesia.

Rome Declaration on World Food Security (1996): World Food Summit, 13-17 November Rome, Italy.

Rosenn, K.S. (1963): "Puerto Rican Land Reform: The History of an Instructive Experiment." 73 *Yale Law Journal*, 334-356.

Ruel, M.T. and Levin, C.E. (2000): "Assessing the Potential for Food-Based Strategies to Reduce Vitamin A and Iron Deficiencies: A Review of Recent Evidence." *FCND Discussion Paper No. 92*, International Food Policy Research Institute.

Schipani, S., van der Haar, F., Sinawat, S. and Maleevong, K. (2002): "Dietary Intake and Nutritional Status of Young Children in Families Practicing Mixed Home Gardening in Northeast Thailand." 23(2) *Food and Nutrition Bulletin*, 175-180.

Senior, C. (1954): "Migration and Economic Development in Puerto Rico." 28(4) *Journal of Educational Sociology*, 151-156.

Shankar, A., Gittelsohn, J., Prahdan, E.K. and others (1998): "Home Gardening and Access to Animals in Households with Xerophthalmic Children in Rural Nepal." 19(1) *Food and Nutrition Bulletin*.

Sloan, N.L., Rosen, D., de la Paz, T. and others (1997): "Identifying Areas with Vitamin A Deficiency: The Validity of a Semiquantitative Food Frequency Method." 87(2) *American Journal of Public Health*, 186-191.

Soemarwoto, O. (1987): "Homegardens: A Traditional Agroforestry System with a Promising Future." In Steppler, H., Nair, P.K.R. (eds.) (1987). *Agroforestry: A Decade of Development*. International Council for Research in Agroforestry, Nairobi, pp. 157-170.

Soemarwoto, O., Soemarwoto, I., Karyono and others (1985): "The Javanese Home Garden as an Integrated Agro-Ecosystem." 7(3) *Food and Nutrition Bulletin*.

Sommers, P. (1978): "Traditional home gardens of selected Philippine households and their potential for improving human nutrition." Master's thesis. University of Philippines, Los Banos.

Stoler, A. (1978): "Garden use and household economy in rural Java." 14, *Bulletin of Indonesian Economic Studies*, 85-101.

Talukder, A., Kiess, L., Huq, N., de Pee, S., Darnton-Hill, I. and Bloem, M. (2000): "Increasing the production and consumption of vitamin A-rich fruits and vegetables: Lessons learned in taking the Bangladesh homestead gardening programme to a national scale." 21 (2) *Food and Nutrition Bulletin*.

Terra, G.J.A. (1954): "Mixed-Garden Horticulture on Java." 3, *The Malayan Journal of Tropical Geography*.

Thaman, R.R., "Mixed Home Gardens in the Pacific Islands: Present Status and Future Prospects." In Landauer, K., and Brazil, M. (eds.) (1990). *Tropical Home Gardens*. United Nations University Press, Tokyo, Japan, pp. 41-68.

tho Seeth, H., Chachnov, S. and Surinov, A. (1998): "Russian Poverty: Muddling Through Economic Transition with Garden Plots." 26(9), *World Development*, 1611-1623.

Vasey, D.E. (1985): "Household Gardens and their Niche in Port Moresby, Papua New Guinea." 7(3), *Food and Nutrition Bulletin*.

Notes

1. The concept of allocating house-and-garden plots to landless labourers as a second-best (and more feasible) alternative to traditional land reform approaches has been gaining ground in India, as evidenced by India's recently adopted Tenth Five-Year Plan:
"Access to even small pieces of land which may not be sufficient for providing income to a family for subsistence can significantly reduce poverty and food insecurity by providing an essential component in a diversified livelihood system." [Government of India (2002): sec. 3.2.57]; and
"Ownership of even a small plot of land enables a family to raise its income, improve its nutritional status, have access to credit facilities and lead a more dignified life.... Horticulture, floriculture and vegetable cultivation on small plots of land, including homestead land, have proved beneficial for the poor. Agricultural labourers, therefore, need to be provided access to land to improve their economic and social well-being." [Government of India (2002): sec. 3.2.71)].
2. Underlying the sustainable livelihoods approach is the theory that people draw on a range of capital assets or poverty reducing factors to further their livelihood objectives (DFID: 2001). Assets are categorized as social, human, natural, physical, financial and political, and may serve both inputs and outcomes (Baumann and others: 2001). Various vulnerability factors over which people have little or no control (such as environmental disasters and political unrest) impact the assets. Assets are also filtered through policies, institutions, and processes that determine the degree to which the people's livelihood objectives are realized (DFID: 2001).
3. On Java, the great majority of homegarden plots are smaller than 200 m^2 (Prosterman and Mitchell: 2002), while on other less densely populated Indonesian islands they average 2,500 m^2 and can reach sizes of 3 hectares (Christanty: 1990). In Bangladesh, homegarden sizes range from 30 to 700 m^2, with an average size of 200 m^2 (Christanty: 1990). (Bloem: 1996). Improved family nutrition and health is a key objective of the sustainable livelihoods framework.
4. Sri Lankan homegardens have been reported to produce 60 per cent of leaf vegetables and 20 per cent of all vegetables consumed by the household (Hoogerbrugge and Fresco: 1993, citing Ensing and others: 1985). Others have reported that homegardens typically produce more than 50 per cent of vegetables, fruits, medicinal plants and herbs consumed by the household (Marsh: 1998). The bulkiness of fresh fruits and vegetables favours their consumption near the production site, especially where the infrastructure is least adequate to allow transport of rural crops to cities (Vasey: 1990).

5. The presence of parasites in the population studied can greatly influence the degree to which people are able to absorb vitamins present in consumed fruits and vegetables (HKI/AP: 2003). It is also difficult to control the influence of socio-economic status. A study in Bangladesh found that children living in households with a homegarden were less likely to have eye diseases associated with lack of vitamin A than children living in households without a homegarden; specifically, children from households without a homegarden were 2.5 times more likely to be nightblind, 2.1 times more likely to have Bitot's spot, 3.4 times more likely to have active corneal lesions and 2.4 times more likely to have corneal scars. However, the study also found that children of poor households were likely to experience similar risk ratios for the same afflictions, making it unclear whether the availability of homegarden produce or other factors, such as foods purchased by the higher earning families, accounted for the observed differences in eye health (Cohen and others: 1985, cited in HKI/AP: 2001).
6. There appears to be a wide range of experience, however, regarding the homegarden's contribution to total calories consumed. For example, researchers in Sri Lanka have reported that homegardens produce over 80 per cent of staples consumed (Hoogerbrugge and Fresco: 1993, citing Ensing and others: 1985), while homegardens in urban Papua New Guinea have been estimated to produce 4-6 per cent of household food energy needs (Vasey: 1985).
7. Vegetables and fruits produce less bio available beta-carotene than previously assumed, and the amount produced varies widely, which means that consumption of vegetables and fruits is not as likely to improve vitamin A status as previously thought; therefore, programmes that promote consumption of dark green leafy vegetables are likely to have a more modest impact on vitamin A deficiency disorders (HKI/AP: 2001, de Pee and others: 1995).
8. Households produced ranging from 180 kg of meat worth US$429 for goats, to 2,700 kg of meat worth $4,286 for pigs; livestock rearing on homegardens was widespread, with 40 per cent of surveyed households raising poultry, 34 per cent raising sheep, 62 per cent raising goats, 10 per cent raising pigs, 6 per cent raising cattle and 2 per cent raising rabbits (Asare and others: 1990).
9. A study of child nutrition among homegardening families in Thailand demonstrates how difficult it is to study causation. Although the study found that children of households that raised fish in addition to growing vegetables were taller and heavier than children of households that only grew vegetables, it was impossible to ascribe this result to the introduction of the fish since the same differences were found even in older children of the households, and it was deemed extremely unlikely that the recent addition of fish to the diet could have reversed earlier nutritional deficits in the older children. In addition, because the families raising fish were slightly

better off financially than the control group, the control group was deemed inadequate (Schipani: 2002).

10. In rural areas studied, home production on plots near the home, the average size of which was approximately 3600 m^2, accounted for 48.3-72.5 per cent of household income, while income from gardening in towns and cities was considerably less, accounting for 0-18.04 per cent of household income (tho Seeth and others: 1998).
11. Although the plots averaged from one-half to one acre initially, subsequent amendments allowed for distribution of parcels smaller than one quarter of an acre near urban areas and more than three acres in unfertile rural areas [Rosenn: 1963, quoting 28 L.P.R.A. (1955), secs. 552, 554].
12. "Now, the workers vote as they want and not as the landowner wants.... The landowners finally realized that labourers had rights that must be respected and the workers realized that selling their votes to the landowners not only was immoral but very bad business. Democracy started to function then and there, a real revolution occurred in the electoral process in Puerto Rico." (Pico: 1964: 153).
13. In a study of a lowland rural village in Central Java, Rajagukguk (1989) found that of 1002 village families, 44 families (4%) owned no pekarangan or household plot, 347 families (35%) owned 50 m^2 or less, 328 families (33%) owned 50-100 m^2, 259 owned (26%) 100-500 m^2 and only 24 families (2%) owned more than 500 m^2. The size of rural pekarangan plots is declining over time, which Arifin (2002) attributes in part to the widespread practice of parents allowing their children to build houses on the pekarangan plots, which are then divided among the children upon the death of the parents.
14. Presumably, mothers who do not work outside the home—whether by choice or because such wage labour is not available—have more time to devote to managing homegarden production.
15. Urban homegardening was found to improve the diet of urban Filipino families by increasing the varieties of fruits and vegetables consumed (Miura: 2003).
16. In a December 2000 survey of 500 rural households in West Bengal, respondents estimated the average cost of non-irrigated arable land to be 46,975 Rupees per acre (Hanstad and Lokesh: 2002), which is equivalent to approximately US$ 1,006 per acre ($US 2,487 per hectare) at December 2000 exchange rates. If one hectare is divided such that 20 families receive 500 m^2 of land each, this represents an average acquisition cost of approximately $124 per family benefited.
17. Brownrigg also notes that school teachers are a particularly poor choice as local homegarden extension agents since they either have no experience in gardening or tend to approach the garden as an academic exercise rather than focusing on actually producing plants and animals.
18. Definition of the general class of beneficiaries is presumed to depend upon the base poverty line, family ownership of various assets, and other measurements already familiar to planners.

19. If "family" is defined as a nuclear family (parents and unmarried children), then it is quite common to find that married children occupy the homesite of their parents, such that several generations, comprised of several nuclear families, occupy the same homesite. In many such situations, the second generation is functionally landless since they own no land of their own and the homesite is not large enough to support the number of occupying nuclear families.
20. Marten (1990) reports that a study of households in West Java revealed that although rice fields controlled by the poorest households produced a higher value of products per square meter as compared to the value produced per square meter on the homegarden, the cost of purchased inputs for homegardening were much less; for the better-off households (defined as any household earning more than $100 per year), the value of homegarden production per square metre equalled the value produced per square metre on rice fields.

4

Access to Land: Some Issues

Srijit Mishra

Introduction

Crisis in Indian agriculture has two dimensions, the agrarian and the agricultural.[1] The former refers to the livelihood crisis of those dependent on agriculture and it includes the tillers of the soil who do not have access to land and other inputs whereas the latter refers to the developmental crisis on account of poor designing of programmes and inappropriate allocation of resources and it includes the increasing allocation of land for non-agricultural purposes. These two are interrelated because what the individual farmer faces has structural underpinnings. In developmental discourse these can be identified with the displacement of people on the one hand and the displacement of ideology on the other hand (Bhaduri: 2008).

Crisis in Indian Agriculture

Agrarian Livelihood Crisis

Some features of the current agrarian livelihood crisis are the following:

- Declining share of agriculture in national income. Growth in gross domestic product (GDP) has been higher than agricultural-GDP, but what is of concern is that agricultural-GDP's growth rate in the 1990s (1993-94 to 2004-05) was lower than that in the 1980 (1983-84 to 1992-93) whereas for overall GDP the growth rate was much higher in the 1990s. The share of agricultural-GDP to total GDP decreased from 30 per cent in 1993-94 to 30 per cent in 2004-05.

- High dependence on agriculture. Employment and unemployment situation based on the National Sample Survey (NSS) indicate that share of employment in agriculture reduced from 64 per cent in 1993-94 to 57 per cent in 2004-05. This is indicative of limited non-farm opportunities in rural India.
- Increasing feminization of agriculture. Using NSS data for rural areas one further observes that the proportion of female headed households increased from 9.7 per cent in 1993-94 to 11.3 per cent in 2004-05. Further, the number of persons employed in agriculture declined from 74.1 per cent to 66.5 per cent for males and from 86.2 per cent to 83.3 per cent for females. This shows that females have even lesser opportunities of going out from agriculture.
- Waning profitability and poor returns. The Situation Assessment Survey (SAS) of farmers carried out as part of NSS in 2003 suggests that per capita per day returns from cultivation is less than eight rupees in 2002-03. Again, from the survey it is said that 40 per cent of the farmers did not want to continue farming largely because of poor returns that makes agriculture a difficult livelihood option.
- Rise in undernourishment among farmers. Using poverty line updating done by the Planning Commission the total number of poor among farmers decreased from 223 million in 1983-84 to 63 million in 2004-05, but while calculating calorie poor using 2400 calorie as a norm then the total number of poor increased from 153 million to 200 million in the same period.
- Marginalization of holdings. Based on NSS of 2004-05, one observes that from all rural households the size-class of land cultivated in 2003-04 is distributed as follows: 62.1 per cent had up to 0.40 hectares (landless and lower marginal holdings), 17.1 per cent had between 0.41-1.00 hectares (upper marginal holdings), 11.3 per cent had between 1.01-2.00 hectares (small holdings) and 9.5 per cent had 2.01 and above hectares.
- Increasing incidence of farmers' suicides in India. Between 1995 and 2008, suicide mortality rate (suicide deaths for 1,00,000 persons) for male farmers increased from 9.7 to 18.0 and that for non-farmers increased from 12.6 to 14.1. For 2006-08, average suicide mortality rate of male farmers is 18.1. Some of the major states where it is higher than this are Kerala (271.6), Maharashtra (59.5), Chhattisgarh (59.2), Andhra Pradesh (37.1),

Karnataka (32.7) and West Bengal (20.9). Being a rare event, relatively higher incidences among farmers is symptomatic of a larger socio-economic malaise. For every single suicide death, hundred thousands of farmers are in distress.

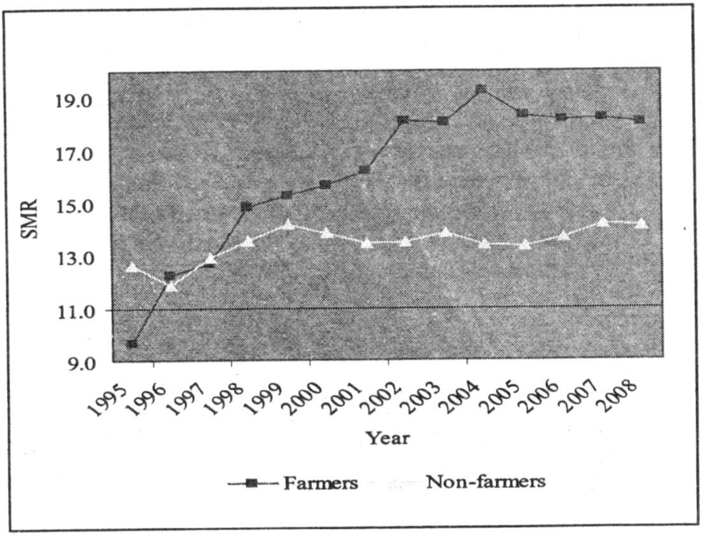

Fig 4.1: Suicide Mortality Rate (SMR) for Males, Farmers and Non-farmers, 1995-2008.

Note: Calculations exclude Tamil Nadu for 1995 and Jharkhand for 2003; method is as elucidated in Mishra (2006).

Source: National Crime Record Bureau (Various Years) and Census of India, 1991 and 2001.

Agricultural Developmental Crisis

Some features of the agricultural developmental crisis are the following:

- Declining production and productivity. Compared to 1980s (Triennium ending, TE 1981-82 to TE 1993-94), the 1990s (TE 1994-95 to TE 2007-08) had trend growth rates that were statistically significantly lower for almost all crop groups with regard to production, productivity and value. For instance, annual trend growth rate for productivity for different crop groups or crops for the two periods respectively are as follows: cereals (3.5 per cent, 1.4 per cent), pulses (1.6 per cent, 0.2 per cent), oilseeds (2.9 per cent, 1.3 per cent), sugarcane (1.8 per cent, –0.2 per cent), cotton (3.0 per cent and 2.0 per cent), coconut (2.9 per cent, –0.1 per cent), and potato (1.8 per cent, 0.7 per cent). What is more, the 1990s growth rate for cereals, pulses and oilseeds is lower than the 1.9 per cent growth rate of population based on censuses.

- Decline of public investment in irrigation and other infrastructure. Public gross fixed capital formation in agriculture as a proportion of agricultural-GDP declined from 5.3 per cent in the Sixth Plan (1980-81 to 1984-85) to 2.1 per cent in the Ninth Plan (1997-98 to 2001-02) and then increased to 3.0 per cent in the Tenth Plan (2002-03 to 2006-07). Further, during 1990-91 to 2006-07, the growth of area under irrigation has been negative for canals and tanks, but positive for borewells/tubewells and other sources. The demise of public systems led to the emergence of private systems, which started a tragedy of commons identified with declining ground water and escalating costs, particularly in Andhra Pradesh.
- Inadequate supply of credit from formal sources. The All India Debt and Investment Survey (AIDIS) of 2003 and comparable earlier surveys indicate that the share of debt for cultivator households from formal sources increased from 32 per cent in 1971 to 66 per cent in 1991 (post-bank nationalization and other developments period), but reduced to 61 per cent in 2002 (post-liberalization period). The loan from informal sources was at a greater interest burden. Further, during the 1990s and till six/seven years ago the number of rural branches declined, the number of agricultural borrowal accounts declined, share of area under marginal holdings increased but the share of credit disbursed to the group decreased. There are no mechanisms to address non-willful default to formal sources arising out of crop failure, price shock or other genuine contingencies faced by the farmer. The recent debt waiver of 2008 is a bookkeeping exercise that will clear the non-performing assets of the banks and to that extent the loan burden of the farmer and makes her/him eligible for further credit. This, however does not address how to improve the poor returns to cultivation.
- Failure of research and extension. The green revolution technology of the 1970s was largely meant for rice and wheat under irrigated conditions. This has meant a neglect of millets

and pulses that are grown under rain-fed and dry land conditions in large parts of the country. Further, research has been compartmentalized to a mono-crop based system that puts practices like intercropping and crop rotation to the background. To add to this, continuing education and further recruitment of extension workers were stopped as part of fiscal discipline adopted by the governments since early 1990s. Moreover, a hierarchy has been put in place with the scientist at the top, extension worker in between and farmer at the bottom such that knowledge production and dissemination, instead of a two-way process, is on top-down basis.

- Changing technology and market conditions has increased uncertainties in product and factor markets. There has been a shift in agricultural practices from the farmers owning their own seeds and using other locally available resources to one where the farmer is increasingly dependent on the market for technology and inputs – seeds, fertilizers and pesticides among others. As a consequence of this, one can observe deskilling among farmers as traditional knowledge becomes redundant, there is supplier-induced-demand, and the farmer is exposed to input shocks in price and quality besides the output shocks in price and quantity.
- Increasing demands for non-agricultural use of land and water. A matter of concern in recent years is that agriculturally prosperous land are increasingly being lost to urbanization and other projects under Special Economic Zones (SEZs) and industrialization among others.

Matrix of Issues

The farmer is exposed to a multitude of risks – vagaries of weather, input prices, spurious inputs, output prices and other exigencies. Some of the concerns can also be summarized under a matrix of issues given in Table 4.1.

Risks: Access to Land and Inputs

One can further elaborate on the risks by comparing alternative practices and the farmer's access to land and inputs. This is done through the prism of choice of technique (Sen: 1960) where an improvement is considered as one where either input reduces or output increases. An illustration is

Table 4.1: Matrix of Issues

Issues	*Demand*	*Supply*
Output, Price, Income	Yield risk: weather, power, pests, spurious inputs; Not profitable; Poor returns	Increased price volatility; High subsidies in the United States and European Union; Low tariff; Minimum Support Prices are not always functional; Futures are a virtual medium that will provide little or no help to the farmer
Input	Supplier-induce-demand; Deskilling; Increasing costs – tragedy of commons	Poor link between research and extension; Unregulated private suppliers; Inadequate public investment
Credit	Formal sources – not timely; Repayment difficult under yield/price shocks; System draws farmers into credit; Consumerism	Decline in branches; Decline in agricultural/net bank credit (direct); Increasing reliance on informal sources at higher interest burden
Land	No access to tiller	Increased demand and land acquisition policies for private entities with public support for non-agricultural usage has reduced supply for agricultural purposes
Other	Dominance of lender/input dealer; Higher family size; Lack of social support	Interlinked markets; Non-farm option is limited; Pub health response (farmers); Pesticide availability

Source: Reserve Bank of India (2006), Mishra (2008).

given in Table 4.2. Under rented land with input intensive technology, the per annum value of input is 2.0 units and output is 4.0 units giving a net return of 2 units and from this consumption is of 1.4 units and savings is 0.6 unit. The cumulative savings at the end of three years would fall short of meeting input costs and providing for consumption at slightly reduced level. Thus, to make ends meet the farmer would have to borrow in a bad year.

In the low external input under rented land, the per annum value of input is 1.2 units and the output is 3.0 units, 3.5 units and 4.0 units in the first, second and third years respectively. As a result the net returns, consumption and cumulative savings keep increasing. In this scenario, the cumulative savings at the end of the third year can compensate for input costs and consumption at a slightly reduced level but still leaves the farmer with some amount of savings.

Low external input under own land will reduce input costs as the farmer does not have to pay rent, but output will remain the same as that of the second scenario. Net returns, consumption and cumulative saving

Table 4.2: Comparing Traditional, Input-intensive and Sustainable Technology: An Illustration

Technology	*Year*	*Input*	*Output*	*Net Return*	*Consumption*	*Cumulative Savings*
Rented land, input intensive, T_0	1	2.0	4.0	2.0	1.4	0.6
	2	2.0	4.0	2.0	1.4	1.2
	3	2.0	4.0	2.0	1.4	1.8
	4	1.5	0.5	-1.0	1.0	-0.2
Rented land, low external input, T_1	1	1.2	3.0	1.8	1.3	0.5
	2	1.2	3.5	2.3	1.5	1.3
	3	1.2	4.0	2.8	1.8	2.3
	4	1.2	1.0	-0.2	1.3	0.8
Own land, low external input, T_2	1	1.0	3.0	2.0	1.4	0.6
	2	1.0	3.5	2.5	1.6	1.5
	3	1.0	4.0	3.0	1.9	2.6
	4	1.0	1.0	0.0	1.4	1.2

Note and Source: Based on similar exercises in Mishra (2008), Mishra and Gopikrishan (2010) and Government of India (2010).

will be relatively higher such that in a bad year it compensates for consumption and there are some savings left that could even help if there is a problem for the second consecutive year. However, the tiller, in large parts of India, does not have access to land. This takes us to the unfinished tasks in land reforms.

Unfinished Tasks in Land and Tenancy Reforms

The important issues under land and tenancy reforms are as follows[2]:

- *Abolition of Intermediaries*. This is with regard to Zamindar and other intermediaries who had a share, sometimes a substantial amount, in the rent. The association of these Zamindars and their intermediaries with the exploitative British made it easier to address this. This has been abolished across India.
- *Tenancy Reforms*. The purpose of this is to regulate contracts between a landowner and a tiller by registration and stipulation of the terms or by abolishing tenancy. Tenancy reform had a lot of resistance at the legislature and also implementation level because of vested interest from the landed gentry. In some cases, where tenancy is abolished in paper, it has gone underground and it is difficult to estimate its nature and extent based on surveys. This is an unfinished task.

- *Land Ceiling*. The agenda here is to distribute surplus land to landless so that land is with the tiller. Such an arrangement also supports efficiency because of the inverse relationship of size-class with productivity. Unfortunately, the failure to implement land ceilings was even more evident. Loopholes in the policies and delays led to its tardy progress. With increasing marginalization of holding, its relevance will reduce, but this continues to be another unfinished task.
- *Consolidation of Holdings*. The idea is to bring all lands of a cultivator together to reduce transaction cost and thereby increase productivity. Consolidation was less enacted and poor land records also made it difficult. This also remains as an unfinished task.

Suggested Measures

Providing greater access to the tiller is appealing from equity as well as efficiency perspective. However, it is not that simple a proposition, as it raises questions on property rights. Nevertheless, some suggestions are as follows[3]:

- The land-lease market should be opened-up in states where it is illegal so that tenancy does not go underground. One should accept tenancy, but guard against reverse tenancy.
- In states where tenants are protected, like West Bengal, there could be provision of making them owners on part of the land while giving up their tenancy rights on the remaining part. This will benefit the landowners as also the tenants.
- Transaction costs in land markets should be reduced. This includes the official fees on account of documentation as also bribes paid in the process.
- Moratorium on land obtained through distribution of ceiling surplus land should be removed after a time lag. This will enable a person or household that has moved out of agriculture to sell the same and move into other activities.
- Efforts should be made to improve administration and enforce transparency so as to do away with rent seeking. Revenue department is known for rent-seeking in public perception.
- Women should be given independent or joint land rights. This will increase their status and bargaining power.

- The government can consider buying land to distribute. If market rates are paid then the existing landowners who do not have much stake in agriculture are likely to sell these.
- While distributing homestead plots the government should consider additional land that could be used as a kitchen garden or for some livelihood options, as is evident from such micro-plot successes.
- People's institutions need to be built and the civil society needs to be strengthened to provide checks and balances.

References

Besley, T. and Burgess, R. (2000): "Land Reform, Poverty Reduction, and Growth Experience: Evidence from India", *The Quarterly Journal of Economics* 115(2): 389-430.

Bhaduri, Amit (2008): "Predatory Growth", *Economic and Political Weekly*, 43 (16), 10-14.

Government of India (2007): *Report of the Expert Group on Agricultural Indebtedness*, Ministry of Finance, New Delhi (Chairperson: R. Radhakrishna).

Government of India (2009): *Report of the Committee on State Agrarian Relations and the Unfinished Task in Land Reforms*, Ministry of Rural Development, New Delhi (Chairperson: C.P. Joshi).

Government of India (2010): *Task Force on Credit Related Issues of Farmers*, Ministry of Agriculture, New Delhi (Chairperson: U.C. Sarangi).

Mearns, R. (1999): *Access to land in rural India*. Policy Research Working Paper 2123. The World Bank: Washington, DC.

Hanstad, T., Haque, T. and Nielsen, R. (2008): Improving Land Access for India's Rural Poor, *Economic and Political Weekly*, 43(10): 49-56.

Hanstad, T. and Nielsen, R. (2009): Land Tenure Reform in India, in R.L. Prosterman, R. Mitchell and T. Hanstad (eds) *One Billion Rising: Law, Land and the Alleviation of Global Poverty*, Leiden University Press.

Mishra, S. (2008): Suicide Mortality Rates across States of India, 1975-2001: A Statistical Note, *Economic and Political Weekly*, 41(16): 1566-1569.

Mishra, S. (2008): Risks, Farmers' Suicides and Agrarian Crisis in India: Is There a Way Out? *Indian Journal of Agricultural Economics*, 63(1): 38-54.

Mishra, S. (2010): Agrarian Crisis and Farmers' Suicides in India, Revised version of a paper presented at a one day international seminar, "Environmental degradation and food crisis – Lessons for India" organized by Greenpeace India on 24 October 2008 at India International Centre, Lodhi Road, New Delhi, India.

Mishra, S. and Gopikrishna, S.R. (2010): *Nutrient Based Subsidy (NBS) and Support Systems for Ecological Fertilization in Indian Agriculture*, Policy Brief, Greenpeace India.

Mishra, S. and Reddy, D.N. (2010): Persistence of Crisis in Indian Agriculture: Need for Technological and Institutional Alternatives, in: D.M. Nachane (ed.) *India Development Report 2010*, Oxford University Press, New Delhi, forthcoming.

Prosterman, R.L. (2009): Concluding Reflections, in R.L. Prosterman, R. Mitchell and T. Hanstad (eds) *One Billion Rising: Law, Land and the Alleviation of Global Poverty*, Leiden University Press.

Reddy, D.N. and Mishra, S. (eds.) (2009): *Agrarian Crisis in India*, Oxford University Press, New Delhi.

Reserve Bank of India (2006): *Report of the Working Group to Suggest Measures to Assist Distressed Farmers*, Rural Planning and Credit Department, Mumbai (Chairperson: S.S. Johl).

Notes

1. The discussion on this, which is elaborated further in the section on 'Crisis in Indian Agriculture', borrows from the author's recent work on related themes, Mishra (2009), Mishra and Reddy (2010) and Reddy and Mishra (2009); also see Government of India (2007).
2. This is largely based on the discussions in Besley and Burgess (2000), Government of India (2009), Hanstad and Nielsen (2009), and Mearns (1999).
3. This is largely based on the discussions in Government of India (2009), Hanstad, Haque and Nielsen (2008), Hanstad and Nielsen (2009) and Prosterman (2009).

5

Structural Changes in Land Distribution and Its Implications for Improving Access to Land

K.N. Nair and *Arindam Banerjee*

On the basis of the pattern of land distribution the emerging agrarian structure in India today presents a grim picture of diminishing size of landholding coupled with the dominance of marginal and unviable small holdings. Typically, these structural features are considered barriers to rapid agricultural transformation. Given this situation, it has become increasingly difficult for the marginal and small holders to earn their livelihood exclusively from agriculture. Consequently, they are compelled to seek livelihood opportunities outside agriculture. Moreover and somewhat paradoxically in the post-reform period from the 1990s, the pace and pattern of growth of the economy in general and of the non-agricultural sector in particular has been of a kind least conducive for absorbing these marginal and small holders productively. Under these circumstances they are left with little option but to continue to cling on to their small pieces of land. It is seen that the process of marginalization of land holding also poses organizational challenges for the efficient utilization of land, water and other inputs and also to the diffusion and adoption of various technologies including agro-mechanical technology. It is in this context that the important issue discussed is the scope for improving the access to land for the landless and marginal farmers (who constitute the bulk of the rural poor) as a means to improve their livelihood. Many prescriptions have been offered by policy making bodies at the national and international levels and though some of the policies have been adopted and implemented yet they have not translated into any noticeable gainful result. We are inclined to argue that for any meaningful intervention in the sphere of land redistribution it is imperative to critically examine the changes in the pattern of land distribution and the factors shaping it. This chapter is a modest attempt in providing some insights to this rather complex issue.

This chapter is organized as follows. Section I examines the changes in the pattern of land distribution. This is followed by an analysis of the factors and processes shaping the distributional changes on Section II. The last section highlights the implications of the analysis for improving the access to land.

I

Changes in the Pattern of Land Distribution

The NSS Land and Livestock Surveys are an important data source on the size distribution of ownership holdings and are available for every ten years from 1961-62. The source also provides data on pure landless category. For the purpose of the present analysis we have refrained from examining the patterns of change by size groups since such has already been commented upon in a number of existing studies and reports.

Estimated number of rural households for the country as a whole in the early sixties was about 64 million and it increased to 148 million by the early two thousand (NSS Land and Livestock Survey, 17th round, Report No. 144 and 59th round, Report No. 491). It is seen that increase in the number of rural households has taken place at a much faster rate in the 1980s and 1990s. The percentage of rural households without land has declined over the last four decades; but in absolute numbers it has increased due to increase in the number of rural households. The average size of ownership holding by excluding the landless declined from 1.99 ha in 1961-62 to 0.73 ha by 2002-03 (*ibid*). In the early sixties, 55 per cent of the holdings belonged to the marginal size group and it increased to nearly 80 per cent by 2002-03 (*ibid*). The distribution of owned area by size category of holdings revealed a trend different from the number of holdings; while the share of the owned area of the medium and large holdings showed a decline, the percentage share of small and semi-medium size classes increased from 33 to 42 and that of marginal from less than 8 to 23 (see Table 5.1).

In order to understand the manner in which the changes noted in the size categories has shaped the distribution we have estimated the share of bottom 60 per cent, middle 30 per cent and top 10 per cent in the total owned area for 1961-62 to 2002-03 for the country as a whole. The figures were obtained by fitting the Lorenz Curve and deriving the respective shares (see Figure 5.1).[1] It is seen that the shares of different groups

Table 5.1: Percentage Distribution of Owned Area by size-classes

Size Class of holding (in hectares)	*1961-62 (17th round)*	*1971-72 (26th round)*	*1981-82 (37th round)*	*1991-92 (48th round)*	*2002-03 59th round*
Marginal	7.6	9.76	12.22	16.93	23.05
Small	12.4	14.68	16.49	18.59	20.38
Semi-medium	20.5	21.92	23.58	24.58	21.98
Medium	31.2	30.73	29.83	26.07	23.08
Large	28.2	22.91	18.07	13.83	11.55
All	**100.0**	**100.0**	**100.0**	**100.0**	**100.0**

Source: NSS Land and Livestock Survey, various rounds.
Note : For the 59th round, the data is for the Kharif season alone.[2]

remained fairly stable from 1961-62 to 1991-92. In the subsequent period, the percentage share in area of the bottom 60 per cent households declined from 10.8 to 8.3. The percentage share of the middle 30 per cent has also marginally reduced from 37.7 to 36.5 whereas the share of the top 10 per cent households increased from 51.5 to 55.2. (see Table 5.2)

Table 5.2: Percentage Distribution of Owned Area by Household Groups: various rounds

Household Group	*1961-62*	*1971-72*	*1981-82*	*1991-92*	*2002-03*
Bottom 60%	10.8	10.8	10.5	11.0	8.3
Middle 30%	37.7	38.2	38.5	38.0	36.5
Top 10%	51.5	51.0	51.0	51.0	55.2

We have also examined the distributional changes at the state level by deriving similar land ownership and operated area shares for two points of time, namely 1961-62 and 2002-03. The results showed (Table 5.3) that in all the states, except Kerala (+7.3), Orissa (+2.1), and Bihar (+1.4), the share of the bottom 60 per cent households has declined; in Kerala, the share has increased at the expense of the middle 30 per cent. However, even in Kerala, we observe an increase in the share of the Top 10 per cent households. In fact, the share of top 10 per cent increased in all the states, except Orissa (moderate decline) and Andhra Pradesh (unchanged). But the latter was reporting one of the highest shares for the Top 10 per cent already in 1961-62. The increase in the share of the Top 10 per cent was particularly large in states like Gujarat (+11.5), Punjab (+10.8) and

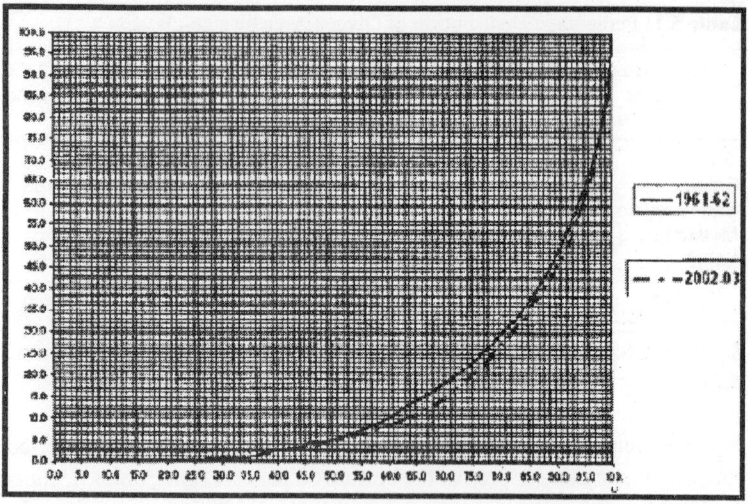

Fig. 5.1: Lorenz Curve for Ownership Distribution of Land, All-India: 1961-62 and 2002-03.

Tamil Nadu (+12.5). In all these states, this concentration of land in the hands of the topmost stratum of households has occurred mainly at the expense of the Bottom 60 per cent. The share of the Bottom 60 per cent in 2002-03 was only 7, 2 and 2 per cent in Gujarat, Punjab and Tamil Nadu respectively. The middle landholding households in Gujarat and Punjab have also lost ownership of land significantly.

Turning to the distribution of operational holdings, it may be seen that the general tendency is for a decline in the share of the Bottom 60 per cent, stagnancy at the Middle 30 per cent, and a slight increase in the share of the Top 10 per cent. There are exceptions to this general trend, with Kerala showing a large increase in the share of the Bottom 60 per cent from 14 per cent to 24 per cent, and Gujarat (-9.0), Punjab (-18.6) and Tamil Nadu (-9.3) showing large declines.

The share of the Top 10 per cent has either stagnated or shown a marginal increase, with the exception of Gujarat, Punjab and Tamil Nadu; the increases in the shares of the Top 10 per cent are 10.4, 21.0 and 8.2 per cent points respectively. It may be observed that the magnitude of the increase in the share of the Top 10 per cent of operated holdings corresponds well with the increase in the share of owned area by this group in these three states.

Table 5.3: State-wise Percentage distribution of Owned and Operated Area by Household Groups: 1961-62 and 2002-03

States	Year	Distribution of Owned Area			Distribution of Operated Area		
		Bottom 60%	Middle 30%	Top 10%	Bottom 60%	Middle 30%	Top 10%
Andhra	1961-62	6.5	34.4	59.1	15.8	36.7	47.5
Pradesh	2002-03	4.5	36.5	59.0	18.0	36.2	45.8
Bihar*	1961-62	10.1	40.4	49.5	19.2	38.8	42.0
	2002-03	11.5	38.0	50.5	17.9	40.6	41.5
Gujarat	1961-62	13.5	43.0	43.5	22.5	39.7	37.8
	2002-03	7.0	38.0	55.0	13.5	38.3	48.2
Karnataka	1961-62	18.0	40.0	42.0	22.5	38.2	39.3
	2002-03	11.5	39.8	48.7	18.9	38.1	43.0
Kerala	1961-62	12.2	33.6	54.2	14.2	36.5	49.3
	2002-03	19.5	23.7	56.8	24.3	27.9	47.8
Madhya	1961-62	16.5	41.0	42.5	22.2	39.9	37.9
Pradesh*	2002-03	15.5	40.0	44.5	20.5	39.2	40.3
Maharashtra	1961-62	12.0	42.5	45.5	19.1	41.3	39.6
	2002-03	10.5	42.0	47.5	20.2	40.3	39.5
Orissa	1961-62	12.0	39.5	48.5	23.5	37.9	38.6
	2002-03	14.1	40.9	45.0	22.5	41.5	36.0
Punjab*	1961-62	6.1	43.4	50.0	23.5	43.0	33.5
	2002-03	2.0	36.7	61.3	49.0	40.6	54.5
Rajasthan	1961-62	16.0	38.5	45.5	20.0	37.5	42.5
	2002-03	13.4	36.9	49.7	15.0	37.0	48.0
Tamil Nadu	1961-62	10.8	39.7	49.5	23.5	38.9	37.6
	2002-03	2.0	36.0	62.0	14.2	40.0	45.8
Uttar	1961-62	15.5	40.5	44.0	23.4	40.2	36.4
Pradesh*	2002-03	14.0	40.7	45.3	18.1	41.6	40.3
West Bengal	1961-62	13.5	42.7	49.8	26.8	39.4	33.8
	2002-03	12.8	38.2	49.0	21.0	40.3	38.7
All-India	1961-62	10.8	37.7	51.5	18.0	37.5	44.5
	2002-03	8.3	36.5	55.2	14.5	37.1	48.4

Source : Estimated by the authors from the Lorenz Curves for each state.

Note : Bihar, Madhya Pradesh, Uttar Pradesh and Punjab include Jharkhand, Chhattisgarh, Uttaranchal and Haryana respectively. While Haryana was formed in 1966, the other three states were formed in 2000.

II

Factors and Processes

The distributional changes noted. for the country as a whole and across the states could be due to the net effect of transfer of land that has taken place through partition of households, distribution of land by the state through land reform measures and through the market process. Coming to the partition of households, one would expect the rate of partition among households to increase with increase in size of landholding. This is because of the positive relation between size of the household and size of landholding under the assumption that the probability of partition will be the same across different size groups. However, there could be some distortion in this pattern if larger holdings tend to delay partitioning due to reasons of economies of scale or other factors that binds them together as joint households (Krishnaji: 1980). Increase in the number of households in the bottom group and decrease in the higher size classes of landholding indicates the consequences of partition on the downward mobility of households across size groups of landholdings.

Changes in land ownership as a result of the implementation of land reforms might have contributed to the changes in land distribution. In this context one of the important measures implemented in the 1950s was the abolition of intermediaries. As a result of this and subsequent struggles by tenant farmers for land rights, some 20 to 25 million tenants in the *Zamindari* areas in West Bengal and the former state of Uttar Pradesh became owners of land. This part of land reforms also resulted in the transfer of ownership of vast tract of forest and waste land to state ownership. Another important measure implemented was the distribution of surplus land above ceiling limit to the landless. During the early years of the implementation of ceiling law, the progress was very slow and it gave opportunities for land owners to circumvent the provisions of the Act and evade their surplus land. Though, there were attempts for the speedy implementation of the Act in recent decades, the extent of land re-distributed so far was about 5.4 million acres distributed to about 5.6 million households (World Bank: 2007).

Regarding the implementation of tenancy laws, the security of tenure and the associated terms and conditions varied across states as per the different legislations. The key provisions of the tenancy laws relating to the circumstances in which land owners can resume leased-out land for own cultivation, the payment of compensation to the land owners in return for the ownership of land has come under lot of litigations in civil courts resulting in delayed or inadequate implementation of land reform measures. A recent study on land policies in India summarized the effectiveness of various land reform measures as follows:

> "Overall, land reform constituted a major effort that resulted in the transfer of almost 10 million hectare—2.5 million ha under

programmes to redistribute of ceiling surplus land, and 7.35 million ha under tenancy legislation. (...) Although the magnitude of accomplishments is put somewhat into perspective if the share of land and population are considered, land reform in India was undoubtedly a major historical event, especially if the amount of land distributed under abolition of intermediaries is added to these figures". (World Bank: 2007: 56-7).

Table 5.4: Share of households and area affected by land reforms across states

	Tenancy		*Ceiling*	
State	*Area (%)*	*Pop (%)*	*Area (%)*	*Pop (%)*
AP	3.48	0.75	8.34	3.81
Bihar	0.00	0.00	4.42	4.00
Gujarat	15.00	11.20	1.95	0.31
HY	0.51	0.01	1.26	0.26
HP	0.16	3.19	0.06	0.05
KA	15.38	5.29	1.71	0.30
KE	8.47	12.49	1.30	1.04
MP	2.15	0.61	2.69	0.71
MA	27.01	1.68	7.74	1.08
OR	0.15	1.43	2.24	1.28
PI	1.89	0.04	1.50	0.25
RI	0.00	0.16	6.63	0.75
TN	3.65	3.23	2.47	1.24
	0.00	0.00	5.81	3.68
WB	6.41	10.80	14.91	19.73
Total	**5.45**	**5.35**	**4.41**	**2.27**

Source: Kaushik (2005), taken from World Bank, 2007.

For the country as a whole, it is estimated that through tenancy reform the amount of land distributed accounted for 5.45 per cent of the land area involving 5.35 per cent of the population. The amount of land transferred through ceiling provision amounted to another 4.41 per cent of the land area and 2.27 per cent of the population (Table 5.4). There are also significant interstate differences in the effectiveness and coverage of land reforms as is evident from the area and population affected. While in a few states, namely West Bengal, Gujarat, Maharashtra, Kerala and Karnataka, the effectiveness of tenancy reforms has been impressive, in other states the effectiveness of implementation has been dismal as is evident from the low coverage in terms of area and population. Coming to the implementation of ceiling laws, Andhra Pradesh, Bihar,

Maharashtra, Uttar Pradesh, Rajasthan and West Bengal are the states, where some progress has been achieved. In both the measures, only West Bengal has made notable achievement. It may also be noted that after the progress made in the 1970s and 1980s, the implementation of both these measures has slowed down in all the states in recent years.

The impact of land reforms on the rural economy has been studied by researchers in the context of specific states. These studies revealed the positive impact of land reforms in reducing poverty and in improving the socio-economic status of the beneficiary households. A nationwide survey spanning the period 1982-99 attempted to assess the impact of land reforms on physical and human capital growth and how their impact evolved overtime.[3] The results of this survey show that the households in those states with higher levels of land reform effort saw their welfare and investment improved more than that of households in states where land reform implementation remained lackluster.

The results from this survey also points out that the benefits of land reforms have been declining overtime. An important reason behind this could be the inability of households to retain their landholding 'due to division among family members or sale and the consequent downward mobility in the size of landholding. The modernization of agriculture through technological changes that required higher investment in capital intensive and labour saving technologies has reduced the viability of small holder farming compared to the medium and large holdings have resulted in the gradual exist of the marginal and small holding from farming. This has resulted in the transfer of land either through sale or by lease from the bottom size of holdings to the medium and larger size of holdings. The increase in the share of land owned and operated by the top 10 per cent of holdings is a reflection of this process. The agrarian crisis that precipitated in the nineties, particularly in export oriented commercial crop cultivation (Patnaik: 2002, Reddy and Mishra: 2007), has also reduced the profitability and the scope of investment in agriculture and emergence of high indebtedness among agricultural households, thereby nullifying some of the earlier gains of land reforms. Whatever be the underlying causes, an in-depth understanding of the process of land transfer is necessary to understand the linkages between the structural changes in land distribution and its implications for development.

As we noted earlier, a constraint in undertaking this task is the absence of reliable data to evaluate the relative importance of the various factors in shaping the land distribution. In the absence of such data, we have attempted to provide some insights into the factors and processes that shaped the distribution of land by drawing on the evidence available from some village level studies.

In the late 1990s, a number of villages in southern India were studied by a village resurvey project coordinated and implemented in the Centre for Development Studies. These were villages studied as part of the 1961 village survey monographs prepared by the Census of India or surveyed by other scholars in the sixties. The principal focus of these studies was to capture the processes of change that have overtaken the rural economy, especially with respect to the caste and economy nexus. These studies have analyzed data on the changes in the distribution of land covering roughly a period of 40 years. While intensive studies of this nature are available for two villages each from Kerala, Tamil Nadu, and Andhra Pradesh, there were also a few more restudies using qualitative research methods from each of these states. Some of the insights from these studies are briefly summarized below.

Though the Kerala case is exceptional as we have indicated earlier yet it may not be out of place to summarise the general trends discernible from the study of the villages resurveyed. These are, (i) a reduction in the concentration of land ownership, (ii) acquisition of land by the bottom size groups through purchase, (iii) significant reduction in landless households due to the allocation of house sites to landless from within the ceiling-surplus land, and (iv) land reforms as such had very little impact in changing the distribution of land.

In all the three villages in Kerala, households belonging to the intermediate castes, namely the Ezhavas and Muslims, increased their share of land and the share of the forward castes declined. Though, this was partly due to the effect of land reforms, the purchase of land by the intermediate castes has also contributed significantly to this in recent years. Migration to the Gulf and the consequent flow of remittances has also facilitated these communities in improving their access to land by purchase. The Scheduled Castes, which constituted the bulk of the agricultural labour did not benefit from the various modes of land transfer.

The study of two villages in Tamil Nadu conducted by Jeyaranjan for Illuppakorai and Rukmini for Arkavadi showed the interrelation between agrarian changes on the one hand and changes in the land distribution on the other (Draft Project Reports, KRPLLD, CDS). In the first village that falls in the Thanjavur delta, agricultural production based predominantly on the paddy crop faced a serious crisis in the 1970s due to water scarcity arising out of the Cauvery water dispute; besides this, there were also problems arising out of the monopoly procurement of paddy by the State and increasing militancy among the agricultural labourers. In response to the severe constraints experienced in the cultivation of paddy, farmers started shifting to crops like sugarcane and banana that did not require flooding the fields. By the 1980s, about 70 per cent of the area in the village was under these two crops. In the subsequent two decades there was some improvement in paddy cultivation because of the improvement in irrigation systems, increase in the mechanization of agriculture and decline in the labour militancy. The expansion of area under sugarcane and banana required more capital to cultivate such crops and the degree of risk and uncertainty involved with regard to marketing arrangements and prices was higher. The impact of these changes on the structure of landholding was mainly the following.

During the period 1975-85, there was a substantial weakening of the smaller size classes of landholders. This was evident from the fact that 18 out of the 59 households who possessed less than 2.5 acres of land became landless during the period. On the other hand, those who owned more than 5 acres of land retained their status. In the subsequent period, there was complete reversal of this process. The number of landed households increased from 40 to 59 and extent of land operated by them increased from 222 acres to 264 acres. The top size-class of landholding who operated more than 10 acres of land and accounted for about 60 per cent of the total owned area in the village, lost about 50 per cent of their area and moved down in the scale of landholding. The latter period also witnessed the strengthening of the medium size of holdings in terms of the number as well as area. However, unlike the medium holdings, the small holdings reduced their average size either due to sale or partitioning. The study also reveals that those who lost land in the first period and acquired it subsequently purchased it in the neighbouring village where the Brahmin households were moving to the urban centres.

Though the above study did not analyze the relationship between caste and land ownership, the second study for Tamil Nadu provides more insights into this aspect. This study shows that there has been consolidation of the landholding position by Vanniars who belonged to the list of backward communities in Tamil Nadu. There was a simultaneous growth of internal stratification within the same community. Vanniars purchased most of their land from Reddiars who belonged to the forward castes. The study traces the risk associated with the development of well irrigation in a low rainfall region as the main reason for the loss of land by some households while some others were able to cope up with the situation and acquire more land.

The village studies conducted in Andhra Pradesh provides ample evidence on the changing pattern of caste and land relationships. It clearly reveals the unleashing of the process of shift in the landownership from upper castes to intermediate castes. As the study observes,

> "(from) 1950s onwards, a slow but a steady process of land transfers from the dominant Reddy and Kamma castes to the less dominant FCs (Kapus) and OBCs such as Yadavas, Goudas, Toorpu Kapus, Koppula Velamas, Kalingas, Munnuru Kapus/ Telagas, Setti Balijas and Pallekapus had taken place. With this came the altered inter-caste economic balance in favour of the hitherto less dominant castes and an overall less iniquitous agrarian structure. However, the process of land transfers from the dominant cultivating castes like the Reddys and Kammas is a much slower process compared to the earlier phase of land transfers from the Brahmins to the Reddys and Kammas". (Rao and Reddy, 2008, p. 50)[4].

The study on occupational change and land ownership in Palanpur for the period 1957-93 also provides insights into the process of land transfer (Dreze: 2002). Unlike other studies cited here, this village was resurveyed four times at periodic intervals, the last one being conducted by Jean Dreze. According to his findings, in the short term, changes in land ownership and distribution are shaped by the process of inheritance and partition. However in the long run, some changes are attributable to land sales and purchases that resulted in inter-caste transfer of land. He also concludes that even in the longer period, land transfer through market process have had only secondary influence in comparison with the process of inheritance and partition.

Implications for Improving Access to Land

From the preceding analysis it can be argued that though land concentration has remained stable over long period in history, land ownership has become more widespread and diffused. However, it is also noted that the large majority of the bottom size groups (the Bottom 60%) account for only a very small percentage of the total owned and operated area, which has declined in almost all the states, except Kerala, over the last four decades. Since access to land has a positive impact on reduction in poverty and improvement in household well-being, the question that confronts us at the policy level is the nature and direction of the reform needed to improve access to land to the poor households.

Prevailing discussion on the subject has outlined, three possibilities in realizing this: (a) proper verification of the lands records, identification of the surplus above the ceiling laws and distribution of the surplus to the land poor, (b) reforming the land lease market by removing the existing restrictions on lease cultivation, and (c) providing financial support to the poor to acquire land. As far as proposition (c) is concerned, it is unlikely to be viable for implementation. Any large scale financing of land for purchase of land may result in pushing up land prices, and if the interest on land is also included in the cost of cultivation, then such farming activity is unlikely to become viable. There is also no experience of this kind available either from India or abroad which can be harnessed to understand the nature of problems involved in the implementation of such a policy. Regarding (a), there may be some scope to recover surplus and public land for redistribution, but this is contingent on the implementation of a programme of verification and modernization of land records by the state governments.

Coming to the prospects for tenancy reform, it is to be noted that the percentage of operated area leased in and the percentage of owned area leased out, and the percentage of households involved in leasing in and leasing out has been declining sharply over the last few decades. This might be due to the underreporting of the incidence of leasing, due to legal restrictions and partly might also be due to the shift in tenancy contract from long duration to seasonal or annual contracts, or leasing in and leasing out of land taking place between close relatives. Certain micro level studies seem to suggest that the incidence of tenancy is much higher than what is reported in the official surveys. In order to assess the extent to which leasing of land modifies the access to land for households with smaller size of holding, we need only look at Table 5.3. It may be noted that in comparison with the share of owned area the share of operated area is higher for the Bottom 60 per cent of the households and this is due to the leasing in of land. The loss in the share of this group over the last four decades has been the gain of the Top 10 per cent as a consequence of the reverse tenancy. Across states, except in Kerala, and Andhra Pradesh, the share of the Bottom 60 per cent has declined, and the share of the Top 10 per cent has increased. In other words, in most parts of the country, the lease market has shifted in favour of the larger size of holding.

The pattern and trends in the distribution of operated area is a clear indication of the fact that any process of reform contributing to removal in the restriction of lease market is likely to work more in favour of the larger size of holdings compared to the smaller size of holdings. This is what one would expect in a rural economy, where there exists indivisibility in the ownership of complementary assets to land like agricultural machinery and equipments, and bullocks, which are crucial to carry out agricultural operations. Analysis of the data from the various rounds of the Land Holding Surveys conducted by the NSSO shows that (a) the ownership of agricultural machinery and implements and cattle tends to increase with increase in landholding and (b) overtime, the proportion of households with these assets has been declining (especially in the bottom size groups) with the downward mobility of landholding and increase in the proportion of marginal and small holdings. In a situation of this kind those who have 'surplus' of such assets would attempt to increase its utilization by increasing the scale of their cultivation by leasing in land. Organizationally, it may be possible to support the marginal holdings to overcome such limitations by various types of group initiatives. However, such attempts in the past have not yielded positive results.

It appears that the policy measures cited above would not contribute to significant improvement in the access to land for a large majority of the Bottom 60 per cent of the rural households. The number and proportion of this category in the structure of landholding is likely to increase in future under the weight of both demographic and non-demographic factors. This process could further accentuate the structural barriers to agricultural growth. From the long term perspective of sustaining agricultural growth, it is necessary to put restrictions in the marginalisation of landholdings. Two policy initiatives need to be initiated in this context: (a) fixing a minimum size of holding based on economic viability of farming, and (b) restrictions on the intergenerational transfer of land from households with cultivation as their main occupation to others. As far as (a) is concerned, this was an idea discussed by agricultural economists in India in the sixties and seventies (Minhas, B.: 1970; Raj, K.N.: 1975), and the idea has now assumed more importance than in the past. Coming to (b), agrarian history of Western Europe and Japan gives examples of such restrictions imposed on the transfer of land.

References

Dreze, Jean 2002: "Palanpur, 1957-93, Occupational Change, Land Ownership and Social Inequality" *In* Madan, Vandana., ed. *The Village in India.* Oxford University Press, 2002, New Delhi.

Jeyaranjan, J. unpublished. Some Aspects of Change in Illuppakkorai Village. Draft Project Report, Kerala Research Programme on Local Level Development, Centre for Development Studies, Trivandrum.

Kaushik, A. 2005: *Nature of Land Market Interventions in India*. Institute of Sustainable Development, Noida.

Krishnaji, N. 1980: "Agrarian Structure and Family Formation: A Tentative Hypothesis", *Economic and Political Weekly*, Vol. 15(13): A38-A43.

Minhas, B.S. 1970: *Rural Poverty, Land Redistribution and Development Strategy: Facts and Policy*, Economic Development Institute, International Bank for Reconstruction and Development, September.

Nair K.N.; Menon, Vineetha., ed. 2007: *Social Change in Kerala: Insights from Micro Level Studies*. Daanish Book, New Delhi.

Patnaik, Utsa. 2002: Deflation and Deja. vu: Indian Agriculture in the World Economy *In* V.K. Ramachandran and M. Swaminathan eds. Agrarian Studies: Essays on Agrarian Relations in Less-Developed Countries.

Tulika, 2002: New Delhi. Raj, K.N. 1975. "Agricultural Development and Distribution of Land Holdings," *Indian Journal of Agricultural Economics*, Vol. 30(1), 1-13.

Rao, G.N.; Reddy, D.N., ed. 2008: *Rural Transformation: Perspectives from Village Studies in Andhra Pradesh*, Daanish Books, New Delhi.

Reddy, D.N.; Mishra Srijit., 2009: Agriculture in the Reforms Regime. *In* D.N. Reddy and Srijit Mishra eds. *Agrarian Crisis in India*, OUP, 2009.

Rukmini, R. unpublished. Socio-economic Changes in the Village of Arkavadi. Draft Project Report, Kerala Research Programme on Local Level Development, Centre for Development Studies, Trivandrum.

World Bank. 2007: *India: Land Policies for Growth and Poverty Reduction.* Agriculture and Rural Development Sector Unit, South Asia Region, The World Bank, OUP.

Notes

1. This does not create much inconsistency as almost all the land operated in the Rabi season is cultivated in the Kharif season also. For instance, in the 48th round, 96 per cent of the operational holdings and 99 per cent of the entire operated area was cultivated in the Kharif season.
2. The disaggregated state-level data on land ownership by size-classes for 1961-62 and 2002-03 have been used for the purpose of drawing the Lorenz curves. For each state, we have a set of Lorenz curves as shown in Figure 5.1 for All-India. The data for both the rounds consisted of 13 intervals. For All-India, the Lorenz curves were constructed for the intermediate rounds (26th, 37th and 48th) also (not shown in Figure 5.1).

3. The nationwide NCAER ARTS-REDS survey covered 5,000 rural households and was conducted in 1982 and 1999.
4. Two volumes synthesizing the results of the studies conducted in Kerala and Andhra Pradesh are available in a published form and the papers prepared for Tamil Nadu are available for reference (Rao and Reddy, 2008; Nair and Menon, 2007).

6

Land Bank: An Institutional Mechanism for Improving Access to Land by the Rural Poor

B.K. Sinha

Introduction

At the time of independence, almost one-third of the country's cultivable wasteland was with the *zamindars*, who while lacking the capacity to bring them under the plough directly would not also provide access to the rural poor as the land market was not sufficiently developed to absorb such large areas.[1] The *zamindars* would only settle those lands when the *salami* (a value computed as the capitalized value of the land charged as the settlement fee at the time of the transactions) was right. At the time of abolition of the *zamindari* interests, almost 36 million hectares of cultivable wasteland accrued to the state which could have been used as land bank to augment the asset base of the rural poor, the landless and the weaker sections of the society. Instead, the emphasis was on putting the wasteland under productive use for augmenting the food production. This in fact, amounted to a national loss. Even the Kumarappa Committee Report did not emphasize the priority claim of the weaker sections to the corpus of land that had passed on to the state as a consequence of the abolition of the intermediary interests.[2]

It was only in the State of Jammu and Kashmir that the rights of the landless and the rural poor were catered to in a significant way. The state imposed a strict ceiling limit of 22.75 acres and the surplus land was used to satisfy the land hunger of the landless and the rural poor. In other states ceiling laws came much later and also where they were imposed, the ceiling was on individual basis and its implementation went by default. This was mostly on account of the contradiction in the class structure of the ruling party. A number of ex-intermediaries had participated in the freedom struggle or were socially linked to the political elite. They even joined the ruling party in large numbers. There was an element of sympathy for the ex-intermediary classes as it was deemed that they were being deprived of their property without adequate compensation. This contributed to weaken the 'land to the tiller' slogan of the Congress in the 1930s. It was on this account and for the lack of a pro-landless vision that the land reforms legislations were allowed to be discussed and passed in a leisurely manner with the result that the intermediaries acquired sufficient opportunity to transfer the lands or to make settlement against fictitious names or to park the land in some safe place for future retrieval. In many instances, the intermediaries did not handover the correct details of the waste and common lands and in many instances they did not submit any account at all to the Government[3].

It was in the 1970s that there was a concerted policy at the national and state levels to enrich the land base of the rural poor. The Congress-I had won the elections in 1970 on the platform of 'Poverty Removal'. Acquisition of ceiling surplus land and providing land to the landless figured prominently in the 20 Points Programme. A Committee of Chief Ministers had formulated the national guidelines for revision of the ceiling laws and lowering of the ceiling limits.[4] It was at this time that the dignity and the empowerment arguments were also getting used for this purpose. However, due to lack of strong will-power, on the part of the states ceiling laws also remained largely ineffective. Besides, there were other categories of land, namely village commons which have been encroached overtime by the rural elites, thereby denying the poor access.[5] Thus encroachment and erosion of government land and village commons, coupled with rapid urbanization and corporatisation of agriculture and ineffective implementation of redistributive land reforms leaves the rural poor low and dry.

Increasing Landlessness and Deepening of Poverty

The situation is characterised by increasing landlessness and deepening of poverty amongst segments of the poor. Poverty is now concentrated amongst the ST/SC, Muslims, landless labour, women and those residing in the remote areas. There is a positive co-relation between poverty and landlessness. It is also to be noticed that while poverty has declined in percentage terms their absolute numbers have continued to rise; while there were 224 million poor in the year 1956 they number 284 million in the year 2004-05. The incidence of poverty is higher amongst the ST

(45%) and SC (37%) as compared to the General (17%). While the incidence of poverty is high amongst the landless, it is comparatively lower amongst the marginal, and the small farmers. Hence, creating a land-base is one of the quickest ways out of the poverty.

There is also a productivity argument in support of providing access to land to the rural poor. The labour productivity on the marginal lands is poor and any improvement in the labour productivity will enhance the labour productivity for agriculture as a whole. The labour productivity on the marginal lands is likely to rise on account of more intense application of labour.

All these produce an inexonerable in favour of providing a greater access to the landless to land particularly the marginal lands including waste, commons, degraded and the surplus lands.

The Objectives of Land Bank

The land bank would have the following objectives:

(i) **To Protect the Existing Corpus of Land of the Rural Poor**: Contrary to the popular perception the rural poor have also their land endowment. The rural poor are normally either landless or belong to the marginal farmer categories and in some cases even the small farmers. It is these categories which are most vulnerable due to their low bargaining strength. These categories also lease in land on tenancy or share cropping basis.

(ii) **To Adopt Measures for Enlargement of the Corpus of Rural Poor**: The core of the rural poor have to be enlarged on grounds of productivity, land use, equity and poverty arguments that has already been detailed earlier in this chapter. The Land Bank is likely to, by providing secure access to the rural landless and the marginal farmers to land, will contribute to greater application of labour, more intense land use, enhanced productivity and hence reduce poverty levels.

(iii) **Mobilisation of the Landless and Marginal Farmers**: Organisation of the rural poor has been a long cherished goal of our national ethos. Mobilising the rural poor around the issue of land is likely to yield rich dividends in other fields as has been demonstrated in the instance of West Bengal which has witnessed higher agricultural growth, raise in literacy, gender empowerment, improvement in the social indicators and political empowerment in the form of intense institutional participation in the local self-government institutions.

(iv) **Cultivation of the Abandoned Holdings**: The Land Bank proposes to takeover and intensively cultivate the abandoned holdings; the long term fallows and other holdings not optimally cultivated thereby adding to the wealth of the rural society. It has been seen that in the rural society all lands are not optimally cultivated. The Land Bank will offer a higher opportunity cost in this regard.

(v) **To Provide Access to the Gender Groups to Land**: The female participation in the agricultural labour force has been continuously on the raise. This is backed by also a pronounced raise in the female headed households on account of factors like outmigration etc. Even otherwise the female make greater labour contribution to the household including the pre- and post-harvest operations. In the last 8 years there has been a remarkable emergence of women's self-help groups in the country and they number 29 lakhs altogether. These SHGs have already proved themselves in their entrepreneurial ventures and it is expected that if provided access to land they will bring the same entrepreneurship and drive to the agricultural operations.

(vi) **To Provide Exit Route to Landowners Disinterested in Continuing Cultivation:** The Land Bank will also enable the landowners disinterested in continuing in agriculture to exit. The Land Bank will have the first right of purchase of any agricultural land in the village and the ceiling laws will not apply to the same.

(vii) **Land Bank to Act as a Leasing Platform:** As discussed leasing is not legal in most parts of the country. The Land Bank will be entitled to lease in and lease out the same at decided rates to the landless. This will enhance the access of the rural poor to land and will also contribute to create a stable lease market.

(viii) **Gainful and Productive Use of Wasteland:** It had been assumed at the time of the abolition of the intermediary interests that the wastelands would be used to supplement the asset base of the weaker sections. To the contrary the wastelands have either been cornered by the rural elite or are being made over to external agencies for different projects. This leads to, particularly in the tribal areas, incursions into the village society and even internal colonisation. By making over the custody over the wastelands, a most productive use of the wastelands will be ensured in addition to its protection against usurpation.

(ix) **To Facilitate Group, Cooperative, Collective Farming at the Village Level:** These solutions were envisaged amongst others by the Kumarappa Committee Report and by Mahatma Gandhi as a part of the solution to the poverty issues. This idea was overtaken by individual based progressive farming. After an experience of almost 60 years of individual based development process the development strategies are coming back to group and community based processes. The Land Bank will promote the use of such process.

(x) **To Add to the Bargaining Strength of the Rural Poor:** The wage structure and land rights of the rural poor have been constrained by their low bargaining strength. The Land Bank is designed to enhance their bargaining strength by providing them with secure access to land and participation in the decision making process.

(xi) **To Fulfill the Demand for Land for Non-Agricultural Uses While Protecting the Rights of the Landless and the Rural Poor**: This is one of the most difficult objectives of Land Bank. This chapter has already discussed the conflicting claims on the rural lands on account of urbanisation, industrialisation, expansion of infrastructure, growth of the service sector and other demands arising out of the globalisation process. The nation cannot afford to take a uni-sectional view promoting one of these sectors while sacrificing the others. This calls for a balanced approach. One of the major drawbacks of the present policy of providing land for the private companies through the land acquisition process, we have seen, lies in the fact that there is no scope of dialogue in this process between the land losers and those are acquiring. For any healthy democratic process to function this dialogue must exist. The Land Bank will also take decisions regarding providing the temporary use of some of the lands for non-agricultural requirements with the consent of the village at large. It is expected that some kind of equilibrium will emerge from this process.

What is Land Bank?

Institution of Resident Landless and Marginal Farmers: The land bank is envisaged as an institution of resident landless and marginal farmers in the rural society directly engaging in personal cultivation and earning major part of the livelihood. Landless labourers or farmers in the village will organise themselves in Self-Help Groups. While making this organisation a certain basic care and caution has to be adopted. The first and the foremost is that the farmers must have social and economic homogeneity. This may prove a difficult task in the context of the Indian villages for the reason that there may be strong social differentiation even within the deprived economic groups. A poor *dalit* may have little in common with a poor upper caste in the same economic bracket and the two may be incapable of having common objectives and pursuing common goals through common action. The methodology followed by the Gramin in creation of their Self-Help Groups is to have economically homogeneous groups. A few core members of the SHGs are identified, adequately briefed and then it is their task to involve other members and assemble a homogeneous group. This may work in the Bangladesh context where the stratification in the society is comparatively less. In India, caste is a dominant factor in the rural societies and though there are signs of weakening in the institution of caste, it continues to override the economic factor. A poor upper caste may find more in common with his richer brethren than he would find with his fellow poor *dalit*. There is also basis for this; the poor upper caste is much better endowed with social capital and his journey out of the poverty on this account is quicker than in the case of his *dalit* counterpart. Hence, the anti-poverty programmes have been more successful in the case of the general categories as compared to SC/STs.

This is, however, not to suggest that caste is the sole factor in the rural society. The Indira Kranthi Patham (IKP) is acknowledgedly the most successful of the SHGs in the country. IKP covers approximately 9 lakh SHGs in the State of Andhra Pradesh pursuing several livelihood activities. It has a strong land component which has been working in two directions—one is securing the poor access to productive lands through facilitating purchase of good quality irrigated lands and the other is facilitating the poor, in convergence with the Revenue Department, to have control over their lands in terms of having secure title, handling their lands locked in courts/disputes, awareness as to the measures taken by the Government to protect the interests of the poor manifested in the form of pro-poor enactments etc. The organisation of the SHGs follows a composite approach. As in the case of Bangladesh while the emphasis is on economic homogeneity there are also mixed social groups as well as exclusive social groups. Assessment of SHGs indicates that socially homogeneous economic groups have performed better.

The SHGs are organised in the form of Village Organisation, Mandal Samakhyas and Zilla Samakhyas. The organisational structure of the SHGs has given them an independent presence with capacity to take up issues. It also gives them the strength to engage the rural elite with much better bargaining strength and draw sustenance from the Government and civil society structures. The Land Bank does not propose to put together a separate structure but envisages utilising the existing SHG structure for its purpose. However, in states where the SHG movement is not focussed on the land issues and these structures do not have the same vitality and success of the IKP, it is proposed that a dedicated SHG structure will have to be created which will integrate into the existing movements and such other land movements that may be in existence.

Organisational Structure

The organisational framework for the Land Bank is of critical importance. There are several options available, such as the Cooperative Societies, SHG under the SGSY programme, Societies Registration Act, Non-Profit Companies under the Companies Act, or set up under special legislation by the State Government. Each of these options has their respective merits. Non-Profit Companies have a strict framework of incorporation, presentation of balance sheet and auditing but require a high degree of legal knowledge which the grass roots organisations will not be capable of. The pattern followed by the Government of Andhra Pradesh in organising their Velugu Movement involves informal organisation at the SHG level and a formal organisation at the higher levels. The SHGs are organised informally but intimation is given to this effect to the concerned offices and to the Banks. They also require certification from the Mandal Development Officer. The Village Organisations which are federations of the SHGs are incorporated under the Andhra Pradesh Mutually Aided Cooperative Societies Act, 1995. This model has worked well in respect to both the livelihood related activities as well as to the land related activities in the State of Andhra Pradesh and present good models for others to replicate.

The Societies Registration Act could have been another model which provides the registered bodies with flexibility and operational freedom. It remains a fact, nevertheless, that this model has not been adequately tried out in any of the States under the SHGs structure. Moreover, this model may not succeed under conditions involving contestation of severe nature over legal rights. Hence, a more secure model needs to be looked for.

The Land Banks are being organised on the pattern of SHGs with the exception that they will have legal rights a legal framework. It is equally recognised that the states follow their own system and the Self-Help Group Movement is in varying stages of development in the states. It is also equally recognised that the work of the Land Bank will be more in legal rights structure and may have to face legal challenges from individuals as well as institutions. Hence, a secure legal framework will also be necessary. The States may decide at their own level that whether they want to continue using the SHG framework or organise the Land Bank under the special legislation which will be necessary in any case. In the case that the States choose to adopt the SHG model still they will be required to make modifications to suit the local conditions.

Transition to Land Bank

A secure proposition for the Land Bank would be to follow the SHG route in matters of organization and incorporation. It has to be clearly recognized that it is not the legal framework that matters as much as the investment in social mobilization which holds the key to the programme. The Land Bank, therefore, may take off as an offshoot of the SHG movement. The existing SHGs may organize land based collaterals and federate at the village level as Land Bank with a strong representation of and linkages with the SHG movement. The main SHG movement definitely represents the most viable social mobilization platform in the country and is in the process of rapid expansion into other related fields apart from livelihood. This mobilization has also institutionally matched other competing institutions like the Panchayats and the mainstream Cooperative Societies. In the State of Kerala the SHG movement operating under the framework of Kudumbashri has an organic linkage with the Panchayats, while it is not the case in Andhra Pradesh. The SHG movement is not to be viewed as merely a livelihood movement or as an organization of the persons availing loans from banks. It is much more; the SHG movement is a genuine organization of the rural poor not only for livelihood but also for assertion of their rights including land and water rights. To view the SHG movement differently is to put fetters on the scope of the movement. The Land Bank is also to be viewed as a movement of the rural poor and the landless and it cannot be viewed any differently from the SHG movement. The Land Bank will require strong support from the SHG movement and from the Panchayat structure in order to survive and grow. In this respect it is to be seen as a part of the SHG movement.

Even in states which have a dedicated legislation and a separate process for the organization of Land Bank it is not to be conceived as an institution existing in isolation. There is a choice here, either the Land Bank could have been a subordinate institution to the Panchayats or an independent institution steering clear of the Panchayats and others or closure to the Self-Help Group movement. In many states the Gram Sabha lands are being managed by the Gram Panchayats; they are being settled by the Panchayats or are being auctioned out to the highest bidder the proceeds of which constitute substantial income. In other States the Panchayats have nothing to do with the revenue administration which remains firmly under the bureaucratic domain. In the former the pattern of allocation of Gram Sabha lands follow the power structure of the village society from which the landless and the rural poor normally find themselves excluded. In the event of the transfer of the management of the Gram Sabha lands to the Land Banks, it may face the hostility of the Panchayats and through it the entrenched power structure.

Besides, the Land Bank will need a high degree of mobilization of the village poor and also institutional support from the SHGs. Hence, while retaining the membership of the SHGs and the VOs, the Land Bank should also have a representation from the Panchayats and it should participate in the meetings of the Gram Sabha and should have a close relationship with the Panchayats. The organization of the rural poor is ordinarily seen in a conflictual mode and any attempt to see the Land Bank in the same light would only diminish its effectiveness and it will be reduced to yet another institution.

Social Mobilization

Social mobilization holds the key to the concept of Land Bank which functions at his best only under the conditions of intense mobilization. Admittedly, there is no one way to social mobilization which may follow myriad paths. The first step in the direction would be to seek out the support institutions which may be in the form of the existing SHGs, or other pre-mobilized institutions. The land based SHGs of Andhra Pradesh for instance could be the best motivators and lead institutions in this regard. Every State will have to select its own mode and its own lead institution and there is no thumb-of-the rule prescription for the same. In fact, differentiated regional strategies may have to be adopted.

There are a number of Civil Society Based Organizations (CBOs) working in the field of land relations and have distinguished themselves on account of their commitment to the subject and their painstaking research work. Such organizations may also be involved subject to their credentials. Likewise, their other organizations also which have an abiding interest in land relations like the Munda-Manki-Pahan Sangh and Consult for Women and Land Rights of Jharkhand which are exclusive organizations of the stakeholders and which have repeatedly agitated on the land issue. The Watershed Groups also have an intimate knowledge of the land situation and command respect in the village society. Such organizations and CBOs need to be associated with the movement for better mobilization.

In Andhra Pradesh, there is a land component of the Indira Kranthi Patham which has taken two directions—one is the land Purchase i.e. securing access of the poor to land assets by facilitating purchase of good quality irrigated lands; the other is land access for facilitating access of the poor to land to which they are otherwise, entitled by making a convergence of the Revenue Department with lawyers, CBOs working in land and concerned individuals over land disputes pending in courts and securing other land rights. The Land Access Movement has been launched so that the land rights of the rural poor are not overtaken by the land purchase movement. It is equally important for the Land Bank that it asserts the right of the landless as per its legal entitlement and does not succumb to the easier option of land purchase.

Training Support

The methodology to be followed will start with identification of the Lead Institutions as already discussed. These Lead Institutions will identify the prospective organizers of the Land Bank SHGs and provide to them initiation in the basics of the SHG Movement, organization, account keeping and other essentials. They will also devise tests to ensure that the prospective organizers have imbibed their training well.

As the next step the prospective lead person will begin the work of Group Formation which is by all means the most onerous of the tasks. In the case that the initial Group Formation is faulty the Land Bank will not have the legs to stand upon. Hence, these two tasks have to be given sufficient time.

The parent SHG group is the support institution which will provide the training and handholding support right through the process. Once adequate number of SHGs have been organized and they have commenced group work in the form of doing community farming, pooling together of their land resources etc., they will qualify to registration as a Land Bank. It is to be emphasized here that the registration of the land based SHGs should not come as an automatic process but have to be earned. Depending upon the regional variations certain tests and benchmarks have to be evolved by the supporting groups to determine that whether a group of land based SHGs have qualified to being declared a Land Bank.

At the Mandal and District levels the organization of the Samakhyas will follow the same pattern. The Land Banks constituted in a particular Mandal will have to qualify in order to be registered. The Revenue Department and the District Administration will have a close integration with the process and will need to support these institutions in a big way. Since, there may be a requirement of borrowing from the financial institution the Lead Bank Officers and other representatives of the banks may also be nominated as members to the Mandal and the District level Lead Bank. In such cases specialized training will be required from the members of the Mandal and the Zila Lead Banks for which a programme has to be drawn up in the participatory mode.

Even after having been registered as a Land Bank, the institution will need continued support for its functioning. In the case of Andhra Pradesh, a support structure has been evolved where retired revenue officials, legal coordinators, community surveyors and paralegals are involved. This organization is however at the Mandal Samakhya and Zilla Samakhya levels.

Instrumentalities of the Land Bank

The Land Bank is a process based institution and will function so long as the processes are in place. The following instrumentalities have been conceived for the Land Bank:

(i) The Land Bank will have the right to use and cultivation in respect of all village waste land and all other village lands which do not have common usages. At present these lands are governed by different sets of laws. The States will have to legislate bringing the lands under the management control of the Land Bank which will include protection against encroachment, assignation to different land based SHGs for cultivation, realisation of the dues, taking steps for improvement, creation of a corpus fund, repayment of loans and dues, tree plantation etc.

(ii) The Land Bank will have the first right of purchase of all the lands being sold in the village. In many States there is a clause of residence which restricts the transfer of land to people outside a certain geographical region. This law had been framed primarily to prevent any large scale incursions into the village society and also to prevent concentration of land. This provision has been given the go-by in most of the States of the origins. The principle governing the Land Bank is that when a person sells the first right of purchase should accrue not to an outsider to the community but to the rural poor of the community. The price of purchase will be the rate governing the registration of the land.

(iii) The Land Bank will also have the right to lease out the land to landless persons and marginal farmers in the village as decided by the Land Bank. Here, there is a conflict between the two positions enunciated. The first of the positions is that the Land Bank will undertake cultivation through its constituent Self Help Groups and the second being that Land Bank can assign such lands for cultivation by their members on individual basis. Ideally, the first auction is the better of the two as it would ensure a pooling of resources for agriculture which will ensure a minimum efficiency. Hence, the second option happens to be a suboptimal one. However, this action cannot be totally ruled out because to arrive at such group dynamics is a difficult proposition and may be left to the different Land Banks or to their Samakhya Groups. They will take the decision in light of their own capabilities, preparedness to take decisions, etc.

(iv) Normally the State Tenancy laws provide for takeover of the abandoned holdings after a certain number of years where the tenant has made no arrangement for their cultivation. Elsewhere there is arrangement for leaving the land fallow either on as current fallow or as long term fallow for a host of purposes including the retention of the fertility of the land. It is an admitted position that given the modern state of scientific advancement leaving land fallow is not a technological necessity and represents a loss to the contrary. The righ to take over the possession of the land is seldom exercised as the administration is urban in character and normally does not have the information relating to this matter. Hence, this right should be made over to the Land Bank or alternatively to the Gram Sabha. Where the Land Bank exercises the right it may be possible that on account of the involvements of its own interests it may take over some holdings where they do not even qualify to action under this clause. Here a provision may be made to the effect that the Land Bank may takeover the management of such abandoned holdings after a period of minimum of two years of abandonment and may make suitable arrangements for their cultivation. The Land Bank may wait for a period of 3 years after which the title in respect of the holding may accrue to the Land Bank subject to the consent of the Gram Sabha. Here, the power has deliberately not been vested into the revenue authorities for the simple reason that this will involve the Land Bank in longer procedures and the issue may get over resolved. However, the powers of appeal may live with a tribunal headed by the Mandal Revenue Officer and by the Collector of the district at the district level.

(v) The Land Bank shall create a corpus of fund for fulfillment of its stated aims and objectives. This corpus of fund will enable the Land Bank to enter the land market in a more aggressive manner and make greater acquisitions for the rural landless. The State Governments may also contribute to this corpus.

(vi) The Land Bank may borrow from the commercial banks and other financial institutions on terms and conditions to be decided by the bank or by the Mandal and the District Samakhya Groups depending upon the size of the transaction.

(vii) The Land Bank can negotiate and lease out land to companies and others subject to unanimous consent of the members of the bank and of the Gram Panchayat as the case may be and on such conditions as they may decide. This provision has been deemed necessary because there are numerous requirements from the State Governments, its engineering departments and from other institutions including the private companies which may be temporary in nature or where support is required for setting up appropriate industries. In such cases the Land Bank will have to seek the approval of the Gram Sabha before the land is finally made over to the leasing party. This clause has been placed in recognition of the fact that the urban requirement of the rural lands is an irrevocable phenomenon and it has to be regularized rather than prevented altogether. As the land belongs to the village community the consent of the Gram Sabha would be required for such leasing.

(viii) The Land Bank shall be subject to annual social audit and audit from such formal institutions including chartered accountants and Panchayat sources as may be provided. The members of the SHGs constituted in the Land Bank will also conduct the social audit in order to determine that whether the Land Bank has deviated from its designed objectives.

(ix) The Land Bank will also take up the problems of the constituent members in the rights mode and will link up with the programmes of the State Governments like the land access programme etc in order to protect the members. For that it is necessary that the Land Bank and their federations maintain close relationship with the State Governments, SHGs, other sister federations and Self Help Groups, Lawyers, etc.

(x) It is the objective that the Land Bank may federate up to the highest level and may set up their own Banking in order to help their constituent members.

(xi) The Government may make such arrangements as it may deem fit, legislatively or otherwise for putting in place administrative and back up support for the Land Banks. It is, however, clearly stated that administrative support does not include administrative control. Land Banks are institutions of the village society of the landless poor and they must not be converted into pockets of the influence of the revenue officials. It is for this very reason that the concept of Land Bank envisages organic linkages with different sectors including the other SHGs, organizations and institutions. Here, the term administrative support may merely imply legal support that the Land Banks may be treated equally as independent institutions in the case of conflicts.

Conclusion

The Land Bank may appear utopian at present but it is a logical extension of the SHG movement that is sweeping the country. This movement has empowered the weaker sections and women and bestowed confidence and dignity to them while reducing their vulnerability to risks. It is natural that the movement should spread to correction of the property relation and towards acquisition of landed assets. It also fits into the pattern as many of the other land-owners are desirous of quitting agriculture for other non-agricultural employment. In fact, the share of women in agricultural employment has been rising as the labour employment in the agriculture sector has been low and the agricultural wages have not been able to keep pace with the wages in other sectors.

A major impact has been made by the introduction of the MGNREGS which has definitely, as per the preliminary indication from studies, contributed to raising the bargaining strength of the rural poor, bestowing a cash surplus to them no matter how insignificant it might be and providing better nutritional, educational and health status. It has also put pressure on agricultural wages thereby reducing the profitability of agricultural operations in such households where the households do not contribute the bulk of the labour. This will hold true particularly in such areas where there is dependence on primary produce and repressed wage rates for their sustainability. Hence, the MGNREGS has already begun to affect the production relations and also the labour relations in a major way in the rural areas which is bound to have a major impact upon the land relations. Only such households contributing the bulk of the labour from their internal resources or the labour by undertaking group operations will be able to survive under the altered production relations. The Land Bank movement helps this new equilibrium to evolve and forecasts the future of the Indian agriculture which is likely to be more group based and much more intensive in nature. The State must go all out to correct the injustices of the yester years which will have a drastic poverty reducing effect upon the rural economy. In addition it will ensure a more optimised utilisation of land resources and hence contribute both to diversification of agriculture and to greater food security.

Notes

1. Appu, P.S.
2. Kumarappa, J.C.
3. Bandopadhyaya (D).
4. Sinha and Pushpendra (1999).
5. Jodha, N.S.

7

Land Reforms in Uttar Pradesh: A View from the Field

Ajit Kumar Singh

I. Introduction

During the recent decades land reforms had receded from the political agenda of the state governments. But today land issues are emerging again all-over the country forcing back attention to the important question of land. Often the debate about land reforms is conducted among economists and policy makers without taking into account the views of the basic stakeholders, i.e., the farmers and the landless agricultural labourers. This chapter presents the views of the farmers on various issues related to land reforms like consolidation of holdings, land ceilings and land lease.

II. Sample Details

This chapter is based on a survey of 280 farmers of Uttar Pradesh selected through a carefully designed survey. One district was selected from each of the main agro-climatic region of the state, namely, Etah (Western U.P.), Hardoi (Central U.P.), Lakhimpur Khiri (U.P. Tarai), Sultanpur (Eastern U.P.) and Jhansi (Bundelkhand). These districts were purposely selected on the criterion of the extent of land distributed to the landless under different programmes. From each district we selected the tehsil where maximum land was distributed. In the next state one block in the tehsil was selected on the same criterion. From each selected block two villages were selected where maximum land was distributed. In the final stages we selected 25 to 30 farming households randomly. The sample details are given in Table 7.1.

Table 7.1: Details about the Respondents

Details	*Lakhimpur*		*Hardoi*		*Sultanpur*		*Etah*		*Jhansi*		*Total*	
	Nos.	*%*	*Nos.*	*%*	*Nos.*	*%*	*Nos.*	*%*	*Nos.*	*%*	*Nos.*	*%*
1. Religion												
Hindu	34	58.6	43	71.7	47	78.3	50	100.0	50	96.2	224	80.0
Muslim	21	36.2	17	28.3	13	21.7			2	3.8	53	18.9
Sikh	3	5.2	-		-		-		-		3	1.1
Total	**58**	**100.0**	**60**	**100.0**	**60**	**100.0**	**50**	**100.0**	**52**	**100.0**	**280**	**100.0**
2. Caste												
SC	13	22.4	7	11.7	1	1.7	4	8.0	1	1.9	26	9.3
OBC	43	74.1	47	78.3	35	58.3	38	76.0	51	98.1	214	76.4
General	2	3.4	6	10.0	24	40.0	8	16.0	-		40	14.3
Total	**58**	**100.0**	**60**	**100.0**	**60**	**100.0**	**50**	**100.0**	**52**	**100.0**	**280**	**100.0**
3. Age Group (Year)												
Below 25	5	8.6	-		-		4	8.0	1	1.9	10	3.6
25-45	24	41.4	21	35.0	16	26.7	16	32.0	19	36.5	96	34.3
45-60	16	27.6	30	50.0	20	33.3	20	40.0	23	44.2	109	38.9
60+	13	22.4	9	15.0	24	40.0	10	20.0	9	17.3	65	23.2
Total	**58**	**100.0**	**60**	**100.0**	**60**	**100.0**	**50**	**100.0**	**52**	**100.0**	**280**	**100.0**
4. Education												
Illiterate	23	39.7	15	25.0	6	10.0	12	24.0	24	46.2	80	28.6
Literate	3	5.2	8	13.3	5	8.3	6	12.0	5	9.6	27	9.6
Up to Elementary	22	37.9	22	36.7	26	43.3	18	36.0	19	36.5	107	38.2
High School/ Intermediate	4	6.9	12	20.0	22	36.7	8	16.0	3	5.8	49	17.5
Graduate and Above	6	10.3	3	5.0	1	1.7	6	12.0	1	1.9	17	6.1
Total	**58**	**100.0**	**60**	**100.0**	**60**	**100.0**	**50**	**100.0**	**52**	**100.0**	**280**	**100.0**

Source: Field Survey.

Out of the 280 sample farmers 80 per cent were Hindus, 19 per cent were Muslims and 1 per cent were Sikhs. About 10 per cent farmers belonged to Scheduled castes, about 75 per cent to OBCs and remaining 14.3 per cent to upper castes. A little less than 75 per cent farmers were in the middle age group (25 to 60 years). 28.6 per cent farmers were illiterate, nearly 50 per cent were literate or educated up to elementary level, while 17.5 per cent were educated up to high school or intermediate and 6 per cent were graduates. The average size of sample households was large-around 8.5.

The average size of land owned was 4.09 acres and varied between 2.53 acres in Hardoi to 5.19 acres in Jhansi (Table 7.2). The size of the operational holding was 3.29 acres. In Lakhimpur Khiri, Sultanpur and Etah the holdings were fully irrigated. Lowest irrigated area was found in Jhansi.

Agriculture was the primary occupation of 76.8 per cent households. 5.7 per cent were engaged in business and 6.7 per cent in service (Table 7.3). About 8.5 per cent worked as labourers and 2 per cent were engaged in animal husbandry. Majority of households were also occupied in some secondary activity mostly animal husbandry and rural labour.

III. Views about Land Consolidation and Its Impact

Land consolidation measures were introduced in the state in the mid-fifties following the U.P. Consolidation of Holdings Act, 1953 which provided for compulsory consolidation. One round of consolidation has been completed in most parts of the state, while second round of consolidation was introduced in selected tehsils since the Sixth Five Year Plan. In all the districts surveyed by us consolidation was done many years ago except in case of Jhansi, which is yet to be covered by consolidation process.

Only 17.1 per cent of our sample farmers reported owning less than two plots before consolidation. About 12 per cent farmers had their holdings spread over 2 to 4 plots and another 12 per cent had holdings spread over 4 to 6 plots. Almost 60 per cent had their land ownership distributed over 6 plots. In nearly one-third cases number of plots owned were over 10, indicating the severe problem of land fragmentation. The situation is reported to have improved dramatically after consolidation (Table 7.4). 52.2 per cent farmers reported that the number of plots was reduced to 1 or 2 and another 21.5 per cent reported the number of plots between 2 and 4. However, about one-fourth farmers reported holdings spread over more than 4 plots. Thus, the problem of fragmentation, though considerably reduced, has not been fully eliminated.

Table 7.2: Size of Land Owned and Operated on Sample Farms

Land Owned/Leased	*Lakhimpur*		*Hardoi*		*Sultanpur*		*Etah*		*Jhansi*		*Total*	
	Total	*Irrigated*	*Total*	*Irrigated*	*Total*	*Irrigated*	*Total*	*Irrigated*	*Total*	*Irrigated*	*Total*	*Irrigated*
Own Land	3.65	3.65	2.91	2.91	3.88	3.48	5.15	5.15	5.19	3.74	4.09	3.74
Leased in	0.03	0.03	0.04	0.04	0.03	0.03	0.06	0.06	0.02	0.02	0.04	0.04
Leased Out	0.34	0.34	0.42	0.42	0.74	0.65	1.16	1.16	0.47	0.47	0.61	0.59
Total Operated Land	3.34	3.34	2.53	2.53	3.17	2.86	4.04	4.04	4.74	3.29	3.52	3.19

Source: Field Survey.

Table 7.3: Primary and Secondary Occupation of Respondents

District	*Primary/ Secondary Occupation*	*Agriculture*	*Animal Husbandry*	*Ag-Labour*	*Non-Ag-Labour*	*Business*	*Service*	*Total*
Lakhimpur	Primary	50	1	2	4	1		58
	%	86.21	1.72	3.45	6.90	1.72	0.00	100.00
	Secondary	5	23	2	7	2		39
	%	12.82	58.97	5.13	17.95	5.13	0.00	100.00
Hardoi	Primary	50	3		1	5	1	60
	%	83.33	5.00	0.00	1.67	8.33	1.67	100.00
	Secondary	9	21		15		1	46
	%	19.57	45.65	0.00	32.61	0.00	2.17	100.00
Sultanpur	Primary	49				1	10	60
	%	81.66	0.00	0.00	0.00	1.67	16.67	100.00
	Secondary	11	14	6	16	6	7	60
	%	18.33	23.33	10.00	26.67	10.00	11.67	100.00
Etah	Primary	44	2			1	3	50
	%	88.00	4.00	0.00	0.00	2.00	6.00	100.00
	Secondary	5	35	1	2	4	1	48
	%	10.42	72.92	2.08	4.17	8.33	2.08	100.00
Jhansi	Primary	22			17	8	5	52
	%	42.31	0.00	0.00	32.69	15.38	9.62	100.00
	Secondary	26	12	1	11		2	52
	%	50.00	23.08	1.92	21.15	0.00	3.85	100.00
Total	Primary	215	6	2	22	16	19	280
	%	76.79	2.14	0.71	7.86	5.71	6.79	100.00
	Secondary	56	105	10	51	12	11	245
	%	22.86	42.85	4.08	20.82	4.90	4.49	100.00

Source: Field Survey.

Table 7.4: Number of Plots Before and After Consolidation

Details	*Lakhimpur*		*Hardoi*		*Sultanpur*		*Etah*		*Total*	
	Nos.	*%*	*Nos.*	*%*	*Nos.*	*%*	*Nos.*	*%*	*Nos.*	*%*
No. of Plots Before Consolidation										
Less Than 2 Acre	24	41.38	15	25.00	–	0.00	–	0.00	39	17.11
2 - 4	16	27.59	10	16.67	1	1.67	–	0.00	27	11.84
4 - 6	9	15.52	12	20.00	5	8.33	1	2.00	27	11.84
6 - 8	4	6.90	6	10.00	14	23.33	5	10.00	29	12.72
8 - 10	1	1.72	10	16.67	13	21.67	11	22.00	35	15.35
10 - 12	4	6.90	5	8.33	11	18.33	5	10.00	25	10.96
12 +		0.00	2	3.33	16	26.67	28	56.00	46	20.18
Total	**58**	**100.00**	**60**	**100.00**	**60**	**100.00**	**50**	**100.00**	**228**	**100.00**
No. of Plots After Consolidation										
Less Than 2 Acre	44	75.86	51	85.00	18	30.00	6	12.00	119	52.19
2 - 4	10	17.24	7	11.67	21	35.00	11	22.00	49	21.49
4 - 6	4	6.90	2	3.33	8	13.33	11	22.00	25	10.96
6 - 8	–	–	–	–	7	11.67	11	22.00	18	7.89
8 +	–	–	–	–	6	10.00	11	22.00	17	7.46
Total	**58**	**100.00**	**60**	**100.00**	**60**	**100.00**	**50**	**100.00**	**228**	**100.00**
Quality of Land Allotted										
Good	18	31.03	26	43.33	23	38.33	12	24.00	79	34.65
Average	32	55.17	26	43.33	35	58.33	25	50.00	118	51.75
Poor	8	13.79	8	13.33	2	3.33	13	26.00	31	13.60
Total	**58**	**100.00**	**60**	**100.00**	**60**	**100.00**	**50**	**100.00**	**228**	**100.00**

Note: Consolidation has not been carried out in Jhansi.
Source: Field Survey.

Majority of farmers were satisfied about the quality of land allotted after consolidation. About one-third reported that the quality of allotted land was good and about half reported it to be satisfactory. Only 13.6 per cent said that poor quality land has been allotted to them. The proportion of farmers reporting poor quality of land was relatively more in Etah district.

Table 7.5 shows the response of the sample farmers about the changes in investment and productivity after consolidation. 86.4 per cent farmers reported improvement in their productivity after consolidation. Out of them about 30 per cent reported an improvement of up to 10 per cent, while 50 per cent reported improvement between 10-20 per cent and the remaining 17 per cent reported an improvement of over 20 per cent. Slightly over 50 per cent reported that their investment on land increased after consolidation mainly on irrigation, mechanization and fertilizers. These findings should be treated as indicative of the gains rather than their exact measurement as they are based on a long recall period. However, these are in line with the findings of the earlier studies as well (Agarwal: 1971, PEO: 1969).

The major gains of consolidation are in terms of reduced work time and reduction in drudgery. Over two-thirds of the farmers reported that consolidation reduced their drudgery and saved time on going to work on farm (Table 7.6). About 60 per cent reported saving in work period by 10 to 20 per cent and about 17 per cent reported saving in time of more than 20 per cent. Only 22.6 per cent felt that saving in time was less than 10 per cent.

40 per cent farmers did not report any type of problem faced during consolidation process (Table 7.7). However, remaining complained of various problems like too much delay in the process (17.5%), discriminatory attitude of revenue officials (7.5%), payment of bribe (5.7%) and running around block and tehsil offices (10.36%). About 10 per cent reported litigation over consolidation involving time and money. Three cases are still pending in spite of a gap of 20 years.

Overall 75 per cent of farmers feel that consolidation is a necessary measure and there is no need to repeal the Act (Table 7.8). However, majority of those who supported consolidation want it to be made voluntary.

The main suggestions given by the farmers for improving the consolidation process are:

(a) There should be no discrimination against anybody by consolidation officials.

Table 7.5: Change in Productivity and Investment after Consolidation

Description	*Lakhimpur*		*Hardoi*		*Sultanpur*		*Etah*		*Total*	
	Nos.	*%*	*Nos.*	*%*	*Nos.*	*%*	*Nos.*	*%*	*Nos.*	*%*
1. Did Productivity Increase After Consolidation										
Yes	50	86.21	52	86.67	58	96.67	37	74.00	197	86.40
No	8	13.79	8	13.33	2	3.33	13	26.00	31	13.60
Total	**58**	**100.00**	**60**	**100.00**	**60**	**100.00**	**50**	**100.00**	**228**	**100.00**
2. If yes, to what extent										
Below 10%	8	16.00	38	73.08	8	13.79	6	16.22	60	30.46
10%–20%	31	62.00	11	21.15	35	60.34	21	56.76	98	49.75
20%–30%	11	22.00	3	5.77	13	22.41	7	18.92	34	17.26
30%	–		–	–	2	3.45	3		5	2.54
3. Has Investment Incresed After Consolidation										
Yes	31	62.00	32	61.54	25	43.10	13	35.14	101	51.27
No	19	38.00	20	38.46	33	56.90	24	64.86	96	48.73
Total	**50**	**100.00**	**52**	**100.00**	**58**	**100.00**	**37**	**100.00**	**197**	**100.00**
4. If yes, on which item										
1. Irrigation	7	22.58	18	56.25	17	68.00	3	23.08	45	44.55
2. Mechanization	12	38.71	6	18.75	2	8.00	7	53.85	27	26.73
3. Fertilizer Seeds	12	38.71	8	25.00	6	24.00	3	23.08	29	28.71

Table 7.6: Impact of Consolidation on Work Time and Drudgery of Work

Description	Lakhimpur		Hardoi		Sultanpur		Etah		Total	
	Nos.	%	Nos.	%	Nos.	%	Nos.	%	Nos.	%
Has consolidation reduced time and drudgery of work										
Yes	46	79.31	51	85.00	37	61.67	21	42.00	155	67.98
No	12	20.69	9	15.00	23	38.33	29	58.00	73	32.02
Total	**58**	**100.00**	**60**	**100.00**	**60**	**100.00**	**50**	**100.00**	**228**	**100.00**
If yes, extent to which work time has reduced										
0-10%	13	28.26	8	15.69	11	29.73	3	14.29	35	22.58
10%-20%	23	50.00	42	82.35	16	43.24	13	61.90	94	60.65
20%-30%	8	17.39	1	1.96	7	18.92	3	14.29	19	12.26
30%	2	4.35	0	0.00	3	8.11	2	9.52	7	4.52
Total	**58**	**100.00**	**60**	**100.00**	**60**	**100.00**	**50**	**100.00**	**228**	**100.00**

Source: Field Survey.

Table 7.7: Problems Faced During Consolidation

Type of Problem	*Lakhimpur*		*Hardoi*		*Sultanpur*		*Etah*		*Total*	
	Nos.	*%*	*Nos.*	*%*	*Nos.*	*%*	*Nos.*	*%*	*Nos.*	*%*
1. No Problem	32	55.17	30	50.00	20	33.33	29	58.00	111	39.64
2. Too much time	17	29.31	9	15.00	12	20.00	11	22.00	49	17.5
3. Discrimination by Revenue Officials	7	12.07	2	3.33	7	11.67	5	10.00	21	7.5
4. Had to pay bribe	2	3.45	3	5.00	9	15.00	2	4.00	16	5.71
5. Frequent Visits to Offices	–	–	16	26.67	10	16.67	3	6.00	29	10.36
6. Other	–	–	–	–	2	3.33	–	–	2	0.71
Did Consolidation lead to litigation										
Yes	1	1.72	1	1.67	17	28.33	9	18.00	28	10.00
No	57	98.28	59	98.33	43	71.67	41	82.00	200	71.43
Total	**58**	**100.00**	**60**	**100.00**	**60**	**100.00**	**50**	**100.00**	**228**	**81.43**
Has the case been decided										
Yes	1	100.00	1	100.00	15	88.24	8	88.89	25	8.929
No	–	–	–		2	11.76	1	11.11	3	1.071
If No, how many years have lapsed										
More than 20 Years	–		–		2	100.00	1	100.00	3	100.00

Source: Field Survey.

Table 7.8: Respondents' Views Regarding Consolidation

Views About Consolidation	*Lakhimpur*		*Hardoi*		*Sultanpur*		*Etah*		*Total*	
	Nos.	*%*	*Nos.*	*%*	*Nos.*	*%*	*Nos.*	*%*	*Nos.*	*%*
Do You Feel Consolidation is Necessary										
Yes	38	65.52	46	76.67	46	76.67	42	84.00	172	75.44
No	20	34.48	14	23.33	14	23.33	8	16.00	56	24.56
Total	**58**	**100.00**	**60**	**100.00**	**60**	**100.00**	**50**	**100.00**	**228**	**100.00**
Suggestion for Improving the Process of Consolidation										
No Discrimination	9	23.68	11	23.91	25	54.35	10	23.81	55	31.98
No Reduction in Plot Size	14	36.84	8	17.39	10	21.74	15	35.71	47	27.33
Consolidation Within Stipulated Time	21	55.26	13	28.26	29	63.04	22	52.38	85	49.42
Consolidation under supervision of officials	13	34.21	21	45.65	13	28.26	14	33.33	61	35.47
Should Consolidation Act be Repealed										
Yes	20	34.48	14	23.33	14	23.33	8	16.00	56	24.56
No	38	65.52	46	76.67	46	76.67	42	84.00	172	75.44
Total	**58**	**100.00**	**60**	**100.00**	**60**	**100.00**	**50**	**100.00**	**228**	**100.00**
If no, should it be made voluntary										
Yes	28	73.68	18	39.13	37	80.43	35	83.33	118	68.60
No	10	26.32	28	60.87	9	19.57	7	16.67	54	31.40

Source: Field Survey.

(b) The total size of land should not be reduced.
(c) Consolidation process should be completed within stipulated time.
(d) Consolidation should be carried out under the supervision of senior officials.

IV. Views About Land Ceilings

Land Ceiling Act in U.P. was first passed in 1960. It provided for a ceiling of 40 acres of "fair quality land" for a family. Not only the Act fixed a very liberal ceiling, it allowed for a large number of exemptions and contained many loopholes, which defeated the purpose of the Act. Hence in 1972 the Land Ceiling Act in U.P. was amended removing some of the exemptions. It also reduced the ceiling to 18 acres of irrigated land for a family. The various loopholes that remained in the Act coupled with its lackadaisical implementation rendered it practically ineffective measure of agrarian change as various studies have shown (Singh: 1989; Government of U.P.: 1971 and 1975). Even the limited land that was declared surplus could not be distributed fully due to various problems.

Only 5 farmers out of 280 surveyed by us reported that their land was taken away under the Ceilings Act. Only one-fifth of the sample farmers were opposed to ceiling legislation on the ground that they have sentimental attachment for ancestral land (29.6%), their holdings will be reduced in size (29.6%), and smaller holdings will reduce their income (Table 7.9). However, it is important to note that almost 80 per cent farmers are in favour of ceiling legislation.

When asked what should be the size of the land ceiling almost half of the respondents felt that it should be below 10 acres of irrigated land (Table 7.10). About 30 per cent favoured a ceiling between 10-12 acres and about 20 per cent a ceiling of over 12 acres of irrigated land. Our respondents feel that for unirrigated land ceiling should be little higher with majority indicating a ceiling between 12 to 16 acres. Around one-fifth would like ceiling to be higher than this.

Thus, it appears that the large majority of farmers think that the present ceiling of 18 acres of irrigated land in the state should be reduced. It needs to be mentioned that majority of our respondents are small and marginal farmers, who think that a reduction in ceiling can make more land available for them.

Table 7.9: Respondents' Views About Abolition of Ceilings Act

Opinion	*Lakhimpur*		*Hardoi*		*Sultanpur*		*Etah*		*Jhansi*		*Total*	
	No.	*%*	*No.*	*%*	*No.*	*%*	*No.*	*%*	*No.*	*%*	*No.*	*%*
Should the Ceiling Act be Repealed												
Yes	7	12.1	5	8.3	13	21.7	6	12.0	23	44.2	54	19.39
No	51	87.9	55	91.7	47	78.3	44	88.0	29	55.7	226	80.7
Total	**58**	**100.0**	**60**	**100.0**	**60**	**100.0**	**50**	**100.0**	**52**	**100.0**	**280**	**100.0**
If Yes, give your reasons												
Ancestral Land has a Sentimental Value	3	42.8	1	20.0	2	15.4	1	16.7	9	39.1	16	29.6
Land Holding Size is reduced	2	28.5	3	60.0	3	23.1	4	66.7	4	17.4	16	29.6
Smaller Land Holding Reduces Income	2	28.5	1	20.0	8	61.5	1	16.7	10	43.5	22	40.7
Total	**7**	**100. 0**	**5**	**100.0**	**13**	**100.0**	**6**	**100.0**	**23**	**100.0**	**54**	**100.0**

Source: Field Survey.

Table 7.10: Respondents' Views About Maximum Landholding Size

Type of Land (Acres)	*Lakhimpur*		*Hardoi*		*Sultanpur*		*Etah*		*Jhansi*		*Total*	
	Nos.	*%*	*Nos.*	*%*	*Nos.*	*%*	*Nos.*	*%*	*Nos.*	*%*	*Nos.*	*%*
Irrigated Land												
Below 8	12	20.69	13	21.67	5	8.33	8	16.00	7	13.46	45	16.07
8 - 10	31	53.45	7	11.67	41	68.33	4	8.00	9	17.31	92	32.86
10 - 12	12	20.69	25	41.67	5	8.33	19	38.00	24	46.15	85	30.36
12 +	3	5.17	15	25.00	9	15.00	19	38.00	12	23.08	58	20.71
Total	**58**	**100.00**	**60**	**100.00**	**60**	**100.00**	**50**	**100.00**	**52**	**100.00**	**280**	**100.00**
Unirrigated Land												
Below 12	9	15.517	12	20.00	5	8.333	11	22.00	7	13.46	44	15.71
12 - 14	29	50.00	26	43.33	20	33.33	19	38.00	9	17.31	103	36.79
14 - 16	13	22.41	14	23.33	21	35.00	7	14.00	24	46.15	79	28.21
16 +	7	12.07	8	13.33	14	23.33	13	26.00	12	23.08	54	19.29
Total	**58**	**100.00**	**60**	**100.00**	**60**	**100.00**	**50**	**100.00**	**52**	**100.00**	**280**	**100.00**

Source: Field Survey.

About 85 per cent farmers felt that the Ceiling Act has been beneficial for the poor as it has improved their income levels as well as their educational and social status (Table 7.11).

It had a beneficial impact on their livelihood through greater access to credit and taking up activities like animal husbandry.

About 62 per cent farmers support land ceiling and redistribution as a poverty eradication strategy. However, over half of them feel that other programmes like loans for self-employment or employment generation programmes like NREGA should be adopted (Table 7.12).

V. Views about Land Leasing

Sub-letting of land is prohibited under the Zamindari Abolition Act of U.P. except for certain categories of handicapped persons like widows, physically handicapped, serving in armed forces, etc. However, in practice leasing out continues as the NSS surveys have shown. About 28 per cent farmers in our survey reported that they had leased out land, while 3 per cent reported leasing in of land (Table 7.13). No case of simultaneous leasing in and leasing out was reported.

Land leasing is being practiced mostly among the small and marginal farmers. In half of the cases the land leased out was small being less than 2 acres, in 36 per cent cases it was 2 to 4 acres and only in 11.5 per cent cases land leased out exceeded 4 acres (Table 7.13A).

The most common term of leasing out is 50-50 per cent sharing of paid out costs (irrigation and fertilizers only) and output. Nearly 80 per cent farmers reported leasing out on sharing of cost and output basis (Table 7.14). However, in about one-fifth cases no sharing in cost was done, while half of the produce was shared. Only 2.5 per cent farmers reported giving land on cash basis. The prevailing rate is around Rs. 4,000 per acre. Incidentally, all the cases where land is leased out on cash basis or without sharing costs were reported from Sultanpur district, where population pressure is more and landlessness is higher.

In nearly 58 per cent cases the terms of tenancy were based on the common practice prevailing in the district (Table 7.15). However, about 18 per cent farmers leasing out reported that they fixed the terms of lease, while in 24 per cent cases the tenant decided the term. The proportion of the latter was relatively higher in Etah and Jhansi districts. Thus, the landowners are not in a definite dominant position as far determination of conditions of lease is concerned.

The main reasons for giving out land on lease were high cost of

Table 7.11: Views of the Respondents' about Benefits of Ceiling Act

Details	*Lakhimpur*		*Hardoi*		*Sultanpur*		*Etah*		*Jhansi*		*Total*	
	Nos.	*%*	*Nos.*	*%*	*Nos.*	*%*	*Nos.*	*%*	*Nos.*	*%*	*Nos.*	*%*
Is Ceiling Act Helpful for the Poor												
Yes	46	79.31	52	86.67	55	91.67	43	86.00	43	82.69	239	85.36
No	12	20.69	8	13.33	5	8.33	7	14.00	9	17.31	41	14.64
Total	**58**	**100.00**	**60**	**100.00**	**60**	**100.00**	**50**	**100.00**	**52**	**100.00**	**280**	**100.00**
If Yes, in which way (multiple response)												
Education has Improved	17	36.96	12	23.08	9	16.36	11	25.58	13	30.23	62	25.94
Social status has gone up	21	45.65	19	36.54	11	20.00	12	27.91	16	37.21	79	33.05
Income has Increased	12	26.09	23	44.23	16	29.09	9	20.93	24	55.81	84	35.15
Taking loan has become easy	6	13.04	2	3.846	9	16.36	14	32.56	7	16.28	38	15.90
Assisted in animal husbandry	9	19.57	17	32.69	5	9.09	8	18.61	3	6.98	42	17.57
Helped to construct houses	7	15.22	6	11.54	2	3.64	13	30.23	4	9.30	32	13.39

Source: Field Survey.

Table 7.12: Respondents' Views About Ceilings and Poverty

Details	*Lakhimpur*		*Hardoi*		*Sultanpur*		*Etah*		*Jhansi*		*Total*	
	Nos.	*%*	*Nos.*	*%*	*Nos.*	*%*	*Nos.*	*%*	*Nos.*	*%*	*Nos.*	*%*
Do you think allotment is a useful method for reducing poverty												
Yes	40	68.97	44	73.33	25	41.67	24	48.00	41	78.85	174	62.14
No	18	31.03	16	26.67	35	58.33	26	52.00	11	21.15	106	37.86
If no, what should be done												
Provide employment	9	50.00	6	37.5	22	62.86	8	30.77	3	27.27	48	45.28
Provide loan for Self-employment	6	33.33	9	56.25	12	34.29	14	53.85	6	54.55	47	44.34
Implement Govt. Programmes such as NREGA	11	61.11	13	81.25	8	22.86	13	50.00	7	63.64	52	49.06

Source: Field Survey.

Table 7.13: Distribution of Respondents by Type of Leasing

Details	*Lakhimpur*		*Hardoi*		*Sultanpur*		*Etah*		*Jhansi*		*Total*	
	No.	*%*	*No.*	*%*	*No.*	*%*	*No.*	*%*	*No.*	*%*	*No.*	*%*
Only Leasing in	1	1.73	2	3.33	2	3.34	2	4.00	1	1.93	8	2.86
Only Leasing out	10	17.24	24	40.00	17	28.33	16	32.00	11	21.15	78	27.85
Neither Leasing In or Leasing out	47	81.03	34	56.67	41	68.33	32	64.00	40	76.92	194	69.29
Total Respondents	**58**	**100.00**	**60**	**100.00**	**60**	**100.00**	**50**	**100.00**	**52**	**100.00**	**280**	**100.00**

Note: No cases of lease in - lease out were reported.

Source: Field Survey.

Table 7.13A: Details Regarding Leasing out of Land

Details about leasing	*Lakhimpur*		*Hardoi*		*Sultanpur*		*Etah*		*Jhansi*		*Total*	
	No.	*%*	*No.*	*%*	*No.*	*%*	*No.*	*%*	*No.*	*%*	*No.*	*%*
Have you leased out land												
Yes	10	17.24	24	40.00	17	28.33	16	32.00	11	21.15	78	27.86
No	48	82.76	36	60.00	43	71.67	34	68.00	41	78.85	202	72.14
Total	**58**	**100.00**	**60**	**100.00**	**60**	**100.00**	**50**	**100.00**	**52**	**100.00**	**280**	**100.00**
If yes, how much (acres)												
Below 1	1	10.00	10	41.67	1	5.88	1	6.25	–		13	16.67
1 to 2	3	30.00	12	50.00	6	35.29	–		7	63.64	28	35.90
2 to 3	3	30.00	2	8.33	5	29.41	5	31.25	1	9.09	16	20.51
3 to 4	3	30.00	–		2	11.76	5	31.25	2	18.18	12	15.38
4+	–		–		3	17.65	5	31.25	1	9.09	9	11.54
Total	**10**	**100.00**	**24**	**100.00**	**17**	**100.00**	**16**	**100.00**	**11**	**100.00**	**78**	**100.00**
Average land leased out	2.00		1.04		2.49		3.63		2.95		2.20	

Source: Field Survey.

Table 7.14: Terms of Leasing Out

Terms for leasing out	*Lakhimpur*		*Hardoi*		*Sultanpur*		*Etah*		*Jhansi*		*Total*	
	No.	*%*	*No.*	*%*	*No.*	*%*	*No.*	*%*	*No.*	*%*	*No.*	*%*
50% Sharing of cost of production & total produce	10	100.00	24	100.00	–		16	100.00	11	100.00	61	78.21
50% Sharing of produce only	–		–		15	88.24	–		–		15	19.23
Fixed amount Rs. 4,000/acre	–		–		2	11.76	–		–		2	2.56
Total	**10**	**100.00**	**24**	**100.00**	**17**	**100.00**	**16**	**100.00**	**11**	**100.00**	**78**	**100.00**

Source: Field Survey.

Table 7.15: Who Fixes the Condition for Leasing

Details	*Lakhimpur*		*Hardoi*		*Sultanpur*		*Etah*		*Jhansi*		*Total*	
	No.	*%*	*No.*	*%*	*No.*	*%*	*No.*	*%*	*No.*	*%*	*No.*	*%*
Self	3	30.00	5	20.83	2	11.76	3	18.75	1	9.09	14	17.95
Tenant	2	20.00	1	4.17	2	11.76	7	43.75	7	63.64	19	24.36
Common Practice	5	50.00	18	75.00	13	76.48	6	37.50	3	27.27	45	57.69
Total	**10**	**100.00**	**24**	**100.00**	**17**	**100.00**	**16**	**100.00**	**11**	**100.00**	**78**	**100.00**

Source: Field Survey.

production (58%) and low incomes from agriculture (20.5%). Only 7.7 per cent reported leasing out due to non-availability of labour. About one-fifth said that they have leased out as they have taken up non-agricultural activity (Table 7.15A).

All the reported cases of tenancy were oral tenancy since leasing out is legally prohibited (Table 7.16). Around one-fourth lessors reported that they engaged their tenants for work without payment or on lower wages, indicating a kind of dependency on the lessors. However, in three-fifths of cases no such dependency was reported, which is suggestive of the fact that leasing is largely a market driven process (Table 7.16).

Over four-fifth farmers reported that there has been no change in the system of tenancy during the past five years. However, about one-fifth reported that practice of advance cash payment is emerging (Table 7.16).

A little over one-third farmers reported that they have not changed in tenants during the last five years. However, about three-fourth respondents reported that they had changed the tenant during the last 5 years (Table 7.17). The main reasons for changing the tenant were low productivity (30.5%), demand of more share in cost by lessee (49.2%), non-utilization of money given for cost sharing (13.6%) and quarrelsome nature of the tenant (6.8%).

About 70 per cent of the farmers leasing out land reported that they have not made any investment on the leased out land. Lack of funds and infertile nature of land were cited as the main reason for not making investment. However, about 30 per cent did report making investment of leased out land in acitivities like leveling of field, bunding and irrigation (Table 7.18).

Over two-thirds of the farmers who reported leasing out land favoured the view that leasing should be legalized (Table 7.19). The main reason for supporting legalization lease was that it will make the process easier (42.3%), will allow retention of land ownership (21.8%) and will allow fair sharing of costs and outputs (35.9%). However, about one-third of lessors were opposed to legalization of lease as they feared that it will result in loss of ownership of land and may lead to corruption and malpractices by revenue officials. About 15 per cent of farmers not leasing out land at present also supported the view that leasing should be legalized. The remaining respondents were apprehensive that it is likely to result in loss of ownership of land. On the whole, 30 per cent farmers surveyed by us were in favour of legalization of tenancy.

About 18 per cent felt that the period of lease should be up to 3 years; while as many as 47 per cent felt it should be five years and 35 per cent

Table 7.15A: Reasons for Leasing Out Land

Reasons	Lakhimpur		Hardoi		Sultanpur		Etah		Jhansi		Total	
	No.	%	No.	%	No.	%	No.	%	No.	%	No.	%
Engaged in non-agricultural activity	3	30.00	4	16.67	1	5.88	0	0.00	0	0.00	8	10.26
High cost of cultivation	5	50.00	9	37.50	7	41.18	3	18.75	5	45.45	29	37.18
Low income from agriculture	2	20.00	9	37.50	8	47.06	11	68.75	6	54.55	36	46.15
Non-availability of agricultural labourers	0	0.00	2	8.33	1	5.88	2	12.50	0	0.00	5	6.41
Total	**10**	**100.00**	**24**	**100.00**	**17**	**100.00**	**16**	**100.00**	**11**	**100.00**	**78**	**100.00**

Source: Field Survey.

Table 7.16: Type of Tenancy and Changes in Tenants

Details of Tenancy	Lakhimpur		Hardoi		Sultanpur		Etah		Jhansi		Total	
	No.	%	No.	%	No.	%	No.	%	No.	%	No.	%
Type of Tenancy												
Oral	10	100.00	24	100.00	17	100.00	16	100.00	11	100.00	78	100.00
In Writing	–		–		–		–		–		–	–
Other work taken from tenant												
Work without payment	–		3	12.50	2	11.76	1	6.25	2	18.18	8	10.26
Work on low wages	2	20.00	3	12.50	1	5.88	4	25.00	1	9.09	11	14.10
No work taken	8	80.00	18	75.00	14	82.35	11	68.75	8	72.73	59	75.64
Change in the leasing system over last 5 years												
No changes	10	100.00	19	79.17	11	64.71	16	100.00	8	72.73	64	82.05
Payment in advance	–		5	20.83	6	35.29	–		3	27.27	14	17.95

Source: Field Survey.

Table 7.17: Details of Change in Tenants Over Last Five Years

Details	*Lakhimpur*		*Hardoi*		*Sultanpur*		*Etah*		*Jhansi*		*Total %*	
	No.	*%*	*No.*	*%*	*No.*	*%*	*No.*	*%*	*No.*	*%*	*No.*	*%*
Change in Tenants												
Yes	7	70.00	15	62.50	12	70.59	12	75.00	7	63.64	53	67.95
No	3	30.00	9	37.50	5	29.41	4	25.00	4	36.36	25	32.05
Total	**10**	**100.00**	**24**	**100.00**	**17**	**100.00**	**16**	**100.00**	**11**	**100.00**	**78**	**100.00**
If, Yes Reasons for Change												
Low Productivity	1	14.29	8	53.33	6	50.00	4	33.33	4	57.14	23	43.40
Not utilizing money for input use	2	28.57	3	20.00	0	0.00	2	16.67	1	14.29	3	15.09
Demanding More Share in cost	3	42.86	3	20.00	6	50.00	4	33.33	2	28.57	18	33.96
Tenant is Quarrelsome	1	14.29	1	6.67	0	0.00	2	16.67	0	0.00	4	7.55
Total	**7**	**100.00**	**15**	**100**	**12**	**100.00**	**12**	**100.00**	**7**	**100**	**53**	**100.00**

Source: Field Survey.

Table 7.18: Investment Made on Leased Out Land

Type of Tenancy	*Lakhimpur*		*Hardoi*		*Sultanpur*		*Etah*		*Jhansi*		*Total*	
	No.	*%*	*No.*	*%*	*No.*	*%*	*No.*	*%*	*No.*	*%*	*No.*	*%*
Investment made on leased out land												
Yes	6	60.00	8	33.33	2	11.76	1	6.25	6	54.55	23	29.49
No	4	40.00	16	66.67	15	88.24	15	93.75	5	45.45	55	70.51
Total	**10**	**100.00**	**24**	**100.00**	**17**	**100.00**	**16**	**100.00**	**11**	**100.00**	**78**	**100.00**
If Yes, in what												
Levelling of land	2	33.33	5	62.50	-		1	100.00	2	33.33	10	43.48
Bunding	1	16.67	-		-		-		3	50.00	4	17.39
Irrigation	3	50.00	3	37.50	2	100.00	-		1	16.67	9	39.13
If no, Reason												
Lack of fund	1	25.00	4	25.00	9	60.00	7	46.67	2	40.00	23	41.82
Land is infertile	3	75.00	12	75.00	6	40.00	8	53.33	3	60.00	32	58.18

Source: Field Survey.

Table 7.19: Perception of Respondents Regarding Tenancy

Details	*Lakhimpur*		*Hardoi*		*Sultanpur*		*Etah*		*Jhansi*		*Total*	
	No.	*%*	*No.*	*%*	*No.*	*%*	*No.*	*%*	*No.*	*%*	*No.*	*%*
Should Tenancy be Legalised												
Yes	7	70.00	16	66.67	11	64.71	10	62.50	9	81.82	53	67.95
No	3	30.00	8	33.33	6	35.29	6	37.50	2	18.18	25	32.05
Total	**10**	**100.00**	**24**	**100.00**	**17**	**100.00**	**16**	**100.00**	**11**	**100.00**	**78**	**100.00**
If yes, Reasons												
Procedure will Become easy	2	28.57	6	37.5	4	36.36	3	30.00	2	22.22	17	32.08
Land will remain with owner	2	28.57	4	25.0	2	18.18	3	30.00	3	33.33	14	26.42
Fair sharing of cost and production	3	42.86	6	37.5	5	45.45	4	40.00	4	44.44	22	41.51
If no, Reasons												
Owners' monopoly will end	1	33.33	5	62.5	4	66.67	3	50.00	1	50.00	14	56.00
Corruption might increase	2	66.67	3	37.5	2	33.33	3	50.00	1	50.00	11	44.00

Source: Field Survey.

Table 7.20: Views About the Term and Conditions of Lease

Details	*Lakhimpur*		*Hardoi*		*Sultanpur*		*Etah*		*Jhansi*		*Total*	
	No.	*%*	*No.*	*%*	*No.*	*%*	*No.*	*%*	*No.*	*%*	*No.*	*%*
What should be the Duration of Tenancy												
Three Year	6	60.00	11	45.83	6	35.29	5	31.25	6	54.55	34	43.59
Five Year	2	20.00	9	37.50	7	41.18	8	50.00	3	27.27	29	37.18
Above Five Year	2	20.00	4	16.67	4	23.53	3	18.75	2	18.18	15	19.23
Total	**10**	**100.00**	**24**	**100.00**	**17**	**100.00**	**16**	**100**	**11**	**100.00**	**78**	**100.00**
What should be the Conditions of Tenancy												
Cash	2	20.00	5	20.83	3	17.65	3	18.75	2	18.18	15	19.23
50% Share of production	3	30.00	3	12.50	4	23.53	3	18.75	4	36.36	17	21.79
Equal sharing of cost and produce	5	50.00	16	66.67	10	58.82	10	62.50	5	45.45	46	58.97
Total	**10**	**100.00**	**24**	**100.00**	**17**	**100.00**	**16**	**100**	**11**	**100.00**	**78**	**100.00**

Source: Field Survey.

felt it should be even longer (Table 7.20). Around 38 per cent who favoured leasing preferred leasing on cash basis, while 62 per cent desired leasing on the basis of sharing of output and costs on 50:50 basis.

VI. Conclusion

Our study based on a field survey of 280 farmers from 5 districts representing different economic and agro-climatic regions shows a general positive response of farmers to the land reform measures introduced in the state. The major findings may be summed up as follows:

(i) Consolidation of holdings is one of the most successful land reform measure. It has reduced time and drudgery of farmers on moving one field to another and enabled them to make investment on development of land. It has a clear favourable impact on agricultural productivity. There is an overwhelming support among the farmers to consolidation measures. However, there are some procedural problems and malpractices associated with the process which need to be removed.

(ii) Land ceilings did not affect the majority of our sample farmers directly. However, by and large they favour land ceiling measures as it has a positive impact on the economic and social status of the land allottees. In fact, a good proportion of farmers feel that the land ceiling size may be further reduced to below 10 acres of irrigated land. The majority of our respondents are small and marginal farmers, who think that a reduction in ceiling can make more land available for them as well as the landless labourers.

(iii) Though sub-letting of land is prohibited under the Zamindari Abolition Act of U.P. except for certain categories of handicapped persons, the practice of oral lease is quite common. Land leasing is being practiced mostly among the small and marginal farmers and is a mechanism to make adjustment in their investment and income requirements. The most common term of leasing out is 50:50 per cent sharing of paid out costs and output. However, the practice of advance cash payment is emerging in some parts. The survey also indicates that the landowners are not in a dominant position in the land lease market which is largely determined by market forces of demand and supply of land. Over two-thirds of the farmers who reported leasing out land favoured the view that leasing should be legalized as they felt that it will make the process easier, will allow retention of land ownership and will allow fair sharing of costs and outputs. Overall our study lends support to the view that land leasing should be allowed as it will increase access of marginal farmers and the landless labourers to land and improve their economic condition.

References

Agarwal, S.K. (1971): *Economics of Land Consolidation in India*, Sultan Chand, New Delhi.

Directorate of Evaluation, U.P. Government (1971): *Impact of Ceilings on Land Holdings Act, 1960: A Case Study in Gorakhpur District*, Lucknow.

P.E.O., Planning Commission, GoI (1969): *Report on the Evaluation of Consolidation of the Holdings Programmes*, New Delhi.

Revenue Department, Government of U.P. (1975): *Report of the Uttar Pradesh Land Settlement Enquiry Committee*, Lucknow.

Singh, Ajit Kumar (1989): "Land Ceiling Legislation in U.P.: An Assessment", in Planning Commission, GOI, *Proceedings and Papers of the Seminar on Land Reforms: A Retrospect and Prospect*, New Delhi.

Part II

TENANCY REFORMS

8

Agricultural Tenancy Reform in India: Policy, Practice and Impact

T. Haque

Introduction

The system of agricultural tenancy in which the landowner does not cultivate his land himself but leases out to others for the purpose of cultivation against payment of rent has been present in all feudalistic or semi-feudalistic agrarian economies. At the same time, it has been traditionally portrayed as an inefficient and exploitative agrarian institution, notwithstanding the fact that empirical evidence and inferences vary widely. At the time of independence, the overwhelming view of India's policy makers was in favour of elimination of all feudal elements and consequently also the system of cultivation by tenants or subtenants. The tenancy reform laws adopted by most Indian states since independence not only aim at conferring ownership or occupancy right on the tenants with respect to tenanted lands, but also either prohibit or restrict leasing and sub-leasing of land in future. However, the question still remains whether agricultural land leasing has either disappeared or reduced as a result of implementation of restrictive tenancy laws and whether or not the overall impact of tenancy reform has been positive on tenurial security, agricultural efficiency and equity. The present chapter analyses the positive and negative aspects of post-independence tenancy reform policy in India and suggests appropriate amendments, keeping in view the emerging agrarian concerns.

Evolution of Tenancy Reform Policy

Prior to independence, the system of land leasing was considered to be an integral part of the feudalistic, albeit unproductive and exploitative agrarian system. Between the landlord and the tenant who actually cultivated the land, there were several layers of rent-receiving intermediaries all of whom had to be supported by the back-breaking labour of the tenant. The system was also marked by rack renting and exploitation and consequently the actual cultivators lacked both incentive and ability to improve the productivity of land (Haque and Sirohi: 1986). The Bengal Tenancy Act, 1885, Bihar Tenancy Act, 1885 and the Orissa Tenancy Act, 1914 granted occupancy rights to tenants who were in possession of land continuously for twelve years. However, sharecroppers and other tenants-at-will were not given any protection. On the whole, the agrarian structure in large parts of the country remained inefficient and unjust based on parasitic rent-receiving relationships and a high degree of sub-infeudation which left the actual cultivators without any incentive and ability to increase agricultural productivity. After independence, the Congress government set up an Agrarian Reforms Committee under the chairmanship of Shri J.C. Kumarappa which submitted its report in 1949. The Committee recommended not only for elimination of all intermediaries, but also abolition of tenancy except in the case of widows, minors and certain other categories of disabled landowners. Several state governments either introduced a new tenancy law or amended the existing law based on this ideology, with some states taking either a more liberal or a strong view on this, depending on the local socio-political conditions. At the central level, various five year plans have made useful recommendations from time to time.

The First Five Year Plan recommended that there should be a distinction between tenants-at-will of large owners and those of small and medium owners. In the case of large landowners, the plan recommended that they can resume land from tenants-at-will for personal cultivation up to the ceiling limit. But tenants of non-resumable land beyond ceiling should be given occupancy rights on payment of a price to be fixed as a multiple of the rental value of the land. In the case of tenants of small and medium landowners, owning land below the ceiling, owners could resume land for personal cultivation within a period of five years. In case the landowner failed to resume land for personal cultivation within five years, tenants would acquire permanent and heritable rights. The Second Five Year Plan specifically pointed out that "the economic circumstances of small owners are not so different from those of tenants that tenancy legislation should operate to their disadvantage. The owner with very small holdings, i.e. those with less than one-third of a family holding should be permitted to resume their entire area, while those owning land between a basic holding and a family holding, should be permitted to resume for personal cultivation one half of the area held by the tenant, but not less than a basic holding. Where tenants were left with land below a basic holding size, ceiling-surplus land should be distributed to them to bring the tenancy to the level of a basic holding". Both First and Second Five Year Plans recommended fair rent in the range of 20 to 25 per cent of the gross produce. The Third Plan made a recommendation that concrete steps should be taken to complete the programme of conferring ownership right on the tenants on non-resumable land. Regarding the tenanted land owned by small owners, it was suggested that the problem should be studied by states in the light of their local conditions with a view to determine the action called for in this direction. The Fourth Five Year Plan, however, took a definitive position on the issue. It suggested that all tenancies should be non-resumable and permanent. The Fifth Five Year Plan made an observation that it is difficult to regulate tenancies, although the logic of development in the field of land reform lies in the right of ownership being conferred on all the tenants. From the Sixth Five Year Plan to Tenth Five Year Plan, there was not much change in the approach to tenancy reform. But mid-term appraisal of the Ninth and Tenth Five Year Plan adopted a very pragmatic approach. The mid-term appraisal of the Ninth Five Year Plan provided the rationale for continuing the ban on leasing-in backward regions, but recommended that tenancy in developed regions should be legalized, as it would increase poor peoples' access to land and also occupational mobility of large landowners. The mid-term appraisal of Tenth Five Year Plan pointed out: "that the Country's tenancy laws are highly restrictive. They were enacted in the context of exploitation of the Peasantry. The current situation however, is very different. The economic and political hold of large landowing classes has been considerably weakened ...... Research studies indicate that restrictive tenancy laws have prevented optimum allocation of land resources and denied the rural poor access to land. They have also led to concealed tenancy, which is widespread. In such cases, the tenant has no protection. The fear of reverse tenancy is also unfounded as small and marginal farmers account for over 80 per cent of the total leased in land in the country. In any case, the existing ceiling laws can be leveraged to prevent reverse tenancy by stipulating that operational holdings could not exceed those prescribed under the ceiling laws" (Govt. of India: 2005).

The Eleventh Five Year Plan (2007-2012) pointed out that security of tenure should not be confused with ownership right. It recommended that tenancy should be legalized in a limited manner. It should provide security to the tenant for the contractual period, which could be long enough to encourage long-term investment by the tenant. It should also protect the rights of the landowner so that he has an incentive to lease out his land instead of keeping it fallow or underutilising it. Instead of prescribed rentals, an upper and lower bound of rents may be prescribed at the state level…. Land owners who would otherwise have to operate small uneconomic holdings should have the opportunity to legally lease out land to other farmers with the assurance of being able to resume possession at the end of the stated period of tenancy. The Twelfth Five Year Plan is in the making now. The Approach Paper of the Twelfth Five Year Plan points out that small and marginal farmers may be better off leasing-out their land to more viable farmers while seeking paid employment themselves (Govt. of India: 2011). This view has also been reinforced in a recent paper by the present Deputy Chairman of the Planning Commission (Ahluwalia: 2011).

Legal Status of Agricultural Tenancy

In view of the fact that laws adopted for abolition of intermediaries in the 1950s could not end all forms of exploitative and unproductive agrarian relations, several state governments either amended their existing tenancy laws or enacted new laws, mainly in the 1960s and 1970s aiming at conferring ownership or occupancy rights to the tenants. While the national land leasing policy, as reflected in Five Year Plans, has undergone a change in recent years, the tenancy laws adopted by various state governments in the fifties, sixties and seventies remain unchanged. The legal status of tenancy in various states is discussed below:

Andhra Pradesh: In Andhra area of Andhra Pradesh, any lease of land to a tenant subsisting on or before 1974 was deemed to be in perpetuity [Andhra Pradesh (Andhra Area) Tenancy Act, 1956, as amended in 1974]. The Lease after 1974 has to be in writing as well as registered and for a minimum period of six years. The Lease is also renewable and heritable, but not transferable except to financial institutions by way of mortgage for loan. The landlord has a continuous right of resumption for personal cultivation, provided he does not hold more than two-thirds of the ceiling area and the tenant is left with no less than one half of the area held by him under lease before such resumption.

In the Telengana area of Andhra Pradesh, however, leasing-out of agricultural land is prohibited except by those whose landholding is equal to or less than three times the family holdings (Andhra Pradesh–Telengana

Area Tenancy Act, 1950, section-7). Such leases will have to be in writing (a copy of which shall be filed before the tehsildar) for a period of five years and renewable for a further period of five years in succession. But such lease may be terminated by giving a notice of one year if the landowner requires the land for personal cultivation. Also the disabled persons such as minor, female, persons with physical and mental infirmity, persons in defence service and those prevented by any sufficient reason from cultivating the land can lease out with the permission of the collector for a period as the collector may fix.

Assam: The Assam (Temporarily Settled Areas) Tenancy Act, 1971 does not explicitly prohibit leasing-out of land by any class of landowners. Sub-letting however is prohibited. The Act classifies all tenants into (a) occupancy tenants and (b) non-occupancy tenants. Occupancy tenants who have held land as tenants for at least three years enjoy security of tenure and can acquire ownership right on payment of compensation at the rate of 50 times the rate of annual revenue payable for such lands. Also, non-occupancy tenant who holds land continuously for three years can also acquire the right of an occupancy right. The *adhiars* (share croppers) have no security of tenure and no right of being recorded. Voluntary surrender of land is not allowed except with the permission of the Deputy Commissioner, who then has to place the land to a landless agriculturist who becomes a non-occupancy tenant of the landlord.

Bihar: The Bihar Land Reforms Act, 1950 which was enacted for abolishing intermediaries, did not regulate conditions of tenancy. However, according to Bihar Tenancy Act, 1885 as amended in 1970, an underraiyat in continuous possession of land for twelve years can acquire the right of occupancy provided the landowners from whom the land is leased in, owns more than five acres of irrigated or ten acres of other land. Besides, the Bihar Land Reforms (Fixation of Ceiling Area and Acquisition of Surplus Land) Act, 1961, Prohibits creation of future tenancy, (section-19) except by disabled persons i.e. minor, widow or unmarried, divorced or separated woman or person suffering from mental or physical disability or a person in the Armed Forces or in public service receiving salary not exceeding Rs. 250 per month.

Gujarat: The Bombay Tenancy and Agricultural Lands Act, 1948, as amended till date, does not explicitly prohibit leasing-out of land by any class of landowner. However, a tenant acquires a right to purchase the land leased within one year of lease period, which means that effectively legal leases will be possible only if the tenant is in no position to exercise his/her right to purchase land due to financial difficulties or otherwise. Besides, the Bombay Tenancy and Agricultural Lands Act, 1948, as amended by Government of Gujarat with effect from 9 March, 1973 (i) prohibits surrender of tenancy right in favour of landlord, (ii) extends the date for all postponed cases for acquiring ownership right (iii) disallows termination of a tenancy of a tenant belonging to SC/STs even on the grounds of personal cultivation or non-agricultural uses, and (iv) allows the agricultural labourers and artisans to purchase the homestead lands belonging to the landlord if the dwelling houses were built by them on such sites, on payment of price to be fixed by the tribunal, not exceeding 20 times the annual rent of the site.

Himachal Pradesh: The Himachal Pradesh Tenancy and Land Reforms Act, 1972 which came into force with effect from 21 February, 1974 provides that all occupancy tenants become owners w.e.f. 21 February, 1974 or from the date they became occupancy tenants and non-occupancy tenants became occupancy tenants from the date notified by the state government in respect of irresumable land of the landowner. While the landowner is permitted to resume his land for personal cultivation, a tenant must be left with not less than half the land cultivated by him. Sub-leasing of land is banned except by disabled persons such as members of armed forces, unmarried/divorced/separated women, a widow, a minor, a person suffering from physical or mental disability and a student of a recognized institution. A tenant cannot surrender the land in favour of the landowner. If he wishes, he can surrender it to the government which then inducts a suitable tenant.

Jammu and Kashmir: The Jammu and Kashmir Agrarian Reforms Act, 1976 provides that all rights, titles and interests in land of any person not cultivating it personally in Kharif 1971 shall be extinguished and vested in the state, free from all encumbrances with effect from 1 May, 1973. The Act provides for conferment of ownership rights on tenants after allowing the resident landlord to resume land for personal cultivation, provided his annual income does not exceed Rs. 500 per month and also the tenant is left with not less than an area of two standard acre Creation of tenancy in future is prohibited under the Act.

Karnataka: According to the Mysore Land Reforms Act, 1961 as amended with effect from 1 March, 1974, no tenancy shall be created or continued nor shall any land be leased out for any period, except by disabled persons, namely a soldier or a seaman and every such lease by such disabled persons shall be in writing. Also, no tenancy existing on 1st March, 1974 shall be terminated by efflux of time. Besides, a tenant is allowed to surrender the land only to the state government and also lands leased out in contravention of the Act shall vest in the state government. The right of the tenant is heritable but not transferable.

Kerala: The Kerala Land Reforms Act, 1963 as amended in 1969, conferred security of tenure and provided for transfer of ownership rights to all tenants, including sharecroppers and *kudikidappukars*. Future leasing is prohibited and any tenancy created after the commencement of the Act is invalid, without any exemption being granted even to defence personnel or the disabled.

Manipur: The Manipur Land Revenue and Land Reforms Act, 1960 as amended in 1975, provides for fixity of tenure for tenants without any right of resumption of land by the landowner. The Act also prohibits any future leasing.

Madhya Pradesh: According to Madhya Pradesh Land Revenue Code, 1959 as amended till date leasing is permitted only by disabled persons, such as widow, unmarried women, deserted women, minor, a person with mental or physical disability, armed personnel, a person in imprisonment, a public or charitable/religious institution or a local authority or a co-operative society. The terms and conditions of lease in such cases are to be as agreed between the lessor and the lessee. If leases are made in contravention of the above restriction or a person cultivates by an agreement for over two years, then it will be an unauthorized lease in the event of which the lessee gets the right of occupancy except when the landowner is a tribal *bhumiswami*.

Maharashtra: According to Bombay Tenancy and Agricultural Lands Act, 1948 applicable to Bombay and Vidarbha regions as well as the Hyderabad Tenancy and Agricultural Lands Act, applicable to former Hyderabad region, there is no explicit restriction on leasing-out of land; however, a tenant has a right to purchase the land leased in by him within one year of creation of tenancy, as no new privileged tenancies could be created after coming into force of tenancy Acts of various regions. There is no minimum or maximum period limit for leasing. But section 4B of Maharashtra (Bombay region) Tenancy Act provides that leases cannot be terminated on the ground that the lease period had expired. A landlord is not permitted to resume land even for self-cultivation, if the tenant happens to be a member of a co-operative farming society or from a tenant belonging to SCs or STs. The tenancy law applicable in Vidarbha region makes provisions for recording the rights of ordinary tenants, while in former Hyderabad region, all leases were to be in writing.

Orissa: According to section-6 of Orissa Land Reforms Act, 1960, as amended in 1973 and 1976, sub-letting of agricultural land, except by a privileged *raiyat* or a person under disability is not legal. A person under disability includes (i) a widow or unmarried/divorced or separated women, (ii) a minor, (iii) a person incapable of cultivating land for reasons of some physical or mental disability, (iv) a serving member of armed forces, and (v) a *raiyat* holding three standard acres or less. A privileged *raiyat* means a co-operative society, Lord Jagannath, any trust or institution declared as a privileged *raiyat* under this Act or any religious or charitable trust of a public nature or any public financial institution. The Orissa Land Reforms Act also prohibits sharecropping or *bhagchas*. However, tenants under a person under disability or a privileged *raiyat* has a right which is heritable, but non-transferable.

Punjab: The Punjab Land Reform Act, 1972 does not specifically prohibit tenancy. However, the landowner having land within the ceiling limit can evict a tenant on ground of personal cultivation, subject to the condition that the minimum area of five standard acres be left with the tenant or an alternative area is provided by the Government. Section 16 of the Act provides that the tenant of a big landowner was entitled to purchase his tenanted land from the landowner, if he had been in continuous possession of the land for a minimum period of six years, preceding the date of his application and if the land was not included within the reserved or ceiling area of the landowner and he had not sub-let the land or any portion of it during the period of his continuous occupation, except when he/she was a disabled person (widow, or unmarried woman or a tenant suffering from physical or mental disability and also the tenant must have land below ceiling.

Haryana: Punjab and Haryana tenancy provisions are more or less similar, excepting that the Haryana ceiling on Landholdings Act, 1972 did not save the old rights of a tenant to purchase ownerships right of landowners under section 18 of the Punjab law or section 21 of the PEPSU law, unlike section 16 of the Punjab Land Reform Act, 1972 which saved it. Consequently, the land of the tenant under ceiling-surplus land of the landowner who did not acquire proprietorship before 24th January, 1971, is vested with the government, although the tenant would be given preference in the allotment of surplus land by the government.

Rajasthan: The Rajasthan Land Reforms and Acquisition of landowners Estate Act, 1963 does not prohibit letting or sub-letting of agriculture land from holders of *khudkasht* lands and *khatedar* tenants and those under disability such as a minor, lunatic, an idiot, a widow or an unmarried/divorced/separated woman, a person in the armed forces or in prison, a student and a person incapable of cultivating due to physical disability.

Tamil Nadu: The Madras Cultivating Tenants Protection Act, as amended in September 1965, does not explicitly prohibit tenancy, but provides that a landlord shall be entitled to resume possession from any cultivating tenant for the purpose of personal cultivation, not exceeding one-half of the land leased out to the cultivating tenant, excepting when the landlord owns exceeding 13 ½ acres of wet land or has been assessed to sales tax or profession tax or income tax. Besides, the Tamil Nadu Public Tenants Act, 1961 provides that no public trust can evict its cultivating tenants except on the above mentioned grounds. Also, according to Tamil Nadu Agricultural Land Records of Tenancy Rights Act, 1969, there is provision for the preparation and maintenance of a complete record of tenancy rights. There is nothing in the law about voluntary surrender of land or acquisition of ownership right by virtue of occupation of land above a certain time period or right of pre-emption or right of purchase the leased in land by the tenant.

Tripura: The Tripura Land Revenue and Land Reforms Act, 1960 provides that subtenants/*underraiyat* will hold land in perpetuity excepting those under persons of disability namely widow, unmarried/separated, divorced woman, minor, a member of the armed forces or a person under physical or mental disability.

Uttar Pradesh: The Uttar Pradesh Zamindari Abolition and Land Reform Act, 1950 provided complete security of tenure to tenants without any right of resumption by the landowner. Leaseing in future is banned under section 156, except by a disabled person and to agriculture-related educational institution. A disabled person is defined as an unmarried/divorced/separated woman, a widow or a woman whose husband is incapable of cultivating due to physical or mental infirmity or a minor whose father suffers from infirmity or a person who is a lunatic or an idiot or blind or a student of a recognized educational institution whose age does not exceed 25 years of age and whose father suffers from infirmity or a serving member of the armed forces or a person under detention or imprisonment. The law does not regulate rent or protect tenants against eviction.

West Bengal: The West Bengal Land Reforms Act, 1955 as amended in 1970, 1971 and 1977 provides that the landowners can resume land for personal cultivation if his total land size does not exceed three hectares subject to the condition that the sharecropper or *bargardar* is left with at least one hectare of land for his personal cultivation. The *bargardars*/sharecroppers' right is heritable, but non-transferable. In West Bengal, only sharecropping tenancy is legally permitted.

Thus, based on legal position of land leasing, various regions of the country can be broadly grouped into five categories. First, the states of Kerala and Jammu and Kashmir legally ban leasing-out of agricultural land without any exception. Second, Telengana area of Andhra Pradesh, Bihar, Jharkhand, Karnataka, Madhya Prdesh, Chattisgarh, Uttar Pradesh, Uttaranchal, Himachal Pradesh and Orissa allow leasing-out of agricultural land only by certain categories of landowners like disabled, widows, minors, defence personnel, etc. Third, the states of Punjab, Haryana, Gujarat, Maharashtra and Assam have not specifically banned leasing, but the tenant acquires the right to purchase the leased land from the owner within a specified period of creation of tenancy. Besides, once tenancy is created, it cannot be terminated in the case of Gujarat and Maharashtra. In Punjab, a landowner having land within ceiling can evict a tenancy but the tenant has to be left with five standard acres of land. Fourth, in Andhra area of Andhra Pradesh, Rajasthan, Tamil Nadu and West Bengal, there are no legal ban on land leasing, although in West Bengal, only sharecropping leases are legally permitted. In both Andhra Area of Andhra Pradesh and in Tamil Nadu, landowner can resume land for personal cultivation only one-half of the land leased out. Fifth, in the Scheduled Tribe areas of Andhra Pradesh, Bihar, Orissa, Madhya Pradesh and Maharashtra, transfer of tribal land to non-tribals even on lease basis can be permitted only by a competent authority. The idea is to prevent alienation of land from tribals to non-tribals. Besides, in many states, the sharecroppers are not explicitly recognized as tenants. These include Andhra Pradesh, Bihar, Jharkhand, Karnataka, Madhya Pradesh, Chhatisgarh, Uttar Pradesh, Uttaranchal, Haryana, Punjab, Rajasthan and Tamil Nadu.

The other restrictions on landleasing-in most states include fixation of fair rent in the range of one-third to one-fifth of the total produce and also no legal permission to evict tenants except on certain specified grounds in states where landleasing is otherwise allowed.

Impact of Tenancy Reform

The agricultural tenancy laws adopted by various state governments since independence do not seem to have the desired impact in terms of either (i) eliminating the system of lease cultivation, or (ii) providing security of tenure to the tenants-at-will.

No doubt the area under tenancy declined from about 35.7 per cent of the total cultivated area at the time of independence to about 6.5 per cent in 2003 (NSSO 8th and 59th Rounds). Till September 2006 about 12.6 million tenants on 16.7 million acres of land were conferred either ownership or occupancy rights (Govt. of India: 2007). But this measure also provoked the landlords to secure mass eviction of tenants, sub-tenants and sharecroppers. While the tenants were given occupancy right over about four per cent of the total land, they were evicted from about 30 per cent of the total land. Nevertheless, concealed/oral tenancy exists in almost all regions of the country. Even the states of Kerala and Jammu and Kashmir where tenancy is legally banned without any exemption granted even to defence personnel or disabled, large scale informal land leasing takes place (Table 8.1). The informal tenants are most insecure, as they either have short duration oral leases or get rotated from plot to plot each year so that they cannot prove continuous possession of any particular piece of land for any specified period which could give them occupancy right according to law of a state. About 57 per cent of the leased in area in Kharif season and 54 per cent in Rabi season were on short term leases i.e. for less than two years (NSSO, 59th round) and did not have any tenurial security or stability. The NSSO data further reveal that share-cropping continues to be the dominant form of leasing-in most places (Table 8.2), although several state laws do not recognize sharecroppers as tenants. The sharecroppers have limited power to negotiate better terms, because the leasing arrangements are informal and outside the law. Also, the available data on agricultural tenancy is generally underreported as tenancy is either banned or highly restricted in most places. For example, some independent studies (Sharma: 2005: and LBSNAA: 1992) show that in Bihar incidence of leasing was as high as 25 to 37 per cent.

Case for Legalisation and Liberalisation of Land Leasing

There are strong socio-economic reasons for farmers leasing-out and leasing-in land, despite legal ban or restrictions. First, due to population pressure on land and inadequate non-farm employment opportunities, farmers of all size groups tend to lease in land to make their operational holdings adequate for family subsistence. The considerations of subsistence and family labour use are the major reasons for leasing-in (Mani and Pandey: 2004). Second, farmers of all categories who have better outside employment opportunities prefer to lease out. This helps them to maximize family income through both rent by way of leasing-out and income by hiring out labour within and outside agriculture. Also leasing of land is used as an adjustment device by many in response to change in family

Table 8.1: Incidence of Land Leasing in Various States (Rural)

States	*% area*	*leased in*	*% households*	*leasing in*
	2003[k]	*1992*	*2003*	*1992*
Andhra Pradesh	8.9	9.6	15.9	15.9
Arunachal Pradesh	12.4	–	9.1	–
Assam	5.3	8.0	7.5	16.7
Bihar	8.9	4.3	12.3	7.0
Chhattisgarh	6.4	–	11.0	–
Gujarat	5.1	4.2	10.1	10.6
Haryana	14.4	41.1	17.0	18.3
Himachal Pradesh	3.0	4.4	21.6	14.4
Jammu & Kashmir	0.3	3.4	3.1	6.4
Jharkhand	2.4	–	6.7	–
Karnataka	3.6	9.3	7.5	13.1
Kerala	4.0	3.2	7.1	9.6
Madhya Pradesh	3.6**	7.4	5.8	12.3
Maharashtra	4.7	5.2	10.7	15.8
Manipur	12.2	–	15.4	–
Meghalaya	6.9	–	18.1	–
Mizoram	3.0	–	5.8	–
Nagaland	0.1	–	7.9	–
Orissa	13.0	11.4	15.7	22.3
Punjab	16.8	18.2	12.1	15.2
Rajasthan	2.8	5.8	4.3	10.9
Sikkim	14.6	–	34.5	–
Tamil Nadu	6.0	12.4	16.3	19.5
Tripura	6.5	–	6.7	–
Uttar Pradesh	10.5	11.8	12.8	16.5
Uttaranchal	4.8	–	11.8	–
West Bengal	9.3	12.0	11.3	–
UTs	10.5	–	30.6	17.8
Total	**6.5**	**9.0**	**11.5**	**14.7**

Source: Government of India (2006), NSSO, 59th round, Report on Operational Holdings.

Note: * = includes Jharkhand

** = Includes Chhattisgarh

k = Kharif season

labour availability, drought power, cash resources, debt situation etc. (Sharma: 2004). Third, lease market transfers land to those having less land available for use, more ability to use land, more assets to invest, a higher adult workforce and fewer off-farm opportunities (Akter *et al.*: 2006). Thus, economic forces drive land leasing, while ban or restrictions have only reduced the extent of land available in the lease market and have reduced the welfare of poor tenants by forcing them to enter into

Table 8.2: Percentage Distribution of area leased out by terms of lease for each state (Rural) Source NSSO, Government of India, 2003

States	*For fixed money*	*For fixed produce*	*For share of produce*	*Other terms*	*All*
Andhra Pradesh	41.74	22.65	21.72	13.89	100
Arunachal Pradesh	0.00	0.00	14.68	85.32	100
Assam	20.47	0.00	38.71	40.82	100
Bihar	8.89	20.99	59.95	10.17	100
Chhattisgarh	7.29	44.36	45.95	2.40	100
Gujarat	14.78	22.55	10.89	51.78	100
Haryana	84.95	2.10	12.75	0.20	100
Himachal Pradesh	31.82	5.77	28.73	33.68	100
Jammu & Kashmir	6.46	0.00	0.00	93.54	100
Jharkhand	0.53	0.00	99.47	0.00	100
Karnataka	34.73	1.77	34.90	28.60	100
Kerala	33.52	0.00	28.64	37.84	100
Madhya Pradesh	10.85	25.15	53.51	10.49	100
Maharashtra	34.44	3.32	53.97	8.27	100
Manipur	5.52	24.61	65.66	4.21	100
Meghalaya	44.07	7.42	43.69	4.82	100
Mizoram	0.00	0.00	100.00	0.00	100
Nagaland	0.00	21.32	37.74	40.94	100
Orissa	11.52	10.65	71.13	6.70	100
Punjab	91.69	1.73	1.98	4.60	100
Rajasthan	17.62	14.79	44.21	23.38	100
Sikkim	31.20	24.47	44.33	0.00	100
Tamil Nadu	47.19	14.70	26.33	11.78	100
Tripura	6.25	7.05	71.71	14.99	100
Uttar Pradesh	14.84	21.18	51.98	12.00	100
Uttaranchal	77.39	5.96	10.29	6.36	100
West Bengal	29.38	48.79	10.55	11.28	100
Group of UTs	33.73	58.12	8.13	0.02	100
All India	31.04	15.30	39.55	14.11	100

Source: Government of India (2006), NSSO, 59th round, Report on Operational Holdings.

informal arrangements in contravention of the rules and also by restricting the poor peoples' access to land through leasing (Haque: 2001, Deininger: 2005). Besides, due to legal restrictions on land leasing, some landowners prefer to keep their land fallow rather than to lease out for fear of losing the land in case they lease out. The lifting of ban on leasing-in such cases will result in better utilization of the available land and labour and also promote both farm and non-farm development by improving the large landowners' incentive and ability to invest. At the same time legalization of tenancy would provide an incentive to the tenants to produce more and enable them to access institutional credit and other services.

The generally cited reasons for imposing either legal ban or restrictions on land leasing-in the sixties and seventies were as follows:

(i) Tenant farms are relatively less efficient than owner-operated farms;
(ii) Tenancy results in exploitation of the marginal and small farmers who lease in land from absentee land owners or large resident farmers; and
(iii) Unrestricted landleasing may encourage reverse tenancy and would result in concentration of operational holdings in a few hands.

In reality however the above mentioned reasons for either prohibiting or restricting landleasing are not valid in the present context. First, the logic that tenant farms are always inefficient, does not seem to be true as resident landowners, small or large, closely supervise/monitor farm activities of their tenants and also provide credit and share inputs in several cases (Haque: 1996). There is no conclusive evidence to suggest that yields under share tenancy are lower than under owner farming or fixed rent tenancy (Vyas: 1970, Bhaumik: 1993). Second, leasing-in and leasing-out takes place in all size groups of farms and in case, a marginal or small farmer leases out land to another small or marginal or even medium farmer, there is no justification for conferring ownership right on the tenants of marginal and small farmers as the laws in some states suggest. In effect this is anti-poor. Third, according to 59th round of NSSO, about 36 per cent of the tenant farmers are landless, while nearly 56 per cent of the tenant households are marginal landowners having less than one hectare land. (Table 8.3) Also, this category of farmers lease in more land than they lease out. Hence, if more land becomes available through leasing, landless and marginal farmers would improve their social status and economic viability. In fact, the very fact that concentration of operational holdings is lower than that of ownership holdings in the country proves that landless and marginal farmers are able to improve their land access through leasing (Haque: 2010). Why, then should there be any restriction on landleasing, if it helps improve equity, along with farm efficiency? Third, the agrarian and political structure has undergone a complete change in India over the past few decades. While the economic and political powers of large farmers and absentee landowners have eroded, the rural poor has become politically more powerful through Panchayat Raj Institutions and other democratic process. As a result, it is no longer true that a formal tenancy relationship would always be exploitative. After all, tenancy results from a voluntary agreement between the landowner

Table 8.3: Share of landless and Semi-landless Households to total household leasing-in land

States	Landless	Less than 0.5 ha	0.5 to 1.0 ha
Andhra Pradesh	53.1	30.4	8.5
Arunachal Pradesh	71.3	20.1	4.1
Assam	34.7	43.6	11.3
Bihar	5.8	87.0	6.0
Chhattisgarh	26.9	43.5	18.6
Gujarat	63.7	18.8	5.5
Haryana	24.0	45.4	8.9
Himachal Pradesh	62.6	32.2	3.7
Jammu & Kashmir	52.0	43.6	0.7
Jharkhand	45.0	42.9	8.1
Karnataka	55.2	28.3	5.3
Kerala	50.0	46.2	3.4
Madhya Pradesh	28.5	39.7	10.9
Maharashtra	60.1	19.8	11.8
Manipur	16.5	67.5	11.8
Meghalaya	35.0	51.5	5.8
Mizoram	29.3	39.7	9.1
Nagaland	97.3	1.5	1.0
Orissa	17.3	71.8	7.9
Punjab	23.8	31.9	13.0
Rajasthan	25.0	22.2	13.3
Sikkim	83.2	16.1	0.3
Tamil Nadu	72.7	21.3	2.3
Tripura	30.0	67.5	1.6
Uttar Pradesh	7.8	69.5	13.2
Uttaranchal	57.8	41.9	0.3
West Bengal	14.1	75.1	8.4
UTs	89.8	10.6	0.1
All India	35.8	47.5	8.2

Source: Government of India (2006), NSSO, 59th round, Report on Operational Holdings.

and the tenant to lease out and lease in for mutual benefit. Finally, the fear that unrestricted landleasing may encourage reverse tenancy and thereby result in concentration of operational holdings in a few hands can be addressed if the law specifically provides that leasing of land is permitted only within the ceiling limits of each state. Moreover, reverse tenancy within the ceiling limit is not always bad. In the context of Punjab, there seems to be a win-win situation for both the lessor and the lessee, as both maximize their income through leasing-out and leasing-in agricultural land and by way of occupational mobility (Haque: 2001). Thus, there is a case for legalization and formalization of landleasing, as it would help promote agricultural efficiency and equity and also reduce poverty. It is interesting to note that in a left dominated state like Kerala where land leasing is legally prohibited, informal tenancy is quite common. Under Kudumbashree project, the Government of Kerala has even rightly permitted leasing of land by women belonging to small and marginal farm households and landless labourers as neighbourhood groups, which seems to have positive impact on socio-economic status of poor women. Also, the Andhra Pradesh government has permitted licensed cultivators and groups of women to engage in lease cultivation in recent years (APLLCR: 2011)

The Way Forward

Considering various pros and cons it may be necessary to review the landleasing laws of various regions and suggest appropriate amendments, which are mindful of both accelerated agricultural growth and equity. The following principles may serve as guidelines for future landleasing policy:

(i) Leasing in and leasing-out of agricultural land for agriculture by an agriculturist within the ceiling limit of each state should be legally allowed. The ceiling on operational holding would take care of any likely problem of increased concentration of land due to legalization of landleasing.

(ii) The land should automatically revert back to the landowner on the expiry of the agreed lease period, unless renewed.

(iii) The lease period should be as agreed upon by the landowner and the tenant. However, in order to encourage the tenants to undertake land improvement measures for productivity improvement, it would be necessary to have at least three to five years of lease agreement with the condition that any capital investment made by the tenant for improving land productivity should be duly compensated by the landowner on the expiry of the lease period.

(iv) All occupancy tenants including recorded *bargardars*/ sharecroppers especially in West Bengal should be conferred ownership right as they are as good as owners, while leasing of all types, including fixed rent tenancies in future should be made legal, as they are more favourable to the poor tenants who do not have to share the results of his extra efforts put in farming.

(v) The clause of adverse possession of land existing in the tenancy laws of most states should be deleted as it discourages the land-owners to lease out any piece of land on continuous basis, thereby causing insecurity of tenure and lower productivity of rotational tenancy.

(vi) The rent and other terms of lease should be as agreed upon by the landowner and the tenant.

(vii) It may not be necessary to record all tenancies in the revenue record, as this would create a fear in the minds of the landowners and may result in eviction of tenants as it happened in the sixties and seventies.

(viii) All tenants, based on their simple lease agreements should be entitled to access institutional credit and marketing facilities.

References

Ahluwalia, Montek S. (2011): "Prospects and Policy Challenges in the Twelfth Plan", *Economic and Political Weekly*, March 21, vol. XLVI, No. 21.

Akter, Shaheen, J. Farrington, Priya Deshinjkar, Pramod Sharma and Laxman Rao (2006): *Land Rental Markets in India: Efficiency and Equity Considerations*—Paper presented at the National Seminar on Land Policy for Accelerated Rural Growth and Poverty Reduction, organized by the World Bank at Hotel Sangrilla, New Delhi.

All India Congress Committee (1949): Report of the Congress Agrarian Reforms Committee, under the chairmanship of J.C. Kumarappa.

Appu, P.S. (1996): *Land Reform in India*, Vikas Publishing House, New Delhi.

Behuria, N.C. (1997): *Land Reforms Legislation in India*, Vikas Publishing House Pvt. Ltd., New Delhi.

Bhaumik, S.K. (1993): *Tenancy Relations and Agrarian Development*: A Study of *West Bengal*, Sage Publications, New Delhi.

Deininger, Klaus, Songqing Jin and Hari K. Nagarajan (2006): *Equity and Efficiency Impacts of Rural Land Rental Restrictions: Evidence from India*—Paper presented at the National Seminar on Land Policy for Accelerated Rural Growth and Poverty Reduction, organized by the World Bank at Hotel Sangrilla, New Delhi.

Govt. of India, Planning Commission, *Faster, Sustainable and More Inclusive Growth – An Approach to the 12th Five Year Plan*, New Delhi.

Govt. of India, Planning Commission. *Eleventh Five Year Plan*, June, 2008

Govt. of India, (2007), Ministry of Rural Development, Annual Report.

Govt. of India, (2006), NSSO, 59th Round, Report on Operational Holdings.

Govt. of India, Planning Commission. Mid-Term Appraisal, *Tenth Five Year Plan*, June, 2005.

Govt. of India, Planning Commission. *Ninth Five Year Plan*, Mid-Term Appraisal, October, 2002.

Govt. of India, Planning Commission. *First Five Year Plan*, 1951.

Govt. of India, Planning Commission, *Second Five Year Plan*, 1956.

Govt. of India, Planning Commission. *Fourth Five Year Plan*, 1969-73.

Haque, T. (2010): 'Improving the Rural Poors', Access to Land and Water in India, Social Development Report, Council for Social Development, Oxford University Press, New Delhi.

Haque, T. (2006): "Land Tenure and Rural Development in Afro-Asian Countries", *Afro-Asian Journal of Rural Development* 39(2), New Delhi.

Haque, T. (2001): *Impact of Tenancy Reforms on Productivity Improvement and Socio-Economic Status of Poor Tenants*, NCAP, Policy Paper No. 13, ICAR, New Delhi.

Haque, T. (2000): "Contractual Arrangements in Land and Labour markets in India", *Indian Journal of Agril. Economics*, Keynote Paper, Conference November, July-Sept.

Haque, T. (1996): *Sustainability of Small Holder Agriculture*, Concept Publishing Co., New Delhi.

Haque, T. and A.S. Sirohi (1986): *Agrarian Reform and Institutional Changes in India*, Concept Publishing Company, New Delhi.

Mani, Gyanendra and V.K. Pandey (2004): "Are Land Market Operations helping towards Poverty removal in Proceedings of the Workshop on Land Markets and Rural Poverty (August 10-11, 2004)", LBSNAA, Mussourie.

LBSNAA (1992): *Land Reforms in India*, Report of IAS Probationers, Land Reform Unit, Mussourie.

Sharma, Alakh N. (2005): "Agrarian Relations and Socio-economic change in Bihar", *Economic and Political Weekly*, March, Mumbai.

Sharma, Naresh (2004): Tenancy, Poverty, Inequality and Land Reforms in Proceedings of the Workshop on Land Reform and Rural Poverty (August 10-11, 2004)", LBSNAA, Mussourie.

Vyas, V.S. (1970): "Tenancy in a Dynamic Setting", *Economic Political Weekly*, Vol. V, No. 26.

9

Legalising Agricultural Land-Leasing in India: An Assessment of Possible Consequences and Some Suggestions

Sankar Kumar Bhaumik

I. Introduction

In a substantial body of economic literature on agrarian development, the system of land-leasing or tenancy has been portrayed as an institution that leads to production inefficiency as compared to the system of ownership cultivation. The land-leasing system is also held responsible for acting as an instrument for exploitation of the land-poor households (the lessees or tenants) by the land-rich (the lessors or landlords). The historical evidences available from large parts of Latin America and Asia confirmed such features of the land-leasing system (Lipton: 2009).

The land-leasing systems in India that evolved over time in different regions have their roots in the land policies pursued by the Mughal and British rulers (Deshpande: 2010). It has been established in the historical literature that the systems of land-leasing that prevailed in different parts of the country during the pre-independence period led to emergence of absentee landlordism, subinfeudation, rack-renting, and high levels of exploitation of poor tenants through interlocking of land-lease contracts with contracts for credit and/or labour. Primarily because of this exploitative feature, the Government of India, after independence, emphasised the need for tenancy reforms in its overall agenda of land reforms. Beginning with implementation of the policy of abolition of intermediaries, the Central and State Governments, through enactment of various Acts, sought to provide protection to the lessees of land or tenants. The ultimate objectives of the Acts were to confer ownership rights on the tenants and some measure of security of tenure, coupled with rent regulation to prevent exploitation of the tenants, and increase agricultural productivity on tenanted lands.

However, as the Indian Constitution puts land-related issues in the 'States List' while the Central government is supposed to extend broad guidelines in this regard, the tenancy reforms Acts passed over the years in different states differed in their spirit, content and thrust. Thus, while some states had put a complete ban on tenancy so as to implement the policy of 'land to the tiller' (e.g. Jammu and Kashmir and Kerala), the states like Rajasthan and Tamil Nadu do not impose any worthwhile restriction on the practice of land-leasing. In between these two extremes, there are states that also put varying degrees of restrictions on the practice of land-leasing while at the same time making special provisions to protect the rights of the tenants. The heterogeneity that is observed among the states with regard to enactment of tenancy Acts is also visible at the level of actual implementation of the Acts passed by them. Here again, while a few states are hailed for better implementing the tenancy reforms Acts and protecting the tenants' rights, most of the states are alleged for their failures in this regard.

In the backdrop of past failures to implement the tenancy reforms measures, the perspective with regard to land-leasing and tenancy reforms has undergone a 'paradigmatic shift' in India in recent years. This seems to have happened particularly since the beginning of the Tenth Five Year Plan (2002-2007). Deviating from past approaches with regard to tenancy reforms and recognising that the attempts to change the ownership distribution of land in favour of poor by imposing ceiling restrictions on land has had very limited success, the Tenth Five Year Plan document observed that the tenants' access to land could in fact be enhanced through withdrawal of restrictions put on land-leasing or tenancy. It is also observed that the ban on tenancy, which was meant for protecting the tenants, has actually ended up in giving rise to 'concealed tenancy' under which the economic interests of the tenants are hurt as they are deprived of recognition as tenants. In such a situation, the tenants are denied of the benefits of laws that provide security of tenure and regulate rent. Thus, the Tenth Plan calls for taking a fresh look at tenancy laws and completely freeing tenancy laws of all conditions that restrict land-leasing (Government of India: 2002, pp. 301, 308).

The Eleventh Five Year Plan (2002-2007) that adopted the strategy of 'Inclusive Growth' seeks to carry forward the above perspective in greater earnest. In order to ensure inclusiveness of growth, the plan seeks to attain a minimum of 9 per cent growth of GDP (which would be possible with attainment of 4 per cent growth of GDP from agriculture and allied sectors) and more effective reduction of poverty through creation of productive employment opportunities and making those opportunities accessible to all, particularly the poor and marginalised. To see these happen, reforms on several fronts have been proposed, land policy front included. As noted by the Plan, "an efficient and corruption free land administration, coupled with a dynamically adaptive land policy, has a vital role in increasing agricultural growth and poverty reduction" (Government of India: 2008, p. 28).

As regards current tenancy situation in the country, the Eleventh Five Year Plan makes following observations (*Ibid.*: pp. 32-33)[2]: (a) In spite of banning or restricting agricultural tenancy in many states, the incidence of some type of tenancy, particularly crop-sharing tenancy, remained substantial; (b) As tenancy is contracted orally, and in most cases in violation of law, the tenant's position becomes precarious and he/she does have incentive to cultivate land efficiently; (c) In several regions, restrictive tenancy legislations compel some landowners to keep their land fallow for fear of losing their rights over land; (d) Tenancy restrictions have been constraining the supply of land in the rental market as is evident from overall decrease in participation in land-lease market, which ultimately hurts the interest of land-poor households; and (e) In majority of the states, the landless and marginal farmers are net leasers-in of land although in some states the medium and large farmers also lease in land giving rise to 'reverse tenancy'.

To protect the interests of tenants and landowners, and also to increase agricultural productivity, the Eleventh Plan makes six important recommendations. First, tenancy should be legalised in a such a manner that the tenant's security of tenure is guaranteed, and the period of lease contract should be long enough to encourage long-term investment by the tenant which would enhance agricultural production, eventually benefiting both the landlord and the tenant. Secondly, legalising tenancy will result in increasing supply of land and will encourage all categories of households to participate in the land-lease market depending upon resource endowment, education, employment prospects, prevailing wage levels, and institutional back-ups available. Thirdly, instead of a prescribed rental, which are violated in informal tenancies, an upper and lower bound of rentals may be prescribed at the state level. Fourthly, as agrarian conditions vary across states, a 'one-size fits all' prescription cannot be advocated. In cases, where the landlords are dominant and there is strong likelihood of the tenants' interests getting hurt, special clause would be necessary to protect them. Fifthly, small landowners who would otherwise have to operate small uneconomic holdings should have the opportunity to legally lease out land to other farmers with the assurance of being able to resume possession at the end of lease contract period. Finally, the marginal and small farmers should be provided adequate institutional support so that they are not compelled to lease out their land. The sharecroppers should have access to credit once they enter into long-term contracts.

The necessity for the liberalisation of the land-lease market has been further reiterated in the Mid Term Review of the Eleventh Five Year Plan, carried out recently by the Planning Commission (2010). According to the Review, "Liberalisation of the land-lease market does not mean abrogation of existing tenancy legislations. These must be suitably amended to permit leasing-in and leasing-out of land, while making ownership rights non-alienable and secure, fixing tenure of lease, recording of lease and allowing landowners to resume land for cultivation after expiry of lease. Reforming tenancy laws would allow all sections to appropriately participate in the lease market depending upon their resource endowment ... Liberalising tenancy also helps in consolidation of holdings as farmers prefer to lease out rather than sell the piece of land that is inconveniently located. Long term tenancy contracts would also help raise agricultural productivity." (*Ibid.*: p. 90)

In the context of evolution of this new perspective with regard to tenancy reforms and legalisation of land-leasing, mention must also be made of the recommendations of the *Committee on State Agrarian Relations and the Unfinished Task in Land Reforms* (Government of India: 2009).[3] This Committee made a detailed review of the land policy issues covering almost all areas of concern (land ceiling, tenancy reforms, homestead rights, tribal land alienation, land acquisition, women's land rights, common property resources, land reforms for North-East, and so on) and came up with concrete recommendations that are under examination at the level of Central and State governments at this moment. There is broad similarity between recommendations of the *Committee on State Agrarian Relations and the Unfinished Task in Land Reforms* and those of the Eleventh Five Year Plan. It also recommended for legalisation of land-leasing so as to help the rural poor with access to land, discourage the land being kept fallow, and enhance occupational mobility in rural areas. However, an important addition is the recommendation that operational holdings be brought under the purview of ceiling restriction, not just ownership holdings, which would disallow land-leasing for agricultural purposes by those who have land above the ceiling limit. Further, to facilitate land-leasing, standard contracts in simple language protecting the rights of both the parties should be devised which should be enforceable at the *Panchayat* level; women farmers' co-operative and other women land-based groups should be encouraged on a preferential basis to lease in land as experiences (from Andhra Pradesh) show that such organisations of women farmers have emerged as the most viable farming units; the fixation of fair rent may be reconsidered in areas with high institutional strength and the market determined rent should be allowed to prevail; and there should be adequate safeguards including adequate institutional support and rural development schemes to overcome poverty and indebtedness.

It clearly emerges from the foregoing discussion that the predominant mood in the official circles is one of adopting a more flexible approach towards tenancy and going in for its legalisation, of course after appropriately balancing the interests of the landowners and tenants. It is in this context, it would be necessary to explore possible consequences of legalisation of land-leasing (tenancy) in India, especially from the perspective of ensuring better land access for the weaker households (landless, marginal and small cultivators). As both agrarian structure and the degree of legal restrictions put on agricultural land-leasing vary across states of India, the implications of further change in existing legal provisions may be understood comparing tenancy situation among the states/groups of states.[4] This chapter attempts to do so to the extent state-level data on land tenancy available from secondary sources permit. Specifically, we seek to obtain some answers to the following questions: What have been the shares of different categories of rural households in 'owned' and 'operational' lands in India and her states? Does the mechanism of land-leasing-in rural India play an 'equalising role' in the sense of helping the land-poor households to access more land for cultivation? What have been the incidences of land-leasing-in rural areas of India and her states? Who are the 'lessors' and 'lessees' in the market for land-lease? What have been the shares of different categories of households in total households reporting leasing-in and also in total area leased-in? What have been the terms and duration of land-lease (tenancy) contracts? Does any variation in these regards exist between different categories of holdings? How far the land distribution patterns as well as the pattern of land-leasing differed between different states that have been subjected to varying degrees of legal restrictions on land-leasing? What would be consequences of legalisation of land-leasing on the landless, marginal and small farmers in rural areas?

This chapter is divided into six sections. Section II provides a brief idea about data sources and the methodology followed to examine possible consequences of legalisation of land-leasing. We discuss the present structures of landholdings in rural India and her states in Section III. Section IV discusses various aspects of tenancy structure like incidence of tenancy, composition of 'lessors' and 'lessees', access to total leased in land by different categories of holdings, and so on. Section V focuses on the terms and duration of tenancy contracts. The final Section VI provides a summary of main findings and discusses their implications for legalisation of tenancy.

II. Data Base and Methodology

The main source of data for the present study has been the reports on landholdings as published by the National Sample Survey Organisation (NSSO). Among all sources of data on landholdings and tenancy, the NSSO data appear to be most comprehensive and are based on a scientific methodology of sampling (Chadha and Sharma: 1992). However, one important limitation of the NSSO data is that they provide gross underestimates of the extent of tenancy. Accepting such a limitation of tenancy data from the NSSO reports, we maintain that such data are still useful to attempt a comparative assessment of tenancy situations in different states/groups of states of the country.

In this study, we seek to understand the possible impact of liberalisation of tenancy by comparing tenancy situations in states/ groups of states that have been subjected to varying degrees of restrictions on tenancy. As already mentioned, the states of India differed widely with regard to the degree of legal restriction put on tenancy in terms of various Tenancy Acts passed by them from time to time. However, depending upon the degree of legal restriction put on tenancy, the states may be grouped to form some broad categories. In this study, we follow the grouping of states suggested by Hanstad, Haque and Nielsen (2008).[5] According to them, there are four groups of states in India considering the degree of restriction put on tenancy. These are: (1) Kerala and Jammu and Kashmir where there is 'complete prohibition' on tenancy. In these states, the tenancy laws appear to be most stringent and do not allow any exceptions whatsoever; (2) Andhra Pradesh (Telangana area), Karnataka, Madhya Pradesh, West Bengal, Bihar,

Himachal Pradesh, Uttar Pradesh and Orissa comprise the group of 'general prohibition/ limited leasing' cases. Most of these states legally prohibit leasing-out of agricultural land except by certain disabled categories like widows, minors, armed personnel, etc. In West Bengal fixed-rent tenancies are disallowed although crop-sharing tenancies are allowed; (3) Assam, Gujarat, Haryana, Maharashtra and Punjab represent the group characterised by 'permissible leasing with ownership potential'. In these states, the tenants acquire the right to purchase the tenanted land after a certain period; and (4) Andhra Pradesh (Andhra area), Rajasthan and Tamil Nadu comprise the group where there is 'no worthwhile prohibition' on land-leasing). It appears that the degree of legal prohibition on tenancy reduces as move down the order—from 1st group to the 4th.

As we are interested about understanding the impact of legal restrictions on the structure of tenancy and also the poor households' access to land available for tenancy cultivation, a cross-sectional comparison involving these four groups of states is likely to throw useful insights in this regard. Thus, we consider below the land distribution patterns and tenancy structures in these four groups of states. Of course, we also consider the situation in individual states (17 major states) so as to understand intra-group differences, if any. Needless to mention, we have computed the group averages by considering the absolute figures on number of holdings, area owned/ operated, area under tenancy, and so on, for individual states rather than their percentages. In this chapter, we mostly depend on land and tenancy data for the latest available year (January-December, 2003) as thrown by the published reports of the National Sample Survey Organisation (NSSO).

III. Structure of Landholdings

Before we examine the tenancy structures in the states/groups of states, it would be worthwhile to look into their structure of landholdings. This is important because the distribution of landholdings not only determines the inequality of land in the rural society but, to a great extent, also shapes the tenancy structure. We consider both the ownership and operational distributions of land. While ownership holdings represent the most secure type of landholdings, the operational distribution by taking cognizance of land-leasing by the households provides a better idea of their relative economic position.

Distribution of Ownership Holdings

Table 9.1 presents data on ownership distribution of land in 17 major states of India as well as in four groups of states. While presenting data on ownership distribution of land, we have categorised all rural households into six notional categories depending upon their owned land area (x): (1) Completely landless ($x = 0.0$ ha); (2) Marginal ($0.0 < x \leq 1.0$ ha); (3) Small ($1.0 < x \leq 2.0$ ha); (4) Semi-medium ($2.0 < x \leq 4.0$ ha); (5) Medium ($4.0 < x \leq 10.0$ ha); and (6) Large ($x > 10.0$ ha).

Some important points to be noted with regard to distribution of ownership holdings in 2003 are as follows:

(1) In 2003, about 7 per cent of rural households in India appear to be completely landless. If we follow the official definition and consider the holdings owning ≤ 0.002 ha as landless, such a figure turns out to be 10.04 per cent.[7] The percentage of completely landless households (owning no land) varied widely among the states. It is the highest in Tamil Nadu (14.44%). The other states reporting more than 10 per cent of households being completely landless are Himachal Pradesh (14.29%), Andhra Pradesh (12.29 per cent) and Maharashtra (11.05%). Among all states, landlessness is the lowest in Rajasthan (1.82%). In five more states the percentages of completely landless households appear to be less than five per cent. These states are Uttar Pradesh (2.22%), Bihar (2.53 per cent), Jammu and Kashmir (2.60%), Punjab (3.43%) and Kerala (4.80%.)

(2) Nearly 73 per cent of the rural households in all-India belonged to the marginal category of ownership holdings. The percentage of marginal holdings also varied widely among the states. It is the highest in Kerala (90.47%), followed by West Bengal (87.71%), Bihar (85.41%), Orissa (79.94%) and Uttar Pradesh (79.50%). There are three states where the percentage of marginal holdings is less than 60 per cent, which are Rajasthan (53.44%), Madhya Pradesh (57.14%), and Maharashtra (57.94%). Taking an overall view, it may be said that rural India has already attained the stage of high degree of marginalisation of ownership holdings in as much as at least one-half of the holdings belonged to the marginal category in all the states in 2003.

(3) At the all-India level, the percentage of small holdings is recorded to be only about 11 per cent. The percentage of the same has

Table 9.1: Percentage Distribution of Ownership Holdings and Owned Area in 2003 (Rural)

Item	*States / Groups of States*	*Size groups of Ownership Holdings (ha)*						
		0.00 (Completely landless)	*> 0.0 but ≤ 1.0 (Marginal)*	*> 1.0 but ≤ 2.0 (Small)*	*> 2.0 but ≤ 4.0 (Semi-medium)*	*> 4.0 but ≤ 10.0 (Medium)*	*> 10.0 (Large)*	*All sizes*
% of Ownership Holdings	Andhra Pradesh	12.29 (14.33)	70.36	9.13	5.22	2.53	0.47	100.00
	Assam	7.88 (8.06)	73.92	14.17	3.58	0.46	0.00	100.00
	Bihar*	2.53 (6.93)	85.41	8.27	2.90	0.73	0.16	100.00
	Gujarat	9.58 (13.60)	63.76	11.88	7.19	6.58	1.02	100.00
	Haryana	5.88 (9.21)	71.21	9.85	7.75	4.87	0.45	100.00
	Himachal Pradesh	14.29 (15.00)	69.45	11.50	4.05	0.60	0.10	100.00
	Jammu & Kashmir	2.60 (3.29)	74.90	14.99	5.65	1.66	0.20	100.00
	Karnataka	5.51 (14.09)	65.49	14.11	8.77	5.39	0.72	100.00
	Kerala	4.80 (4.80)	90.47	3.55	0.88	0.30	0.00	100.00
	Madhya Pradesh*	5.36 (12.06)	57.14	18.09	12.10	6.37	0.94	100.00
	Maharashtra	11.05 (17.66)	57.94	13.12	11.98	5.10	0.80	100.00
	Orissa	5.65 (9.56)	79.94	9.73	3.72	0.89	0.07	100.00
	Punjab	3.43 (4.57)	73.05	9.52	7.90	5.15	0.95	100.00
	Rajasthan	1.82 (5.65)	53.44	16.53	14.04	10.10	4.07	100.00
	Tamil Nadu	14.44 (16.55)	75.76	5.68	2.87	1.23	0.04	100.00
	Uttar Pradesh*	2.22 (4.17)	79.50	11.99	4.60	1.60	0.08	100.00
	West Bengal	5.02 (6.15)	87.71	5.65	1.40	0.22	0.00	100.00
	All India	6.59 (10.04)	73.08	10.82	6.02	2.97	0.53	100.00
	Group 1	4.42 (4.54)	87.78	5.52	1.70	0.54	0.03	100.00
	Group 2	4.00 (7.82)	76.87	11.28	5.30	2.27	0.28	100.00
	Group 3	8.89 (13.05)	64.61	12.26	8.80	4.73	0.71	100.00
	Group 4	9.53 (12.31)	67.08	9.90	7.21	4.68	1.60	100.00

(Contd.)

(*Contd.*)

Item	*States / Groups of States*	*Size groups of Ownership Holdings (ha)*						
		0.00 (Completely landless)	*> 0.0 but ≤ 1.0 (Marginal)*	*> 1.0 but ≤ 2.0 (Small)*	*> 2.0 but ≤ 4.0 (Semi-medium)*	*> 4.0 but ≤ 10.0 (Medium)*	*> 10.0 (Large)*	*All sizes*
% of Owned Area	Andhra Pradesh	0.00	21.89	19.96	21.17	22.93	14.06	100.00
	Assam	0.00	44.44	34.88	16.36	4.32	0.00	100.00
	Bihar*	0.00	42.02	26.27	17.92	9.13	4.66	100.00
	Gujarat	0.00	13.60	16.05	18.96	39.12	12.28	100.00
	Haryana	0.00	13.15	15.83	24.62	34.14	12.26	100.00
	Himachal Pradesh	0.00	43.77	28.00	19.76	6.44	2.03	100.00
	Jammu & Kashmir	0.00	36.26	25.50	19.54	11.11	7.59	100.00
	Karnataka	0.00	16.65	19.45	23.18	29.52	11.20	100.00
	Kerala	0.00	60.85	21.18	10.81	7.17	0.00	100.00
	Madhya Pradesh*	0.00	13.41	19.61	26.46	29.48	11.04	100.00
	Maharashtra	0.00	12.39	17.57	30.89	27.36	11.79	100.00
	Orissa	0.00	41.50	27.04	19.71	9.97	1.78	100.00
	Punjab	0.00	9.17	15.65	25.32	34.53	15.32	100.00
	Rajasthan	0.00	9.26	11.19	18.61	28.41	32.53	100.00
	Tamil Nadu	0.00	33.14	23.05	22.05	20.53	1.23	100.00
	Uttar Pradesh*	0.00	35.92	27.15	20.27	14.34	2.32	100.00
	West Bengal	0.00	58.32	25.75	11.90	4.03	0.00	100.00
	All India	0.00	23.04	20.37	21.97	23.07	11.55	100.00
	Group 1	0.00	49.70	23.14	14.77	8.96	3.44	100.00
	Group 2	0.00	29.01	23.52	21.69	19.41	6.37	100.00
	Group 3	0.00	15.27	18.36	25.50	29.61	11.26	100.00
	Group 4	0.00	14.13	13.61	19.31	26.80	26.15	100.00

(*Contd.*)

Average size of Ownership Holdings (ha)	Andhra Pradesh	0.00	0.19	1.35	2.51	5.62	18.71	0.62
	Assam	0.00	0.33	1.36	2.52	5.21	-	0.55
	Bihar*	0.00	0.21	1.34	2.60	5.22	12.43	0.42
	Gujarat	0.00	0.22	1.37	2.68	6.04	12.17	1.02
	Haryana	0.00	0.15	1.34	2.64	5.84	22.78	0.83
	Himachal Pradesh	0.00	0.35	1.36	2.73	5.99	11.31	0.56
	Jammu & Kashmir	0.00	0.45	1.58	3.21	6.22	35.01	0.93
	Karnataka	0.00	0.25	1.35	2.59	5.36	15.16	0.98
	Kerala	0.00	0.16	1.39	2.88	5.51	-	0.23
	Madhya Pradesh*	0.00	0.29	1.35	2.72	5.74	14.53	1.24
	Maharashtra	0.00	0.22	1.37	2.63	5.47	15.00	1.02
	Orissa	0.00	0.25	1.34	2.56	5.43	12.10	0.48
	Punjab	0.00	0.11	1.38	2.69	5.61	13.54	0.84
	Rajasthan	0.00	0.36	1.41	2.75	5.84	16.61	2.08
	Tamil Nadu	0.00	0.15	1.37	2.60	5.66	11.20	0.34
	Uttar Pradesh*	0.00	0.27	1.37	2.67	5.43	16.66	0.61
	West Bengal	0.00	0.20	1.34	2.50	5.40	-	0.29
	All India	0.00	0.23	1.37	2.65	5.64	15.83	0.73
	Group 1	0.00	0.20	1.48	3.07	5.89	35.01	0.35
	Group 2	0.00	0.24	1.35	2.65	5.55	14.56	0.65
	Group 3	0.00	0.22	1.36	2.64	5.70	14.44	0.91
	Group 4	0.00	0.21	1.39	2.72	5.81	16.54	1.01

Notes: (1) Figures in parentheses indicate percentage holdings whose landownership is 0.002 ha; and

(2) * Bihar includes Jharkhand, Madhya Pradesh includes Chhattisgarh and Uttar Pradesh includes Uttaranchal.

Source: Author's calculation using data available from NSSO (2006a, Table 1R).

been the highest in Madhya Pradesh (18.09%) and lowest in Kerala (3.55%). The other states where the percentage of the small holdings turned out to be less than 10 per cent are West Bengal (5.65%), Tamil Nadu (5.68%), Bihar (8.27%), Orissa (9.73%), Punjab (9.52%), Andhra Pradesh (9.13%) and Haryana (9.85%).

(4) When we consider the percentages of semi-medium, medium and large holdings, it is found that, at the all-India level, these categories together represent about one-tenth of all ownership holdings. The share of these categories put together in total ownership holdings is the highest in Rajasthan (28.20%) and lowest in Kerala (1.18%). The other states where the same exceeded 10 per cent are Rajasthan, Madhya Pradesh, Maharashtra, Karnataka, Gujarat, Punjab and Haryana. On the other hand, the states where semi-medium, medium and large holdings together represent less than 5 per cent of total holdings are West Bengal, Assam, Bihar, Assam, Tamil Nadu and Himachal Pradesh.

(5) Comparison of four groups of states brings out several interesting points: (a) The percentage of 'completely landless' households is much high for the 3rd and 4th groups of states where tenancy is permitted more liberally. In fact, it is highest for the 4th fourth group of states where there is virtually no legal prohibition on tenancy. However, it needs to be noted that this group comprises two most contrasting states. While Rajasthan recorded the lowest percentage of 'completely landless' households, the same has been highest in Tamil Nadu. Overall, it appears that the degree of landlessness is low in the states where tenancy is restricted (fully/largely) (1st and 2nd groups) as compared to where it is more liberally permitted. This is reflective of some success with regard to implementation of land reforms measures in the 1st and 2nd groups of states. (b) As regards the percentage of marginal holdings, the groups of states that experienced more legal restrictions on tenancy seem to report higher incidence of marginal holdings as compared to where tenancy is not restricted by law. (c) Expectedly, in the groups of states experiencing fewer legal restrictions on tenancy, the percentages of semi-medium, medium and large holdings are more as compared to two other groups where tenancy gets restricted (fully/partially).

(6) At the all-India level, the share of marginal holdings in total owned lands stands out to be only 23.04 per cent although their

share in total holdings was found to be 73.08 per cent. On the other hand, in all-India, the semi-medium, medium and large holdings together captured nearly 57 per cent of all owned lands although their share in total holdings was reported to be just about 10 per cent. This is indicative of severe inequality in the distribution of owned land.

(7) There exists wide variation between the states of India as regards shares of marginal holdings in total owned lands. Nearly three-fourths of total owned lands belonged to the marginal holdings in Kerala and West Bengal. The other states where the shares of this category in total owned lands exceeded 40 per cent are Assam (44.44%), Himachal Pradesh (43.77%), Bihar (42.02 %) and Orissa (41.50%). Some of the states reporting very low percentages of owned lands belonging to the marginal holdings are Punjab (9.17%), Rajasthan (9.26%), Maharashtra (12.39%), Haryana (13.15%), Madhya Pradseh (13.41%) and Gujarat (13.60%). In all these states, the semi-medium, medium and large holdings together accounted for a very high percentage of owned lands; the highest being in Rajasthan (79.55%), followed by Punjab (75.18%), Haryana (71.02%), Gujarat (70.35%), Maharashtra (70.04%) and Madhya Pradseh (66.98). Even in Andhra Pradesh these categories together represented 58.16 per cent of owned lands.

(8) The share of marginal holdings in total owned land is significantly high in the groups of states that were subjected to more legal restriction of tenancy. The shares of marginal holdings in total owned land or the 1st, 2nd, 3rd and 4th groups turned out to be 49, 29, 15 and 14 per cent respectively.

There is a clear negative relationship between the share of marginal holdings in total owned lands and the degree of legal restriction on tenancy. This observation, in conjunction with our earlier observation that the problem of landlessness is less severe in first two groups of states, indicates the positive contributions of land reforms measures (of which tenancy reforms is a part) in bringing down landlessness as well as enhancing access to land by the marginal landholders.

(9) Table 9.1 also presents data on average size of ownership holdings (in ha) for different groups of holdings in all-India, individual states and groups of states. At the all-India level, the average size of all ownership holdings is found to be 0.73 ha.

Overall, it appears that the average size of ownership holdings is high in the states where the share of higher categories of holdings in total land area is high and the same for lower categories is low. Moreover, the average size of ownership holdings goes up significantly as we move from a more-restrictive tenancy regime (from legal point of view) to a less-restrictive tenancy regime. In the groups of states reporting more restrictive tenancy regimes, the lower value of average size of ownership holdings seems to be an outcome of fairer distribution of land in them.

Distribution of Operational Holdings

Let us now turn our attention to the distribution of operational holdings. To understand the patterns of distribution of operational holdings and operated areas in rural all-India and the states, we formed five groups – marginal, small, semi-medium, medium and large. The data on percentages of operational holdings and operated areas for these five groups are presented in Table 9.2. The following points need to be noted specifically:

(1) In 2003, at the all-India level, nearly 70 per cent of all operational holdings belonged to the marginal category. Among the states, the percentage of marginal holdings is extremely high in Kerala (92%), West Bengal (89%) and Bihar (83%). The percentage of marginal holdings remained above the all-India level (70%) in five other states (Himachal Pradesh, Tamil Nadu, Orissa, Uttar Pradesh and Jammu and Kashmir). Among the states, it is relatively low in Rajasthan (49%), Maharashtra (50%), Madhya Pradesh (51%) and Gujarat (60%).

(2) The shares of higher categories in total operational holdings appear to be quite high in the states of Rajasthan, Maharshtra, Madhya Pradesh, Punjab, Gujarat, Haryana and Karnataka. Among these states, the semi-medium, medium and large holdings together account for 32 per cent of all holdings in Rajasthan; the same exceeds 20 per cent in others states. Kerala, West Bengal, Bihar, Assam and Orissa are the states where the combined share of these categories in all operational holdings remained very low (within 5% or so).

(3) When we look at the percentages of different categories of holdings in four groups of states, it appears that the degree of

Table 9.2: Percentage Distribution of Operational Holdings and Operated Area in 2003 (Rural)

Item	*States / Groups of States*	*Size groups of Operational Holdings (ha)*					
		≤ 1.0 Marginal)	*> 1.0 but ≤ 2.0 (Small)*	*> 2.0 but ≤ 4.0 (Semi-medium)*	*> 4.0 but ≤ 10.0 (Medium)*	*> 10.0 (Large)*	*All sizes*
% of Operational holdings	Andhra Pradesh	60.73	20.67	11.97	5.51	1.12	100.00
	Assam	76.23	18.44	4.71	0.63	0.00	100.00
	Bihar*	82.62	12.23	4.01	0.98	0.15	100.00
	Gujarat	59.98	17.28	11.13	9.77	1.84	100.00
	Haryana	66.28	12.77	12.28	7.81	0.86	100.00
	Himachal Pradesh	79.47	14.52	5.12	0.77	0.12	100.00
	Jammu & Kashmir	76.18	15.82	6.06	1.73	0.21	100.00
	Karnataka	58.19	20.42	13.23	7.11	1.05	100.00
	Kerala	91.76	6.19	1.51	0.54	0.00	100.00
	Madhya Pradesh*	51.16	23.27	16.68	7.67	1.23	100.00
	Maharashtra	49.52	21.41	19.09	8.78	1.20	100.00
	Orissa	78.38	15.24	5.20	1.13	0.05	100.00
	Punjab	66.30	11.16	12.86	7.83	1.86	100.00
	Rajasthan	49.41	18.50	15.92	11.52	4.66	100.00
	Tamil Nadu	77.07	13.41	6.73	2.70	0.09	100.00
	Uttar Pradesh*	76.68	15.91	5.59	1.70	0.13	100.00
	West Bengal	88.75	8.88	2.15	0.22	0.00	100.00
	All India (Kharif)	69.76	16.25	9.00	4.20	0.79	100.00
	All India (Rabi)	69.96	15.91	8.95	4.35	0.84	100.00
	Group 1	87.99	8.52	2.61	0.83	0.05	100.00
	Group 2	70.85	17.04	8.37	3.29	0.45	100.00
	Group 3	60.23	17.93	13.37	7.32	1.16	100.00
	Group 4	61.54	16.27	11.89	7.65	2.65	100.00

(Contd.)

(*Contd.*)

Item	States / Groups of States	Size groups of Operational Holdings (ha)					
		≤ 1.0 Marginal)	> 1.0 but ≤ 2.0 (Small)	> 2.0 but ≤ 4.0 (Semi-medium)	> 4.0 but ≤ 10.0 (Medium)	> 10.0 (Large)	All sizes
% of Operated Area	Andhra Pradesh	18.57	21.13	22.77	22.09	15.45	100.00
	Assam	41.99	36.03	17 08	4.90	0.00	100.00
	Bihar*	43.02	27.43	17.61	8.69	3.24	100.00
	Gujarat	13.07	15.02	18.95	37.34	15.62	100.00
	Haryana	10.42	13.53	26.00	34.98	15.07	100.00
	Himachal Pradesh	44.48	27.60	19.56	6.40	1.96	100.00
	Jammu & Kashmir	31.00	21.74	16.76	9.24	21.27	100.00
	Karnataka	16.22	20.04	24.84	27.76	11.14	100.00
	Kerala	57.77	23.35	11.68	7.20	0.00	100.00
	Madhya Pradesh*	13.09	20.27	28.48	26.96	11.19	100.00
	Maharashtra	11.96	17.75	30.41	29.23	10.66	100.00
	Orissa	42.99	28.67	18.81	8.61	0.92	100.00
	Punjab	7.25	11.74	26.20	36.35	18.46	100.00
	Rajasthan	8.97	10.92	18.62	28.39	33.10	100.00
	Tamil Nadu	30.94	24.15	23.00	20.39	1.51	100.00
	Uttar Pradesh*	35.72	29.20	19.84	12.46	2.77	100.00
	West Bengal	58.34	26.74	12.23	2.69	0.00	100.00
	All India (Kharif)	22.57	20.89	22.46	22.23	11.85	100.00
	All India (Rabi)	21.75	20.31	22.34	23.08	12.52	100.00
	Group 1	45.42	22.60	14.02	8.14	9.81	100.00
	Group 2	26.38	24.44	23.34	18.95	6.89	100.00
	Group 3	14.30	17.68	25.53	30.33	12.16	100.00
	Group 4	13.37	13.57	19.50	26.79	26.78	100.00

Avearge size of Operational Holdings (ha)	Andhra Pradesh	0.41	1.36	2.54	5.34	18.44	1.33
	Assam	0.38	1.36	2.52	5.40	-	0.69
	Bihar*	0.31	1.33	2.60	5.24	12.60	0.59
	Gujarat	0.34	1.37	2.69	6.03	13.39	1.58
	Haryana	0.20	1.36	2.72	5.76	22.53	1.29
	Himachal Pradesh	0.40	1.36	2.73	5.95	11.31	0.72
	Jammu & Kashmir	0.40	1.36	2.74	5.29	99.90	0.99
	Kamataka	0.38	1.35	2.59	5.39	14.57	1.38
	Kerala	0.23	1.39	2.84	4.92	-	0.37
	Madhya Pradesh*	0.40	1.38	2.70	5.56	14.44	1.58
	Maharashtra	0.40	1.37	2.63	5.50	14.62	1.65
	Orissa	0.39	1.35	2.60	5.48	12.69	0.72
	Punjab	0.14	1.38	2.67	6.08	13.00	1.31
	Rajasthan	0.43	1.41	2.79	5.87	16.95	2.38
	Tamil Nadu	0.31	1.38	2.61	5.77	12.58	0.76
	Uttar Pradesh*	0.35	1.38	2.66	5.50	16.25	0.75
	West Bengal	0.29	1.33	2.53	5.49	-	0.44
	All India (Kharif)	0.34	1.37	2.65	5.63	15.96	1.06
	All India (Rabi)	0.33	1.37	2.67	5.69	15.96	1.07
	Group 1	0.27	1.38	2.78	5.11	99.90	0.52
	Group 2	0.35	1.37	2.66	5.50	14.52	0.95
	Group 3	0.33	1.37	2.65	5.74	14.58	1.39
	Group 4	0.36	1.40	2.74	5.86	16.88	1.67

Notes: (1) * Bihar includes Jharkhand, Madhya Pradesh includes Chhattisgarh and Uttar Pradesh includes Uttaranchal; and (2) Information for individual states refer to 'Kharif' season.

Source: Author's calculation using data available from NSSO (2006b, Table S2-R).

marginalisation of holdings is more pronounced in areas characterised by restrictive tenancy legislations. Thus, the percentage of marginal holdings goes down as we move from more-restrictive tenancy regimes to less-restrictive tenancy regimes. As is evident from Table 9.2, the percentages of marginal holdings are 88, 71, 60 and 62 in four groups of states, respectively.

(4) As regards distribution of operated land, it is found that the marginal holdings, at the all-India level, represented only 22.57 per cent of total operated area (during Kharif season) although their share in total holdings was reported to be nearly 70 per cent. There are three states (Punjab, Haryana and Rajasthan) where the marginal holdings represented only 10 per cent or less of total operated land. The share of marginal holdings in total operated area turned to about 58 per cent in West Bengal and Kerala. In the states of Assam, Bihar, Himachal Pradesh and Orissa, the same exceeded 40 per cent. On the other hand, semi-medium, medium and large holdings together controlled 70 per cent or more of operated areas in Punjab, Rajasthan, Haryana, Gujarat and Maharashtra, which is quite high as compared to their share at the all-India level (57%).

(5) Comparing the percentage share of marginal category in total operated land for four groups of states, it is observed that the same goes up steadily as we move from less-restrictive tenancy situation (3rd and 4th groups) to more-restrictive tenancy situation (1st and 2nd groups). The marginal holdings controlled as high as 45 per cent of operated lands in the 1st group of states as against only 13 per cent in the 4th group. The situation is just opposite with regard to share of semi-medium, medium and large holdings in total operated land. They together represented as high as 73 per cent of total operated land in 4th group of states as against only 32 per cent in the 1st group. Thus, it is clear that, like the ownership distribution of land, the distribution of operational holdings is also relatively more equal in the states where the degree of legal restriction on tenancy is greater.

(6) At the all-India level, the average size of all operational holdings is found to be 1.06 ha for Kharif season which does not differ much in the Rabi season (1.07 ha). Interestingly, among the states of India, the average operated area for the marginal category is

very low in two green-revolution states of Punjab (0.14 ha) and Haryana (0.20 ha). There are two other states (Kerala and West Bengal) where the same turned out to be less than 0.30 ha. As expected, the average size of operational holding for higher categories is high both in green-revolution states as well as in states like Jammu and Kashmir, Rajasthan, Madhya Pradesh, Gujarat, Maharashtra, Uttar Pradesh and Tamil Nadu. Further, among four groups of states, the average size of all operational holdings is the highest for the 4th group and lowest for 1st group.

The above observations together imply that the degree of marginalisation of holdings has been greater in areas experiencing more restrictive tenancy system. However, in such areas, the landholdings structure has been relatively more favourable to the lower categories of holdings (particularly marginal ones) as revealed by their higher share in both total holdings and total operated area. This seems to have happened due to better land and tenancy reforms in such areas.

Inequality of Land Distribution

Our discussion above indicated existence of high inequality in the distribution of land in most of the states, particularly where land and tenancy reforms have been weak. We have computed the Gini coefficients to better assess the degree of inequality in land distributions (owned and operational).[8]

Table 9.3 shows that the value of Gini coefficient with regard to distribution of owned lands in rural all-India increased between 1970-71 and 2003. During the most recent period of 1991-92 to 2003, it remained almost unchanged. The fact that the value of Gini coefficient with regard to distribution of owned land is 0.738 in 2003 is indicative of very high degree of inequality in the distribution of owned land. Following the trend of inequality in the distribution of owned lands, the inequality in the distribution of operational lands also increased between 1970-71 and 2003. No significant change in the distribution of operational land is visible in rural all-India for the recent period of 1991-92 and 2003.

However, comparing the Gini inequality coefficients for the two distributions – owned and operated – it is found that the degree of inequality for the latter has always been somewhat lower than the former both in all-India and the states (except in Jammu and Kashmir[9]). This indicates contributory role of tenancy to equalise land distribution in rural

Table 9.3: Inequality of Land Distribution in Rural India During 1960-61 to 2003

Year	Gini inequality coefficient of	
	Ownership Holdings	Operational Holdings
1960-61	0.738	0.624
1970-71	0.654	0.579
1981-82	0.697	0.589
1991-92	0.743	0.630
2003	0.738	0.624[a] 0.637[b]

Note: 'a' and 'b' represent Kharif and Rabi seasons respectively.
Source: Author's calculation using data available from NSSO Report Nos. 144 (for 1960-61), 144 & 215 (for 1970-71), 330 & 331 (for 1981-82), 309 & 407 (for 1991-92) and 491 & 492 (for 2003).

areas, though not very significantly as revealed by continuance of high inequality in the distribution of operational holdings.

The Gini coefficient values for individual states as well as groups of states for both the distributions have been presented in Table 9.4. A close perusal of the computed Gini coefficient values lead to following observations:

(1) Besides two green-revolution states of Punjab and Haryana, the Gini coefficient values for ownership distribution of land are very high in Tamil Nadu, Andhra Pradesh and Gujarat. In all these states, such values are above the all-India level. This value is lowest in Jammu and Kashmir. The other states to have relatively low value of Gini coefficient for ownership distribution of land are Assam, Himachal Pradesh, Kerala and Uttar Pradesh.

(2) Comparison between four groups of states shows that Gini inequality in the distribution of owned land has been the lowest for 1st group that experienced complete ban on tenancy and highest for 4th group where tenancy is liberally allowed. In fact the value of Gini coefficient for ownership distribution goes up monotonically as we move from legally restricted-tenancy areas to liberal-tenancy areas.

(3) As in the case of ownership distribution of land, the operated lands are also most unequally distributed in the states of Punjab and Haryana. Two other states where the Gini coefficient values for operational distribution of lands lay above all-India level

Table 9.4: Land Distribution Inequality in the States in 2003 (Rural)

States / Groups of States	*Gini inequality coefficient of*	
	Ownership Holdings	*Operational Holdings***
Andhra Pradesh	0.792	0.588
Assam	0.558	0.480
Bihar*	0.695	0.565
Gujarat	0.758	0.647
Haryana	0.792	0.727
Himachal Pradesh	0.586	0.472
Jammu & Kashmir	0.548	0.588
Karnataka	0.709	0.587
Kerala	0.620	0.519
Madhya Pradesh*	0.663	0.561
Maharashtra	0.725	0.558
Orissa	0.661	0.493
Punjab	0.809	0.748
Rajasthan	0.671	0.633
Tamil Nadu	0.815	0.619
Uttar Pradesh*	0.636	0.554
West Bengal	0.667	0.519
All India	0.738	0.624
Group 1	0.654	0.579
Group 2	0.697	0.589
Group 3	0.743	0.630
Group 4	0.789	0.667

Notes: (1) * Bihar includes Jharkhand, Madhya Pradesh includes Chhattisgarh and Uttar Pradesh includes Uttaranchal;
(2) ** representing Kharif season.

Source: Author's calculation using data available from NSSO (2006a,b).

are Gujarat and Rajasthan. On the other hand, the same appears to be quite low in the states of Himachal Pradesh (reporting lowest value) Assam, Orissa, Kerala and West Bengal.

(4) As regards inequality in the distribution of operational lands in four groups of states, we find that the value of Gini coefficient is the lowest for 1st group (representing most restrictive tenancy situation) and highest for 4th group (where tenancy is liberally allowed). Moreover, the degree of inequality in the distribution of operational land also increases monotonically, as we go down the order, from more-restrictive tenancy situation to less-restrictive tenancy situation. This is not solely due to unequal distribution of tenanted lands in areas experiencing more liberal tenancy regimes. In fact, the high degree of inequality in the distribution of operated land in areas experiencing relatively

less restriction on tenancy results also from high inequality in the distribution of owned land in those areas. This is confirmed by the fact there exists high correlation between the Gini inequality coefficients of owned and operational holdings.[10]

IV. Structure of Tenancy

In this section, we discuss the structure of tenancy in terms of incidence of tenancy, composition of 'lessors' and 'lessees', and access to tenanted land by different categories of holdings.

Incidence of Tenancy

To understand the incidence of tenancy, we consider three aspects: percentage of holdings reporting leasing into total operational holdings, percentage of leased-in area to total operated area, and average leased-in area (ha) per holding under different size groups of holdings. Table 9.5 presents information on these aspects for the year 2003.

It is found that, at the all-India level, nearly one-tenth of all operational holdings are reported to have leased-in land. The percentage of holdings reporting leasing-in is the highest for Orissa and lowest for Jammu and Kashmir. Apart from Orissa, in seven other states, the percentage of holdings reporting leasing-in is found to be more than 10 per cent. These states are Andhra Pradesh, Bihar, Orissa, Punjab, Tamil Nadu, Uttar Pradesh and West Bengal. As expected, among four groups of states, percentage of holdings reporting leasing-in is lowest for the 1st group which puts in place strongest possible restriction on land-leasing. However, the percentage of holdings reporting leasing-in is highest for 2nd group that also includes states which are known to have quite a bit of restriction on land-leasing. This indicates greater inclination among the rural households in the states of this group to engage in land-leasing notwithstanding legal restrictions.

Another important point to note is that, at the all-India level, among all size groups of holdings, the percentage of holdings reporting leasing-in is highest for the large holdings (14.36%). What is more, there are four states where a very high percentage of large holdings are engaged in leasing-in, which are Haryana (53.72%), Punjab (32.71%), Andhra Pradesh (30.30%) and Uttar Pradesh (29.16%). On the other hand, a relatively high percentage of marginal and small holdings have been leasing-in in the eastern Indian states of Bihar, Orissa and West Bengal.

Table 9.5: Incidence of Tenancy in the States in 2003 (Rural)

Item	*States / Groups of States*	Size groups of Operational Holdings (in ha)					
		≤ 1.0 (Marginal)	*> 1.0 but ≤ 2.0 (Small)*	*> 2.0 but ≤ 4.0 (Semi-medium)*	*> 4.0 but ≤ 10.0 (Medium)*	*> 10.0 (Large)*	*All sizes*
% of operational holdings reporting leased-in	Andhra Pradesh	12.41	19.97	17.81	9.63	30.30	14.66
	Assam	6.98	11.68	12.47	11.72	-	8.13
	Bihar*	13.63	12.82	10.40	1.98	0.00	13.27
	Gujarat	5.23	5.26	4.55	2.96	27.41	5.34
	Haryana	4.57	13.02	32.96	22.25	53.72	10.94
	Himachal Pradesh	10.67	3.08	4.98	0.00	0.00	9.18
	Jammu & Kashmir	2.44	0.84	0.51	0.00	0.00	2.02
	Karnataka	5.70	3.72	6.69	3.49	19.35	5.41
	Kerala	5.29	6.66	9.52	0.00	-	5.41
	Madhya Pradesh*	5.90	8.75	7.32	5.02	6.07	6.73
	Maharashtra	6.19	5.91	9.80	10.86	3.07	7.19
	Orissa	18.70	20.29	12.48	0.00	0.00	18.40
	Punjab	4.47	14.62	31.05	38.13	32.71	12.18
	Rajasthan	1.32	3.76	4.38	5.61	8.12	3.07
	Tamil Nadu	9.57	12.94	12.82	3.16	4.64	10.06
	Uttar Pradesh*	11.90	18.23	15.61	9.78	29.16	13.10
	West Bengal	14.27	12.46	12.17	10.89	-	14.05
	All India (Kharif)	10.14	11.73	11.12	8.14	14.36	10.43
	All India (Rabi)	10.36	12.61	10.96	8.69	14.51	10.73
	Group 1	4.70	4.05	4.46	0.00	0.00	4.59
	Group 2	12.16	13.52	10.14	5.38	11.23	11.99
	Group 3	5.76	7.99	13.59	13.09	20.69	7.92
	Group 4	5.85	7.08	6.48	5.23	8.06	6.13

(*Contd.*)

(*Contd.*)

Item	States / Groups of States	Size groups of Operational Holdings (in ha)					
		≤ 1.0 (Marginal)	> 1.0 but ≤ 2.0 (Small)	> 2.0 but ≤ 4.0 (Semi-medium)	> 4.0 but ≤ 10.0 (Medium)	> 10.0 (Large)	All sizes
% of leased in area to operated area	Andhra Pradesh	11.93	11.05	10.10	4.39	7.31	8.95
	Assam	4.00	6.45	6.14	5.30	-	5.31
	Bihar*	12.41	8.59	6.40	0.48	0.00	8.87
	Gujarat	2.42	2.76	3.13	1.37	21.00	5.12
	Haryana	7.22	7.01	20.36	12.65	19.76	14.40
	Himachal Pradesh	5.05	1.56	0.61	0.00	0.00	2.80
	Jammu & Kashmir	0.53	0.43	0.37	0.00	0.00	0.32
	Karnataka	4.69	2.78	3.81	1.96	6.61	3.54
	Kerala	3.54	4.45	7.91	0.00	-	4.00
	Madhya Pradesh*	5.42	4.59	3.94	1.95	2.27	3.54
	Maharashtra	4.16	3.61	5.25	6.19	1.06	4.66
	Orissa	17.02	13.50	9.64	0.00	0.00	13.00
	Punjab	6.04	10.07	18.94	21.54	13.12	16.83
	Rajasthan	0.98	1.69	2.74	3.46	3.04	2.77
	Tamil Nadu	8.67	7.90	4.56	0.79	15.50	6.04
	Uttar Pradesh*	10.80	10.25	8.28	3.61	17.92	9.44
	West Bengal	11.02	6.90	6.72	6.65	-	9.27
	All India (Kharif)	8.64	6.82	6.26	4.23	6.12	6.45
	All India (Rabi)	8.82	7.44	5.96	4.49	7.55	6.74
	Group 1	2.59	2.67	3.75	0.00	0.00	2.31
	Group 2	10.87	8.00	5.74	2.16	5.08	6.92
	Group 3	4.06	4.66	7.94	7.36	11.57	7.07
	Group 4	4.54	3.91	3.17	3.05	3.18	3.43

(*Contd.*)

(*Contd.*)

Area leased-in per operational holding reporting leasing-in (ha)	Andhra Pradesh	0.39	0.75	1.44	2.43	4.45	0.81
	Assam	0.22	0.75	1.24	2.44	-	0.45
	Bihar*	0.28	0.89	1.60	1.27	-	0.40
	Gujarat	0.16	0.72	1.85	2.79	10.26	1.51
	Haryana	0.32	0.73	1.68	3.27	8.29	1.69
	Himachal Pradesh	0.19	0.69	0.33	-	-	0.22
	Jammu & Kashmir	0.09	0.69	2.00	-	-	0.16
	Karnataka	0.32	1.01	1.47	3.03	4.98	0.90
	Kerala	0.15	0.93	2.36	-	-	0.27
	Madhya Pradesh*	0.37	0.72	1.45	2.16	5.39	0.83
	Maharashtra	0.27	0.84	1.41	3.14	5.04	1.07
	Orissa	0.36	0.90	2.01	-	-	0.51
	Punjab	0.19	0.95	1.63	3.43	5.21	1.81
	Rajasthan	0.32	0.63	1.74	3.63	6.36	2.15
	Tamil Nadu	0.28	0.84	0.93	1.45	42.00	0.46
	Uttar Pradesh*	0.32	0.77	1.41	2.03	9.99	0.54
	West Bengal	0.23	0.74	1.39	3.35	-	0.29
	All India (Kharif)	0.29	0.80	1.49	2.92	6.80	0.66
	All India (Rabi)	0.28	0.81	1.45	2.94	8.30	0.67
	Group 1	0.15	0.91	2.34	-	-	0.26
	Group 2	0.32	0.81	1.50	2.21	6.57	0.55
	Group 3	0.23	0.80	1.55	3.23	8.15	1.24
	Group 4	0.28	0.77	1.34	3.42	6.67	0.93

Note: (1) *Bihar includes Jharkhand, Madhya Pradesh includes Chhattisgarh and Uttar Pradesh includes Uttaranchal,
(2) Information for individual states refer to Kharif season.

Source: Author's calculation using data available from NSSO (2006b, Tables 2R and 3R).

Overall, it appears that, among large holdings, the propensity to lease-in is relatively greater in the developed states, particularly the green-revolution ones, while such a propensity is more among marginal and small holdings in the backwards states.

As regards the percentage of leased-in area to total operated area, the following points need to be noted specifically (Table 9.5):

(a) As per the NSSO data, at the all-India level, only about 6 to 7 per cent of the operated areas were leased-in areas (6.45 per cent during Kharif season and 6.74 per cent during Rabi season). Of course, this is an underestimate due to concealment of tenancy in large parts of the country. However, these data on percentage of leased-in areas may be used to make an interstate comparison to understand their relative position as regards incidence of tenancy.

(b) It is found that the states of India vary widely as regards percentage of leased-in area to total operated area. Among the states, this is highest in Punjab (16.83%), followed by Haryana (14.40%), Orissa (13.00%), West Bengal (9.27%), Andhra Pradesh (8.95%) and Bihar (8.87%). In all remaining states, the percentage of leased in area to total operated area is lower than the same for all-India. Jammu and Kashmir, where tenancy is banned, recorded the lowest figure (0.32%).

(c) Comparison between the four groups of states shows that the percentage of leased-in area to total operated area is expectedly the lowest for the 1st group, but it is second lowest for the 4th group where no worthwhile prohibition on tenancy exists. On the other hand, the percentage of leased-in area to total operated area is relatively high for 2nd and 3rd groups, which also differ in terms of degree of legal restriction on tenancy. All these facts seem to suggest that the incidence of tenancy (percentage of leased-in area to total operated area) in the states of India does not depend solely on the provision of restrictive tenancy legislations.[11] Thus, removal of restriction on tenancy might not necessarily lead to expansion of tenanted area, as exemplified by the 4th group of states where tenancy is allowed rather liberally and yet the incidence of tenancy has been very low.

(d) Another noteworthy point is that the percentage of leased-in area to operated area for the large holdings remained above 15 per cent in Gujarat, Haryana, Uttar Pradesh, Tamil Nadu and Punjab, which is much high as compared to the same at the all-India level (6.45%). On the other hand, in Assam, Bihar, Himachal Pradesh, Jammu and Kashmir, Kerala, Orissa and West Bengal, the large holdings are virtually non-existent as lessees. For the marginal holdings, the percentage of leased-in area to operated area, at the all-India level, is found to be 8.64 per cent and it remained above this level in six states only, which are Orissa, Bihar, Andhra Pradesh, West Bengal, Uttar Pradesh and Tamil Nadu.

Table 9.5 further shows that the average leased-in area per holding reporting leasing-in is only 0.66 ha in all-India. However, the same is quite high in Punjab (2.15 ha), Rajasthan (1.81 ha), Haryana (1.69 ha) and Gujarat (1.51 ha). The average leased-in area per holding reporting leasing-in is very low (less than 0.66 ha as observed for all-India) in Jammu and Kashmir, Himachal Pradesh, Kerala, West Bengal, Bihar, Assam, Tamil Nadu, Orissa and Uttar Pradesh. Among four groups of states, it is lowest for 1st group but highest for 3rd group that includes the two green-revolution states.

Composition of Lessors and Lessees

We now focus our attention on the issue of composition of 'lessors' and 'lessees'. Specifically, our objective is to know which category of landholdings they belong to.

The Lessors

To know the background of 'lessors', we have computed percentage distribution of their numbers as well as leased out areas according to size class of ownership holdings for the states/groups of states (Table 9.6). Some important observations here are the following:

(1) At the all-India level, nearly 84 per cent of the 'lessors' belonged to the category of marginal and small landholding groups. There are 10 states where the percentage of lessors from marginal and small landholding groups exceeded the same for all-India. These states, in descending order of percentage of lessors belonging to the marginal and small category, are Jammu and Kashmir,

Table 9.6: Percentage Shares in Total 'Lessors' (Those Reporting Leasing out) and Total Leased-out Land by Different Groups of Ownership Holdings in 2003 (Rural)

States / Groups of States	% share in total lessons by			% share in total leased-out land by		
	≤ 2.0 ha (Mar + Small)	> 2.0 but ≤ 4.0 ha (Semi-Med)	> 4.0 ha (Med + Large)	≤ 2.0 ha (Mar + Small)	> 2.0 but ≤ 4.0 ha (Semi-Med)	> 4.0 ha (Med + Large)
Andhra Pradesh	91.20	2.80	6.00	70.46	9.70	19.84
Assam	91.50	8.50	0.00	76.91	23.09	0.00
Bihar*	90.10	8.74	1.16	65.26	17.02	17.73
Gujarat	84.80	3.50	11.70	61.96	3.36	34.68
Haryana	82.14	11.58	6.29	36.92	32.65	30.43
Himachal Pradesh	95.70	3.00	1.30	91.57	7.20	1.23
Jammu & Kashmir	97.90	0.00	2.10	93.55	0.00	6.45
Karnataka	67.30	10.60	22.10	22.77	14.97	62.26
Kerala	93.21	5.19	1.60	64.13	14.47	21.40
Madhya Pradesh*	67.69	16.01	16.30	37.00	14.11	48.89
Maharashtra	67.30	24.40	8.30	37.83	37.01	25.16
Orissa	88.70	8.70	2.60	62.44	22.33	15.23
Punjab	60.20	27.50	12.30	25.71	40.72	33.57
Rajasthan	59.06	11.51	29.43	23.12	13.71	63.17
Tamil Nadu	90.49	6.51	3.00	76.44	12.75	10.81
Uttar Pradesh*	86.57	10.68	2.75	58.18	22.47	19.35
West Bengal	92.40	5.10	2.50	65.26	8.03	26.72
All India	83.51	10.17	6.32	47.78	20.85	31.37
Group 1	93.25	5.15	1.60	64.66	14.21	21.13
Group 2	84.80	9.98	5.22	50.19	18.60	31.21
Group 3	74.34	17.03	8.63	48.66	31.52	29.77
Group 4	81.80	7.89	10.31	40.86	13.39	45.75

Note: * Bihar includes Jharkhand, Madhya Pradesh includes Chhattisgarh and Uttar Pradesh includes Uttaranchal.
Source: Author's calculation using data available from NSSO (2006a, Table 2R).

Himachal Pradesh, Kerala, West Bengal, Assam, Andhra Pradesh, Tamil Nadu, Bihar, Orissa, Uttar Pradesh and Gujarat. On the other hand, there are six states where the percentage of lessors from marginal and small category remained below the same observed for all-India. These states, in ascending order of percentage of lessors belonging to the marginal and small category, are Rajasthan, Punjab, Maharashtra, Karnataka, Madhya Pradesh and Haryana.

(2) Comparison between four groups of states shows that as high as 93 per cent of lessors from 1st group of states belonged to the marginal and small landholding category. On the other hand, 82 per cent of lessors belonged to the same category in 4th group of states. It is clearly observed that the percentage of lessors drawn from the marginal and small landholding category goes down as we move from more restrictive tenancy regime to less restrictive tenancy regime. In other words, where tenancy is liberally allowed, the propensity of large landowners leasing-out land becomes greater.

(3) Table 9.6 also shows that, at the all-India level, only about 48 per cent of leased out land came from the marginal and small landowners although their share in total number of lessors was 84 per cent. On the other hand, the shares of semi-medium, and medium and large landowners in total leased out land were 21 and 31 per cent respectively (their shares in total number of lessors are 10 and 6% respectively). Clearly the dominance of the lessors from higher landownership categories gets reflected more in terms of their contribution in total leased out area.

(4) There are 11 states where the percentage share of marginal and small landowners in total leased out land exceeded the same for all-India level. These states, in descending order of percentage share of marginal and small landowners in total leased out land, are Jammu and Kashmir, Himachal Pradesh, Assam, Tamil Nadu, Andhra Pradesh, Bihar, West Bengal, Kerala, Orissa, Gujarat ...desh. On the other hand, the percentage share of ...ll landowners in total leased out land was ...low the same for all-India level) in six states ...taka, Rajasthan, Punjab, Haryana, Madhya ...harashtra. Obviously, in these six states, the ...otal leased out land by the lessors belonging to ...ership categories has been relatively high.

(5) Comparison between the four groups of states shows that the percentage contribution to total leased out land by the landowners from higher categories goes up as we move from more-restricted tenancy areas to less-restricted tenancy area. Thus, while the combined share of semi-medium, medium and large landowners to total leased out land has been 35 per cent in the 1st group of states, the corresponding figure for the 4th group is found to be 59 per cent.

While discussing the composition of the lessors, it might also be interesting to gather some idea about their occupational background. The NSSO report for 59th Round provides some information in this regard. We have data on percentage distribution of tenanted land (excluding homestead) according to background of the lessors. Such data presented in Table 9.7 lead to the following observations: (a) At the all-India level, nearly 26 per cent of all tenanted land belonged to the lessors who are engaged in non-farming activities (this includes non-farm work within rural areas, urban work and other works)[12]; (b) There are two states (Kerala and Tamil Nadu) where more than 50 per cent of tenanted land belonged to the lessors who are engaged in non-farming activities. In eight other states (Rajasthan, Punjab, Maharshtra, Haryana, Karnataka, Andhra Pradesh, Himachal Pradesh and West Bengal) the percentage of tenanted land belonging to lessors engaged in non-farming activities is higher than the same for all-India. In states where this is less than the all-India figure are Jammu and Kashmir, Gujarat, Madhya Pradesh, Uttar Pradesh, Orissa, Assam and Bihar; and (c) When we compare between the four groups of states, it is found that as high as 55 per cent of tenanted land has been drawn from the lessors who are engaged in non-farming activities in the 1st group of states.[13] The same for the 4th group of states also appears to be quite high (48%). For the 2nd and 3rd groups corresponding figures are 20 and 30 per cent respectively.

The Lessees

Let us now find out which group of landholdings the lessees belong to and how total leased-in land is distributed among the lessees belongin to different groups of holdings. As regards the distribution of lessees various groups of holdings, the following points may be noted (Tabl

(1) At the all-India level, nearly 86.05 per cent of all belonged to the 'landless+marginal+small' cate

Table 9.7: Percentage Distribution of Tenanted (Leased-in) Land (excluding homestead) According to Background of Lessors

States / Groups of States	*% of tenanted land from lessons engaged in*	
	Farming activities	*Non-farming activities***
Andhra Pradesh	72.54	27.46
Assam	77.70	22.30
Bihar*	77.04	22.96
Gujarat	85.58	14.42
Haryana	71.29	28.71
Himachal Pradesh	72.91	27.09
Jammu & Kashmir	100.00	0.00
Karnataka	72.23	27.77
Kerala	41.95	58.05
Madhya Pradesh*	83.78	16.22
Maharashtra	69.65	30.35
Orissa	79.51	20.49
Punjab	55.69	44.31
Rajasthan	55.34	44.66
Tamil Nadu	46.51	53.49
Uttar Pradesh*	83.72	16.28
West Bengal	73.19	26.81
All India	73.96	26.04
Group 1	45.10	54.90
Group 2	80.31	19.69
Group 3	69.81	30.19
Group 4	52.33	47.67

Notes: (1) * Bihar includes Jharkhand, Madhya Pradesh includes Chhattisgarh and Uttar Pradesh includes Uttaranchal;
(2) ** includes rural non-farm work, urban work and other works.

Source: Author's calculation using data available from NSSO (2006a, Table 14R).

corresponding percentages for semi-medium and large holdings are 9.59 and 3.27 respectively.

(2) In half of the states, the share of the 'landless+marginal+small' holdings in all lessees exceeded 90 per cent. These states are Jammu and Kashmir, West Bengal, Kerala, Himachal Pradesh, Bihar, Orissa, Assam, Uttar Pradesh and Tamil Nadu. In three states, namely Gujarat, Karnataka and Madhya Pradesh, the corresponding percentages remained around 75 per cent. In Maharashtra, nearly 60 per cent of the lessees belonged to this category. On the other hand, two green-revolution states of

Table 9.8: Percentage Shares in Total 'Lessees' (Those Reporting Leasing-in) and Total Leased-in Land by Different Groups of Operational Holdings in 2003 (Rural)

States / Groups of States	% share in total number of 'lessees' by				% share in total leased-in land by			
	≤ 2.0 ha (Landless$ + Mar + Small)	> 2.0 but ≤ 4.0 ha (Sem-Med)	> 4.0 but ≤ 10.0 ha (Med)	≥ 10.0 ha (Large)	≤ 2.0 ha (Landless$ + Mar + Small)	> 2.0 but ≤ 4.0 ha (Sem-Med)	> 4.0 but ≤ 10.0 ha (Med)	≥ 10.0 ha (Large)
Andhra Pradesh	79.53	14.53	3.63	2.31	50.85	25.69	10.84	12.63
Assam	91.84	6.02	2.14	0.00	75.42	19.78	4.80	0.00
Bihar*	96.70	3.15	0.15	0.00	86.87	12.71	0.42	0.00
Gujarat	75.69	9.47	5.41	9.43	14.29	11.60	10.04	64.08
Haryana	42.88	37.00	15.89	4.22	11.80	36.74	30.78	20.67
Himachal Pradesh	97.22	2.78	0.00	0.00	95.74	4.26	0.00	0.00
Jammu & Kashmir	98.48	1.52	0.00	0.00	80.65	19.35	0.00	0.00
Karnataka	75.28	16.34	4.62	3.76	37.19	26.66	15.39	20.77
Kerala	97.28	2.66	0.06	0.00	76.94	23.06	0.00	0.00
Madhya Pradesh*	75.08	18.13	5.68	1.10	46.30	31.67	14.87	7.16
Maharashtra	60.24	26.01	13.23	0.51	24.45	34.29	38.84	2.42
Orissa	96.47	3.53	0.00	0.00	86.07	13.95	0.00	0.00
Punjab	37.69	32.78	24.54	4.99	9.63	29.48	46.51	14.38
Rajasthan	43.92	22.75	21.01	12.32	9.85	18.38	35.43	36.34
Tamil Nadu	90.55	8.58	0.83	0.04	76.09	17.39	2.64	3.89
Uttar Pradesh*	91.79	6.66	1.26	0.28	72.57	17.40	4.77	5.26
West Bengal	97.99	1.86	0.15	0.00	89.21	8.86	1.93	0.00
All India	86.05	9.59	3.27	1.09	52.38	21.80	14.57	11.25
Group 1	97.41	2.54	0.05	0.00	77.18	22.82	0.00	0.00
Group 2	92.30	6.14	1.22	0.34	71.88	18.17	5.45	4.50
Group 3	61.93	22.73	12.32	3.02	19.86	28.65	31.60	19.89
Group 4	77.46	12.56	6.49	3.49	33.20	18.03	23.87	24.90

Notes: (1) * Bihar includes Jharkhand, Madhya Pradesh includes Chhattisgarh and Uttar Pradesh includes Uttaranchal. (2) Data refer to Kharif season. (3) $ refers to those who operated some land, though they did not own any land.

Source: Author's calculation using data available from NSSO (2006b, Tables 4R and 5R).

Punjab (38%) and Haryana (43%) and Rajasthan (44%) are reported to have relatively low percentage of lessess under the 'landless+marginal+small' category.

(3) The overwhelming presence of lessees belonging to the 'semi-medium+medium' category is visible in Punjab and Haryana. In both these states, more than one-half of all lessees belonged to this category. In two other states, namely Rajasthan and Maharashtra, this category of lessees represented nearly two-fifths of all lessees. While in Andhra Pradesh the corresponding figure is nearly 18 per cent, it is 15 per cent in Gujarat. In the remaining states, it varied between 1.52 per cent (Jammu and Kashmir) to 8.16 per cent (Assam).

(4) The five states where the percentage of lessees from large landholdings category is relatively high are Rajasthan (12.32%), Gujarat (9.43%), Punjab (4.99%), Haryana (4.22%), Karnataka (3.76%) and Andhra Pradesh (2.31%). In the remaining states, the presence of such lessees has been negligible.

(5) When we compare between four groups of states, it is found that the areas experiencing higher legal restrictions on tenancy (1st and 2nd groups) have extremely high persence of lessees from 'landless+marginal+small' category. On the other hand, the percentage of lessees from higher categories of holdings increases in areas having more liberal tenancy regimes. Thus, it is clear that while in areas experiencing more tenancy reforms the land-lease market is dominated by the lessees from lower categories of holdings, the lessees from higher categories make quite a substantial presence in the land-lease market when tenancy is liberally allowed. Clearly, the land-lease market has a tendency to become more favourable to the lessees from lower categories of holdings ('landless+marginal+small') when tenancy reforms measures are put in place.

Distribution of Total Leased-in Land

The actual control over the land-lease market by different categories of tenants could not be properly understood in terms of their numbers alone. It is also necessary to look into the distribution of all leased-in (tenanted) lands between the tenants drawn from various categories of operational holdings. The data on distribution of total leased-in lands between tenants from different categories of holdings in the states/groups of states have been also presented in Table 9.8 (also see Figure 9.1). Our main observations here are the following:

(1) At the all-India level, 52 per cent of all leased-in land belonged to the tenants from 'landless+marginal+small' category. On the other hand, the shares of tenants from 'semi-medium+medium' and large categories in total leased-in land are found to be 36 and 11 per cent respectively.

(2) There are eight states where 75 per cent or more of leased-in land belonged to the tenants from 'landless+marginal+small' category. These states are Himachal Pradesh (95.7%), West Bengal (89.2%), Bihar (86.9%), Orissa (86.1%), Jammu and Kashmir (80.7%), Kerala (76.9%), Tamil Nadu (76.1%) and Assam (75.4%). In Andhra Pradesh, nearly half of total leased-in land belonged to this category. The same in Karnataka and Maharashtra are 37.2 and 24.5 per cent respectively. The four states where the tenants from 'landless+marginal+small' category have very low share in total leased-in land are Punjab (9.7%), Rajasthan (9.9%), Haryana (11.8%) and Gujarat (14.3%).

(3) The four states where the land-lease markets are largely dominated by the tenants from 'semi-medium+medium' category are Punjab (having a share of 76% in total leased-in land), Maharashtra (73%), Haryana (68%) and Rajasthan (54%). The share of this category in total leased-in land varied between 37 to 46 per cent in three states (Andhra Pradesh, Karnataka and Madhya Pradesh). In the remaining states of Jammu and Kashmir, Tamil Nadu, Gujarat, Uttar Pradesh, Kerala and Assam, the same varied between 19 and 25 per cent. Himachal Pradesh, West Bengal, Bihar and Orissa are the states where the 'semi-medium+medium' category does not capture much of leased-in land (the share varied between 4 to 14%).

(4) The tenants from large holdings captured an extremely high proportion of leased-in area in Gujarat (64.1%). The same is also quite substantial in Rajasthan (36.3%), Karnataka (20.8%), Haryana (20.7%), Punjab (14.4%) and Andhra Pradesh (12.6%). In seven states, the share of large holdings in leased-in land has been zero. These states are West Bengal, Orissa, Kerala, Jammu and Kashmir, Himachal Pradesh, Bihar and Assam.

(5) Comparing the four groups of states, we observe a highly contrasting picture as regards shares of different categories of

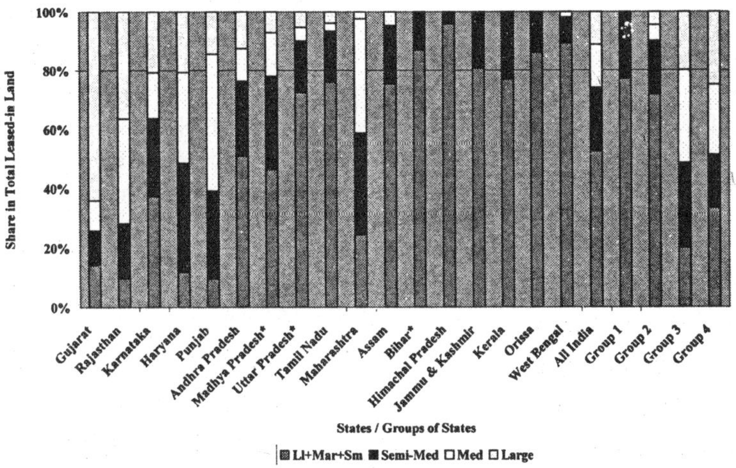

Fig. 9.1: Shares of Different Categories of Operational Holdings in Total Leased-in Land in 2003.

holdings in total leased in land. In areas where tenancy is restricted (1st and 2nd groups of states), the share of lower categories of holdings ('landless+marginal+small') in total leased-in land is very high, and hence the same for upper categories of holdings (particularly for medium and large categories) is very low. The situation is just opposite in areas (3rd and 4th groups) known for fewer restrictions on tenancy. Table 9.8 shows that the shares of 'landless+marginal+small' category in total leased-in land have been 77, 72, 20 and 33 in 1st, 2nd, 3rd and 4th groups of states, respectively. On the other hand, the medium and large categories together accounted for 0, 10, 51 and 49 per cent of total leased-in land in these four groups of states respectively.

It appear from above discussion that, in a regime of restrictive-tenancy, the lower categories of holdings are able to exercise much better control over the land-lease market in the sense of capturing a larger share of tenanted lands as opposed to the regime of liberal-tenancy. To put it differently, in the regime of liberal-tenancy, not only much larger number of tenants from higher categories of landholdings compete with the smaller tenants in the land-lease market but also they succeed in cornering much high proportion of tenanted lands.

V. Terms and Duration of Tenancy Contracts

Terms of Tenancy Contracts

The NSSO report for 2003 provides data on distribution of total holdings and total leased-in land under different terms: fixed money, fixed produce, share produce, relatives under no specified terms, and so on. In this study, we have considered data on distribution of total leased-in land under different terms of tenancy in order to identify the predominant forms of tenancy contracts among different categories of holdings in the states/ groups of states. As regards the terms of tenancy contracts, we have clubbed the first two terms (fixed money and fixed produce) and called it 'fixed-rent' tenancy. The second category is that of 'share produce', and the final category of 'others' included all remaining terms.

The data on terms of tenancy for the states/groups of states for the year 2003 have been presented in Table 9.9. Some important observations in this regard are the following:

(1) At the all-India level, both during Kharif and Rabi seasons, nearly 50 per cent of all tenanted lands have been under fixed-rent contract. The shares of share-produce contract and other terms in total leased-in land have been nearly 40 per cent and 10 per cent respectively.

(2) In nine out of 17 major states, the percentages of leased-in land under fixed rent tenancy have been more than 50 per cent. The presence of such a form of tenancy has been overwhelming in the two green revolution states of Punjab and Haryana. In both these states as high as 81 per cent of all tenanted lands are under fixed-rent contract. The seven other states where the percentage of tenanted areas under fixed-rent tenancy exceeded 50 per cent are Karnataka (74%), Andhra Pradesh (70%), Tamil Nadu (70%), Gujarat (57%), Rajasthan (53%), West Bengal (52%) and Madhya Pradesh (51%). Even in Kerala the share of fixed-rent tenancy in total leased-in land is found to be 47 per cent. In three states (Himachal Pradesh, Uttar Pradesh and Maharashtra), one-third or slightly more of tenanted lands are found under fixed-rent tenancy. On the other hand, the percentages of tenanted land under fixed-rent tenancy have been quite low in Jammu and Kashmir (12%), Orissa (19%), Assam (19%) and Bihar (30%).

(3) The states which are largely predominated by share-produce tenancy are Orissa (73%), Bihar (67%), Jammu and Kashmir (64%), Assam

Table 9.9: Terms of Tenancy under Different Categories of Holdings in 2003 (Rural).

(Percentage of total leased-in area)

% of leased in area under	States/Groups of States	≤ 1.0 (Marginal)	> 1.0 but ≤ 2.0 (Small)	> 2.0 but ≤ 4.0 (Semi–medium)	> 4.0 but ≤ 10.0 (Medium)	> 10.0 (Large)	All sizes
		Size groups of Operational Holdings (ha)					
Fixed rent (Fixed money + Fixed produce)	Andhra Pradesh	72.81	75.05	80.32	77.97	22.87	69.58
	Assam	7.51	27.12	10.71	59.68	–	19.27
	Bihar*	34.07	21.97	21.52	100.00	–	29.58
	Gujarat	12.27	76.73	20.05	44.51	67.42	56.98
	Haryana	79.08	90.49	70.57	79.89	98.46	80.96
	Himachal Pradesh	39.24	27.51	46.68	–	–	37.74
	Jammu & Kashmir	10.54	20.97	0.00	–	–	11.53
	Karnataka	67.47	73.26	69.39	53.87	100.00	73.57
	Kerala	37.21	89.43	21.39	–	–	47.11
	Madhya Pradesh*	48.70	35.16	52.83	55.38	94.74	50.74
	Maharashtra	17.98	45.10	28.48	38.30	100.00	35.19
	Orissa	16.03	19.55	28.65	–	–	18.84
	Punjab	75.50	70.08	90.23	79.60	70.43	80.64
	Rajasthan	21.95	39.55	29.35	55.04	66.69	52.47
	Tamil Nadu	70.44	70.98	43.99	23.40	0.00	62.02
	Uttar Pradesh*	28.73	37.57	54.87	43.54	25.87	36.64
	West Bengal	50.82	46.35	74.93	57.77	–	52.20
	All India (Kharif)	35.80	45.29	54.05	60.19	64.61	48.67
	All India (Rabi)	41.35	44.76	57.55	60.02	56.90	50.37
	Group 1	35.48	84.54	20.22	–	–	44.82
	Group 2	34.03	34.13	50.20	51.94	65.85	39.40
	Group 3	27.08	53.45	53.73	62.05	75.95	58.56
	Group 4	64.82	62.20	34.33	53.78	63.02	55.83

(Contd.)

(*Contd.*)

% of leased in area under	States/Groups of States	Size groups of Operational Holdings (ha)					
		≤ 1.0 (Marginal)	> 1.0 but ≤ 2.0 (Small)	> 2.0 but ≤ 4.0 (Semi–medium)	> 4.0 but ≤ 10.0 (Medium)	> 10.0 (Large)	All sizes
Share produce	Andhra Pradesh	17.81	17.83	12.62	18.18	77.11	24.01
	Assam	67.91	58.12	31.59	36.78	–	54.93
	Bihar*	60.88	76.99	76.64	0.00	–	66.88
	Gujarat	61.95	22.08	55.46	55.19	31.90	38.02
	Haryana	20.07	6.42	26.13	15.59	0.00	15.87
	Himachal Pradesh	34.09	58.21	0.00	–	–	36.36
	Jammu & Kashmir	45.34	76.89	96.63	–	–	64.43
	Karnataka	29.77	26.58	26.69	46.13	0.00	24.77
	Kerala	18.33	10.11	0.00	–	–	11.97
	Madhya Pradesh*	47.13	55.24	26.19	42.88	5.18	38.99
	Maharashtra	49.87	35.97	29.90	44.01	0.00	37.62
	Orissa	77.02	72.07	59.26	–	–	73.07
	Punjab	16.65	29.88	0.00	18.78	28.37	15.35
	Rajasthan	51.88	47.81	70.50	35.13	24.85	39.27
	Tamil Nadu	19.24	10.25	44.04	1.85	99.95	23.37
	Uttar Pradesh*	64.38	56.92	39.85	46.87	0.00	53.52
	West Bengal	38.76	39.25	2.00	5.89	–	34.97
	All India (Kharif)	51.27	47.04	31.26	31.87	26.73	40.39
	All India (Rabi)	49.16	46.51	32.75	30.93	36.81	40.79
	Group 1	20.09	14.88	5.27	–	–	15.35
	Group 2	59.08	59.60	39.10	42.90	1.33	52.12
	Group 3	49.65	35.49	23.37	29.78	23.13	28.92
	Group 4	23.02	20.74	61.50	33.81	28.98	33.67

Other terms	Andhra Pradesh	9.37	7.12	7.07	3.85	0.02	6.41
	Assam	24.58	14.76	57.70	3.54	–	25.80
	Bihar*	5.05	1.04	1.84	0.00	–	3.54
	Gujarat	25.78	1.19	24.48	0.29	0.68	5.00
	Haryana	0.85	3.08	3.30	4.53	1.54	3.17
	Himachal Pradesh	26.67	14.29	53.32	–	–	25.89
	Jammu & Kashmir	44.12	2.14	3.37	–	–	24.07
	Karnataka	2.77	0.16	3.91	0.00	0.00	1.66
	Kerala	44.45	0.46	78.61	–	–	40.92
	Madhya Pradesh*	4.18	9.60	20.98	1.74	0.08	10.27
	Maharashtra	32.14	18.94	41.63	17.70	0.00	27.18
	Orissa	6.95	8.38	12.09	–	–	8.09
	Punjab	7.85	0.04	9.77	1.62	1.19	4.01
	Rajasthan	26.17	12.65	0.15	9.83	8.46	8.26
	Tamil Nadu	10.32	18.77	11.97	74.75	0.05	14.61
	Uttar Pradesh*	6.89	5.51	5.27	9.59	74.13	9.84
	West Bengal	10.42	14.40	23.06	36.33	–	12.83
	All India (Kharif)	12.93	7.67	14.69	7.94	8.67	10.94
	All India (Rabi)	9.48	8.73	9.69	9.05	6.29	8.84
	Group 1	44.43	0.58	74.51	–	–	39.83
	Group 2	6.88	6.27	10.70	5.16	32.83	8.48
	Group 3	23.27	11.06	22.90	8.17	0.92	12.53
	Group 4	12.16	17.06	4.17	12.40	7.99	10.50

Notes: (1) * Bihar includes Jharkhand, Madhya Pradesh includes Chhattisgarh and Uttar Pradesh includes Uttaranchal;
(2) Information for individual states refer to Kharif season.

Source: Author's calculation using data available from NSSO (2006b, Table 5R).

(55%) and Uttar Pradesh (54%). In West Bengal (where share-tenancy is allowed, not fixed-rent tenancy), Himachal Pradesh, Maharashtra, Gujarat, Madhya Pradesh and Rajasthan, the share of share-produce tenancy in total leased-in land varied between 35 to 39 per cent. In Kerala, Punjab, Haryana, Tamil Nadu, Andhra Pradesh and Karnataka this share fell below 25 per cent.

(4) When we compare the terms of tenancy for different categories of holdings, it is observed that, at the all-India level, there is a general tendency of marginal and small holdings leasing-in relatively higher proportion of land under share-produce tenancy, both during Kharif and Rabi seasons. However, there are some states where there is predominance of fixed-rent tenancy for all landholdings groups. These states are Punjab, Haryana, Andhra Pradseh, Tamil Nadu, Karnataka and West Bengal. On the other hand, share-tenancy seems to have dominated for all categories of holdings in Assam, Bihar, Orissa and Jammu and Kashmir. The remaining states seem to have followed the general pattern which is that the lower categories of holdings preferring share-tenancy while fixed-rent tenancy being preferred by higher categories of holdings.

(5) Comparison between four groups of states shows that the proportion of leased-in area under fixed-rent tenancy goes up as we move from areas governed by more restrictions on tenancy to less-restrictive tenancy areas. Thus, while in 1st and 2nd groups of states, the percentages of leased-in area under fixed-rent tenancy are 43 and 39 respectively, the corresponding percentages in 3rd and 4th groups of states are found to be 59 and 56 respectively. Consequently, share-tenancy seems to emerge as the predominant form of tenancy in areas experiencing more legal restriction on tenancy, while fixed-rent tenancy seems to be important in areas with less restriction on tenancy.

Duration of Tenancy Contracts

To understand the duration of land-tenancy contracts, we have considered data on percentage distribution of total leased-in land under different contract periods – less than one year, one year to less than two years, two years to less than five years, and five years and above. Such data for the tenants belonging to different categories of operational holdings have been presented in Table 9.10. Our main observations here are the following:

(1) At the all-India level, during Kharif season, nearly 33 per cent of

leased-in areas have been contracted for a period of one to less than two years. While the share of long-period contracts (five years or more) has been nearly 23 per cent, the share of very short-period contract (less than one year) has been 24 per cent. On the other hand, 20 per cent of leased-in areas are under two to less than five years contract. If we consider all land-lease contracts for less than five years, the corresponding figure comes to around 53 per cent for the Kharif season and 54 per cent for the Rabi season. Thus, the pattern of distribution of land-lease contracts according to contract duration does not differ much between two seasons.

(2) Another important point to note is that there is a tendency among the smaller tenants to get involved in shorter duration lease contracts while the tenants from higher categories being engaged more in longer duration lease contracts. Table 9.10 shows that while the percentage of leased-in land of tenants from marginal holdings under lease contract of less than one year has been 27 per cent, the corresponding figure for tenants belonging to large holdings has been 6 per cent only. On the other hand, while only 18 per cent of leased-in areas of tenants from marginal holdings are under a contract period of five years and above, the same for large holding is about 46 per cent.

(3) There is also wide variation among the states as regards duration of land-lease contracts. In nine out of 17 major states (Haryana, Rajasthan, West Bengal, Assam, Orissa, Uttar Pradesh, Madhya Pradesh, Jammu and Kashmir and Maharashtra), 20 to 35 per cent of leased-in areas are found to be under very short duration contracts (less than one year). In six states (Uttar Pradesh, Karnataka, Bihar, Rajasthan, Punjab and Haryana), the percentage of leased-in areas under contracts for one year to less than two years ranged between 35 to 52 per cent. In three states (Karnataka, Kerala and Jammu and Kashmir) the contracts of two years to less than five years covered one-third or more of leased-in areas. In another set of six states (Kerala, Assam, West Bengal, Gujarat, Tamil Nadu, and Himachal Pradesh), the percentage of leased-in land under long duration contracts (five years or more) ranged between 34 and 57.

(4) It seems that the land-lease contracts of areas experiencing fewer restrictions on tenancy (3rd and 4th groups of states) are dominated more by shorter duration contracts while those of restrictive-tenancy areas being more of longer duration type. This conclusion follows from the fact that the contract period for 72 per cent of

Table 9.10: Duration of Tenancy Contracts Under Different Categories of Holdings in 2003 (Rural)

(Percentage of total leased in area)

Duration of lease	States / Groups of States	Size groups of Operational Holdings (ha)					
		≤ 1.0 (Marginal)	> 1.0 but ≤ 2.0 (Small)	> 2.0 but ≤ 4.0 (Semi-medium)	> 4.0 but ≤ 10.0 (Medium)	> 10.0 (Large)	All sizes
Less than one year	Andhra Pradesh	24.34	16.08	22.90	18.48	0.00	18.11
	Assam	21.81	23.54	25.34	2.25	-	22.31
	Bihar*	15.24	22.63	22.14	85.71	-	18.42
	Gujarat	43.08	20.08	9.44	62.59	0.00	11.64
	Haryana	13.45	11.51	34.64	17.92	0.00	19.70
	Himachal Pradesh	9.30	3.73	0.00	-	-	8.04
	Jammu & Kashmir	49.77	22.55	0.00	-	-	32.23
	Karnataka	6.78	23.95	22.53	27.81	5.19	16.58
	Kerala	21.55	0.00	5.05	-	-	12.15
	Madhya Pradesh*	39.36	43.96	23.09	15.04	0.00	28.98
	Maharashtra	38.19	31.14	49.94	25.71	0.04	35.48
	Orissa	32.23	21.61	1.06	-	-	24.72
	Punjab	37.15	9.60	16.11	11.01	4.50	12.16
	Rajasthan	29.12	42.78	15.92	12.43	25.00	20.19
	Tamil Nadu	4.80	6.74	0.23	0.00	0.00	4.30
	Uttar Pradesh*	29.61	31.88	29.88	20.65	0.00	28.39
	West Bengal	27.66	7.88	3.68	20.96	-	21.47
	All India (Kharif)	27.36	25.28	29.82	21.02	5.64	24.07
	All India (Rabi)	24.07	22.17	26.17	23.02	3.90	21.07
	Group 1	23.38	1.61	4.77	-	-	13.45
	Group 2	26.27	28.75	22.29	20.46	1.56	24.80
	Group 3	31.18	22.39	33.37	20.71	0.79	21.43
	Group 4	7.62	16.80	10.59	11.94	23.62	14.59

(Contd.)

(*Contd.*)

One year to less than two years	Andhra Pradesh	25.48	27.91	43.99	13.27	4.25	26.87
	Assam	32.31	23.41	17.87	37.47	-	25.81
	Bihar*	48.44	44.50	41.93	0.00	-	46.34
	Gujarat	48.61	67.93	3.96	24.52	0.00	11.43
	Haryana	42.97	68.51	42.39	57.82	56.03	51.70
	Himachai Pradesh	17.00	2.77	0.00	-	-	14.08
	Jammu & Kashmir	0.00	3.98	99.50	-	-	20.41
	Karnataka	63.05	21.68	18.43	14.32	53.20	35.11
	Kerala	14.33	15.53	14.41	-	-	14.66
	Madhya Pradesh*	32.70	26.72	6.10	31.81	70.49	25.28
	Maharashtra	8.34	32.01	20.16	20.20	0.00	20.06
	Orissa	28.46	24.24	54.63	-	-	30.85
	Punjab	31.35	40.34	53.65	55.12	26.83	48.96
	Rajasthan	26.53	38.61	73.13	58.65	30.90	48.87
	Tamil Nadu	25.50	17.95	20.49	0.00	0.00	20.57
	Uttar Pradesh*	36.00	38.11	30.30	27.75	24.24	34.66
	West Bengal	22.85	28.19	15.47	35.80	-	23.51
	All India (Kharif)	33.52	33.23	33.02	37.99	23.43	32.86
	All India (Rabi)	38.07	36.95	33.91	35.73	11.87	32.97
	Group 1	13.39	14.71	19.06	-	-	15.03
	Group 2	36.04	34.07	25.37	26.94	44.81	33.46
	Group 3	26.35	39.79	33.21	40.72	16.94	32.55
	Group 4	25.62	23.72	55.23	56.32	29.20	38.89

(*Contd.*)

(*Contd.*)

Duration of lease	States / Groups of States	Size groups of Operational Holdings (ha)					
		≤ 1.0 (Marginal)	> 1.0 but ≤ 2.0 (Small)	> 2.0 but ≤ 4.0 (Semi-medium)	> 4.0 but ≤ 10.0 (Medium)	≥ 10.0 (Large)	All sizes
Two years to less than five years	Andhra Pradesh	18.28	49.02	13.41	21.92	11.63	24.60
	Assam	8.42	25.35	7.73	0.00	-	15.28
	Bihar*	20.98	14.37	26.52	0.00	-	19.83
	Gujarat	6.01	10.43	56.49	12.42	32.57	29.90
	Haryana	31.60	19.88	12.06	23.70	27.70	20.41
	Himachal Pradesh	12.42	69.91	0.00	-	-	20.76
	Jammu & Kashmir	49.76	56.43	0.00	-	-	42.07
	Karnataka	22.43	8.70	51.13	31.00	41.61	33.22
	Kerala	40.86	68.34	1.90	-	-	39.00
	Madhya Pradesh*	15.21	8.52	17.56	29.81	3.88	15.55
	Maharashtra	31.64	12.47	5.53	16.09	65.87	14.84
	Orissa	26.68	23.32	34.04	-	-	26.71
	Punjab	5.62	19.95	15.05	10.23	7.11	11.77
	Rajasthan	35.25	6.07	2.28	26.98	37.20	25.02
	Tamil Nadu	26.72	8.71	34.08	76.33	0.00	22.60
	Uttar Pradesh*	22.01	14.94	17.39	21.30	0.00	17.77
	West Bengal	14.25	9.48	44.48	16.38	-	16.02
	All India (Kharif)	21.13	19.95	18.11	18.93	24.47	20.27
	All India (Rabi)	18.74	19.24	23.09	20.00	27.07	21.07
	Group 1	41.44	67.49	1.79	-	-	39.20
	Group 2	20.77	14.55	24.62	26.32	13.54	19.74
	Group 3	20.18	17.46	13.40	15.20	28.30	17.96
	Group 4	27.71	7.97	13.09	28.94	35.16	24.17

(*Contd.*)

(*Contd.*)

Five years and above	Andhra Pradesh	31.89	6.99	19.70	46.33	84.12	30.42
	Assam	37.47	27.70	49.06	60.28	-	36.60
	Bihar*	15.34	18.50	9.41	14.29	-	15.42
	Gujarat	2.30	1.56	30.11	0.46	67.43	47.04
	Haryana	11.97	0.10	10.92	0.56	16.26	8.18
	Himachal Pradesh	61.28	23.59	100.00	-	-	57.12
	Jammu & Kashmir	0.47	17.04	0.50	-	-	5.29
	Karnataka	7.75	45.67	7.90	26.87	0.00	15.08
	Kerala	23.27	16.13	78.64	-	-	34.18
	Madhya Pradesh*	12.73	20.81	53.25	23.34	25.63	30.19
	Maharashtra	21.83	24.38	24.36	38.00	34.09	29.63
	Orissa	12.63	30.83	10.28	-	-	17.72
	Punjab	25.88	30.11	15.19	23.64	61.57	27.11
	Rajasthan	9.10	12.54	8.67	1.94	6.90	5.91
	Tamil Nadu	42.98	66.60	45.19	23.67	100.00	52.53
	Uttar Pradesh*	12.38	15.07	22.43	30.31	75.76	19.17
	West Bengal	35.24	54.45	36.37	26.87	-	39.00
	All India (Kharif)	17.99	21.54	19.05	22.06	46.45	22.80
	All India (Rabi)	19.12	21.64	16.83	21.24	57.16	24.89
	Group 1	21.78	16.20	74.38	-	-	32.32
	Group 2	16.91	22.62	27.72	26.28	40.09	22.00
	Group 3	22.29	20.36	20.02	23.37	53.98	28.06
	Group 4	39.06	51.50	21.09	2.80	12.02	22.35

Notes: (1) * Bihar includes Jharkhand, Madhya Pradesh includes Chhattisgarh and Uttar Pradesh includes Uttaranchal; (2) Information for imdividual states refer to Kharif season.

Source: Author's calculation using data available from NSSO (2006b, Table 6R).

tenanted lands has been two years or more in 1st group of states, which becomes 47 per cent for the 4th group of states.

VI. Summary of Findings and Implications

Our study shows that in rural India, both ownership and operational distributions of land have been highly unequal. In 2003, nearly 73 per cent of ownership holdings in rural India belonged to the marginal category (owning less or equal to 1.0 ha) while their share in total owned land has been only 23 per cent. The distribution of operational holdings shows an almost similar pattern such that the marginal category represented 70 per cent of all operational holdings, their share in total operated area being 23 per cent. Clearly, rural India has already attained a very high degree of marginalisation of holdings. It also needs mention here that although the large holdings (owning/operating more than 10.0 ha) represented only 0.53 per cent of all ownership holdings in rural India, their share in total owned land has been nearly 12 per cent. In the same way, 0.79 per cent of all operational holdings belonging to large category captured 12 per cent of total operated land. The prevailing situation is representative of high degree of inequality in the distribution of owned/operated lands in rural India. The Gini inequality coefficients for ownership and operational distributions of land (computed as 0.738 and 0.624 respectively for 2003) reinforce such a conclusion.

As expected, the land distribution patterns varied widely between the states of India. The degree of marginalisation of holdings (both ownership and operational) has been high in Kerala, West Bengal, Bihar, Orissa, Uttar Pradesh, Tamil Nadu, Jammu and Kashmir and Assam. On the other hand, some states like Punjab, Haryana, Gujarat, Karnataka, Rajasthan, Madhya Pradesh and Maharashtra display relatively high presence of holdings from upper categories. The Gini inequality coefficient of land distribution also shows wide interstate variation. Considering the Gini inequality coefficients for both the distributions (ownership and operational), it seems that the land distribution has been more unequal in the states of Punjab, Haryana, Tamil Nadu, Gujarat, Andhra Pradesh, Rajasthan, and Karnataka. On the other hand, Assam, Jammu and Kashmir, Kerala and Orissa are the states where such inequalities have been relatively low.

We compared land distribution patterns of four groups of states that have been formed considering the degree of legal restrictions put on tenancy in them. Several interesting findings emerge in this context. First, the groups of states that have more legal restrictions on tenancy have lower degree of landlessness as compared to those experiencing fewer restrictions on tenancy. Secondly, the shares of marginal holdings in total holdings as well as total owned/operated lands are high for groups of states that experienced more restrictions on tenancy. There is clear indication that the landholdings structures have been more favourable for the lower categories of holdings in such states. Thirdly, the average sizes of ownership and operational holdings are much high in areas experiencing fewer restriction on tenancy. Fourthly, the inequality of land distribution (as measured by Gini coefficient values) is also high in areas having less restrictive tenancy legislations. In fact, we observed that the degree of inequality in land distribution (both ownership and operational) increases monotonically as we move from more legally-restrictive tenancy areas to more-liberal tenancy areas. All these findings clearly indicate the contributory roles of land and tenancy reforms measures not only to strengthen the land-base of lower categories of holdings but also to establish a fairer land distribution system.

In 2003, about one-tenth of all operational holdings have been engaged in land-leasing in rural India. But the percentage of leased-in land to total operated area appeared to be just 6 to 7 per cent. Of course, this is an underestimate due to concealment of tenancy in most of the states. There is wide interstate variation as regards percentage of holdings reporting leasing-in as well as percentage of leased-in area to total operated area. The states for which percentage of holdings reporting leasing-in is above the all-India level are Orissa, Andhra Pradesh, Bihar, Orissa, Punjab, Tamil Nadu, Uttar Pradesh and West Bengal. On the other hand, percentage of leased-in area to operated area (incidence of tenancy) has been highest in Punjab (16.8%), followed by Haryana (14.4%), Orissa (13.0%), West Bengal (9.3), Andhra Pradesh (9.0) and Bihar (8.9). In remaining states, the percentage of leased-in area to total operated area has been lower than the same for all-India.

The incidence of tenancy varied between different categories of operational holdings. At the all-India level, although incidence of tenancy seems to follow a declining trend with increase in size of operational holdings, the incidence of tenancy for large holdings appear quite high (second highest among all landholding groups). The incidence of tenancy for large holdings has been much above the level observed for all-India level (6.45% during Kharif season) in Gujarat, Haryana, Uttar Pradesh, Tamil Nadu and Punjab. On the other hand, in Assam, Bihar, Himachal Pradesh, Jammu and Kashmir, Kerala, Orisssa and West Bengal, the large holdings are virtually non-existent as lessees.

Comparison between different groups of states revealed that, as expected, the percentage of holdings reporting leasing-in is lowest for the group that experienced complete prohibition on land-leasing (1st group). However, the percentage of holdings reporting leasing-in is highest for 2nd group that includes states which also have imposed quite a bit of restriction on land-leasing. Obviously, many rural households in the states of this group are eager to engage in land-leasing notwithstanding legal restrictions. It is also interesting to note that although the incidence of tenancy (percentage of leased-in area to operated area) is lowest for the 1st group of states where there is complete prohibition on tenancy, it is second lowest for the group where tenancy is most liberally allowed (4th group). All these observations seem to suggest that the incidence of tenancy in the states of India does not depend much on the provision of restrictive tenancy legislations.

As regards the background of lessors and lessees, we found that, at the all-India level, an overwhelming proportion of them belonged to the lower categories of holdings. In all-India, while 84 per cent of lessors are drawn from marginal and small landholdings groups, 86 per cent of the lessees belonged to the landless, marginal and small landholdings groups. This has been a general feature in the states of India, although there are some exceptions. For example, more than 50 per cent of the lessees are drawn from semi-medium and medium categories of holdings in Punjab and Haryana. The same category represented nearly 40 per cent of the lessees in Maharashtra and Rajasthan, 18 per cent in Andhra Pradesh, and 15 per cent in Gujarat. The percentage of lessees from the large landholdings category has also been quite high in these states.

There seems to be a tendency of percentage of lessors from higher landholdings groups increasing as we move from more-restrictive tenancy regime to a less restrictive tenancy regime. The same is also true about percentage contribution to total leased out land by the lessors from higher landholdings categories.[14]

At the all-India level, nearly 26 per cent of all tenanted land (excluding homestead) belonged to the lessors who are engaged in non-farming activities. However, there is wide interstate variation as regards contribution to total tenanted land by the lessors engaged in non-farming activities. Some of the states where the contribution to total tenanted land by the lessors engaged in non-farming activities is relatively high are Kerala, Tamil Nadu, Rajasthan, Punjab, Maharshtra, Haryana, Karnataka, Andhra Pradesh, Himachal Pradesh and West Bengal. Comparison between different groups of states shows that the contribution to total tenanted land by the lessors with non-farming background is also very high for two extreme groups – 1st group (55%) where there is complete prohibition on tenancy and 4th group (48%) where no worthwhile legal prohibition on tenancy exists. Again, within the 1st group, while in one state (Jammu and Kashmir) all tenanted land has been drawn from lessors with farming background, in another state (Kerala) as high as 58 per cent of tenanted land has been obtained from lessors with non-farming background. Thus, it may be concluded that, in the land-lease market, there is no clear relationship between amount of land to be contributed by the lessors with non-farming background and the degree of legal prohibition on land-leasing.[15]

In order to understand access to land-lease market by different categories of holdings, it is not adequate to look into their numbers among the households engaged in land-leasing. Numerical preponderance of smaller categories of holdings in the land-lease market does not necessarily mean that such a market has been favourable to them. A better indicator of access to land-lease market by different categories of holdings is their shares in all tenanted lands in relation to their shares in total holdings. In this respect, we find that, at the all-India level, only about 52 per cent of all tenanted land belonged to the landless, marginal and small holdings although they represented as high as 86 per cent of all lessees. On the other hand, while only about 10 and 4 per cent of lessees are drawn from semi-medium and medium and large landholding groups, they captured as high as 22 and 26 per cent of all tenanted land. Clearly, the functioning of the land-lease market in rural India has been highly unequal from the point of view of distribution of land available for tenant-cultivation.

There also exists wide variation among the states as regards distribution of tenanted lands across different categories of holdings. We found that the tenants from landless, marginal and small landholdings categories together obtained 75 per cent or more of all tenanted lands in eight out of 17 major states, which is more or less consistent with their share in total number of lessees. These states are Himachal Pradesh, West Bengal, Bihar, Orissa, Jammu and Kashmir, Kerala, Tamil Nadu and Assam. In Andhra Pradseh, nearly one-half of total leased-in land belonged to these categories. The same in Karnataka and Maharashtra are 37 and 25 per cent respectively. However, the four states where the lessees from landless, marginal and small landholding categories obtained very low share of total leased-in land are Punjab (9.7%), Rajasthan (9.9%), Haryana (11.8%) and Gujarat (14.3%). Actually, the land-lease markets in Punjab, Maharashtra, Haryana and Rajasthan are largely dominated by the lessees from semi-medium and medium categories. In Gujarat, more than 60 per cent of all tenanted land belonged to the large category of holdings. The presence of large category among the lessees is also quite substantial in Rajasthan, Karnataka, Haryana, Punjab and Andhra Pradesh.

We have another interesting observation as regards access to land-lease market by the lower landholding categories. We find that in areas where there is legal prohibition of tenancy (1st and 2nd groups of states), the share of landless, marginal and small landholding categories in all tenanted land is very high, and hence the same for upper categories of holdings is very low. The situation is just opposite in areas known for fewer restrictions on tenancy (3rd and 4th groups of states). Our computations reveal that the lessees from landless, marginal and small landholding categories together enjoyed 77, 72, 20 and 33 per cent of all tenanted lands in 1st 2nd, 3rd and 4th groups of states, respectively. On the other hand, the medium and large categories among the lessees together accounted for 0, 10, 51 and 49 per cent of total leased-in land in these four groups of states, respectively. The obvious conclusion here is that in a situation of high legal restriction on tenancy, the lower categories of holdings (landless, marginal and small) are able to exercise much better control over the land-lease market in the sense of capturing a larger share of lands available for tenancy cultivation as opposed to the situation where tenancy is liberally permitted. To put it differently, with fewer restrictions on tenancy, not only much larger number of tenants from higher categories of landholdings competes with the smaller tenants in the market for land-lease, but also they succeed in cornering high proportion of tenanted lands.

As regards the terms of tenancy, we find that nearly 50 per cent of tenanted lands have been under fixed-rent contract at the all-India level; the shares of share-produce contract and other terms being 40 and 10 per cent respectively. What is more, in 9 out of 17 major states, percentages of tenanted lands under fixed-rent rent tenancy exceeded 50 per cent. It seems that fixed-rent contract has already emerged as the most important contractual form in the land-lease markets in India. However, the form of tenancy seems to vary between different groups of operational holdings. The more frequently observed pattern is that of lower categories of holdings getting engaged more in share-produce tenancy while fixed-rent tenancy being preferred by higher categories of holdings. However, there are some exceptions. In some states (e.g. Punjab, Haryana, Andhra Pradseh, Tamil Nadu, Karnataka and West Bengal) there is predominance of fixed rent tenancy all along farm size continuum. Another interesting finding regarding the form of tenancy is that, in most of the states that have more legal restrictions on land-leasing, share-produce tenancy emerges as the most important form of tenancy while fixed-rent tenancy obtains prevalence in states with fewer restrictions on land-leasing.

We also looked into the duration of land-lease contracts. It is found that, at the all-India level, during Kharif season, the highest percentage of leased-in lands (33%) have been contracted for a period of one to less than two years. The percentages of leased-in lands contracted for less than one year, two to five years, and five years or more have been 24, 20 and 23 respectively. The distribution pattern of tenanted lands under different periods of contract does not differ much during the Rabi season. However, there is a clear tendency of tenants from lower landholding groups getting more of shorter-duration lease contracts while those from higher landholding groups being engaged in longer-duration lease contracts. There is also wide variation among the states as regards duration of lease contracts. Comparison between different groups of states showed that the areas experiencing fewer legal restrictions on tenancy are dominated by shorter-duration land-lease contracts while those of restrictive-tenancy areas being dominated by longer-duration contracts.

Implications for Legalisation of Land-Leasing

An important conclusion of this study is that the land and tenancy reforms measures adopted in India did contribute, though not so satisfactorily in all states, in equalising land distribution patterns and ensuring better access to tenanted lands by the lower categories of holdings (landless, marginal and small). Comparison of tenancy reforms-outcomes of the states of India that have been subjected to varying degrees of legal prohibition on land-tenancy leads to such a conclusion.

From the perspective of the rural poor, the effect of liberalisation of land-leasing might not be a happy one in all areas although supply of land for tenancy cultivation might improve. Our analysis of state-level data has highlighted at least two important features of land-lease markets in rural India: First, the extent of tenancy or quantum of land available for tenancy cultivation is not dependent upon the degree of legal prohibition/relaxation on land-leasing; it seems to be related to other 'non-legal' factors.[16] Secondly, there is no guarantee that the landless, marginal and small holdings would gain with relaxation of land-leasing laws inasmuch as they have failed to establish adequate control over the land-lease market, especially in those states which have fewer restriction on tenancy. In fact, the state-level data available from secondary sources indicate to the contrary: the poorer sections (tenants from lower categories of holdings) appear to have enjoyed better access to tenanted lands where tenancy is legally prohibited/ restricted.

Under the circumstances, one may ponder about some possible

consequences of liberalisation of land-leasing. First, it is possible that many households from higher categories, who until now abstained from the land-lease market due to legal prohibition on land-leasing, but were eager to augment their scale of farming further, would feel provoked to engage in land-leasing-in a competitive situation created after withdrawal of restrictions on land-leasing. This type of development would hurt the interests of the tenants from lower categories of holdings unless safeguard measures are adopted to protect them. In fact, such type of development has been taking place in many states which have fewer restrictions on tenancy (e.g. Rajasthan, Punjab, Haryana, Gujarat, Maharashtra, Karnataka and Madhya Pradesh). Secondly, liberalisation of land-leasing might result in emergence of a new class of landlords (absentee or resident) who harp on maximising their earnings through multiple channels, including leasing-out of land. Thirdly, the process of equalisation of landholdings among the rural households through 'sale-of-land-by-big-and-purchase-by-small', which is already weak, might get shrunk further with liberalisation of land-leasing. The landholders who are getting disinterested in farming would prefer leasing-out their lands rather than selling off if land-leasing is legally approved and their ownership rights over leased out land protected. Due to all these reasons, legalisation of land-leasing might eventually end up by producing an adverse distributional consequence and further widening inequality among rural households.

The above observations do not necessarily imply that various routes for augmenting land access by the poorer sections in rural areas should not be explored. In the context of land-leasing, we would like to emphasise that all possible gains of liberalisation of land-leasing be weighed against its possible adverse consequences on the land-poor households before taking any drastic decision in this regard. Of course, the gains/losses for the poorer sections following liberalisation of land-leasing would not be uniform across states/regions. Thus, state-specific detailed investigation on the subject is a necessity. Our limited exercise here indicates that, broadly speaking, there are two sets of states upon which the consequence of liberalisation of land-leasing is likely to be different. In the first set of states (Himachal Pradesh, West Bengal, Bihar, Orissa, Jammu and Kashmir, Kerala, Tamil Nadu, Assam and Uttar Pradesh) the landless, marginal and small holdings have already established higher access over the tenanted lands. So the effect of liberalisation of land-leasing might not be so severe for them in these states. However, for the second set of states (Punjab, Rajasthan, Haryana, Gujarat, Maharashtra, Karnataka and Madhya Pradesh), as the holdings from upper categories dominated the land-lease market, the lower categories are unlikely to benefit much following liberalisation of land-leasing.[17] Interestingly, both sets include the states which presently prohibit land-leasing (fully/partially) and which don't.

Suggestions

In the light of our study, a few suggestions may be put forward to safeguard the interest of the landless, marginal and small landholdings groups in the market for land-lease:

- First, and foremost, any effort towards liberalisation of land-leasing should be preceded by imposition of a ceiling on the size of the operational holding in order to combat its possible adverse distributional consequence.[18] In this context, one suggestion is that it be fixed at 10 ha. However, our computation (Table 9.8) shows that, at the all-India level, such a ceiling would help recover only about 11 per cent of tenanted lands from the holdings operating more than 10 ha.[19] Obviously, the ceiling limit on operational holdings should be set much below 10 ha if the landless, marginal and small holdings are to benefit. Setting a ceiling limit of 2 ha would render land-leasing an exclusive business by the landless, marginal and small farmers. However, as such a low limit would exclude the middle farmers from land-lease market, which might not be feasible for political reasons, this may be fixed at 4.0 ha which would take the semi-medium farmers on board. This may even go up a bit further, say 5.0 ha, in those states where higher categories of holdings have controlled high proportion of tenanted lands (e.g. Gujarat, Rajasthan, Punjab, Haryana, Maharashtra and Karnataka). In any case, the size of operational holdings should be fixed well below 10 ha to attain any perceptible improvement in land access by the rural poor in India.
- The duration of land-lease contracts needs to be appropriately specified. Longer-duration contracts would encourage the lessees to undertake investment on leased-in lands which would raise agricultural productivity and production. However, in the event of no legal stipulation as regards duration of lease-contract, the lessors might insist on shorter-duration contracts in their efforts to exercise control over the lessees' behaviour. At present, such a tendency is visible in areas with fewer restrictions on tenancy. On the basis of this study, we suggest 'minimum 2 years or more'

as the duration of land-lease contract which may be acceptable to the majority of lessors and lessees.

- The term of lease is another important issue. As fixed-rent tenancy has been emerging as the dominant form of tenancy, more so in regions known for fewer restriction on tenancy as also in agriculturally developed areas, the preference for such tenancies would increase further following liberalisation of land-leasing. Our suggestion here is that while it would be inappropriate to legally prohibit this form of tenancy[20], an upper limit to fixed-rental should be worked out to prevent excessive rent-exploitation of the lessees. As the fixed-rent tenancy puts entire burden of production risk on the tenant, the value of fixed-rent should not exceed 25 per cent of value of prospective output from tenanted land.[21] This might be computed for each village/ groups of villages by the *panchayat* representatives in consultation with block-level agricultural development officers, and should be publicised at the beginning of crop year. As regards crop-sharing contracts, the shares of landlord and tenant in crop output should depend on lease-market conditions as also whether the landlord shares in any cost of cultivation. In a situation of zero cost-sharing by the landlord, the landlord's share in output should be bounded between one-fourth to one-third of crop output (main product + by-product) which may go up to 50 per cent when the landlord bears 50 per cent of costs owing to all material inputs and ploughing.[22]
- The landowners' right of resumption of leased out land after the expiry of lease duration may be recognised to motivate more landowners to lease out land. In most areas, such a step would protect the interest of a large number of 'micro-landlords' (lessors from marginal and small landholdings groups) although the lessors from higher categories of holdings would also benefit.[23]
- The recording of all land-lease contracts along with their terms and conditions should be made compulsory. The local *panchayats* may be asked to keep record of such contracts apart from devising means for contract enforcement by both the lessors and lessees. In the event of contract violation by either party, the matter may be amicably settled locally with the help of local *panchayats* instead of pushing the matter to legal court.[24] Authentication of the contract document by the *panchayats* should be made compulsory to obtain loans from institutional

agencies. Further, there should be a separate scheme to advance preferential loans to those tenants who engage themselves in formal contract with the lessors. This would encourage formalisation of land-lease contracts.

- Finally, we would like to emphasise that any attempt to provide access to land by the poorer sections (through legalisation of land-leasing or otherwise) should be accompanied by ensuring their access to complementary inputs. Mere access to land without these complementary inputs and a favourable context is not likely to yield satisfactory result towards better agricultural growth and poverty reduction (De Janvry and Sadoulet, 2008).

Acknowledgement: The author is thankful to his daughter *Jayini* for her help in data processing.

Notes

1. To obtain some brief idea about the approaches of land and tenancy reforms pursued under different Five Year Plans and the interstate variation that existed with regard to legal provisions for land reforms, one may refer to Hanstad, Nielsen and Brown (2004) and Deshpande (2010).
2. The validity or otherwise of these observations (hypotheses) could not be tested rigorously using secondary (NSSO) data which are highly aggregative, and misses out many important aspects of tenurial relations. It is essential to undertake more detailed micro-level investigations covering diverse agrarian situations.
3. It may be noted that the emergence of this new perspective on land and tenancy reforms policies is also guided by the findings of several academic works carried out in recent years in India, China, and other countries (see, for example, Deninger, 2003; Deininger and Jin, 2005; Jin and Deininger, 2009; Haque, 2001; World Bank, 2007, Ch. 7; Hanstad, Haque and Nielsen, 2008).
4. The recent studies on tenancy situation in India as well as the official documents mentioned above (including the report of the *Committee on State Agrarian Relations and the Unfinished Task in Land Reforms)* do not appear to have adequately analysed all available state-level secondary data on tenancy. This becomes clear from the simple fact that instead of considering the shares of landless, marginal and small farmers in total leased-in land, the policy favouring liberalisation of land-leasing has been drawn largely on the basis of information on their share in total holdings. Considering the former alone would lead to a different policy conclusion

as would be clear from this study. Incidentally, in a recent study, Sharma (2010) considers this later aspect, but uses data on 'average leased-in area per reporting household under ownership holdings' rather than 'average leased-in area per reporting household under operational holdings', which leads to overstatement of shares of lower categories of holdings in total leased-in area.

5. We have dropped Andhra Pradesh and considered 16 major states while compiling data for these four groups. This is because the data for Andhra Pradesh, as available from published NSSO reports, could not be bifurcated between Andhra and Telangana areas which have quite divergent legal provisions with regard to tenancy.
6. How the land and tenancy markets changed over time in the states/groups of states that have been subjected to varying degrees of land and tenancy reforms is also an important issue that deserves detailed investigation. But we do not consider the same in this chapter due to paucity of time and space.
7. Using an alternative definition that considers landless households as those 'who neither own any land other than homestead nor cultivate on own homestead land', Rawal (2008) observed that the degree of landlessness is much high in rural India (31.12%).
8. We have computed the Gini coefficient (G) values by considering data for all size classes as provided by the NSSO reports. The formula is $G = 1 - \left[\sum \left(P_j - P_{j-1}\right)\left(Q_j - q_{j-1}\right)\right]/10^4$ where P_j and Q_j are respectively the cumulative percentage of number of owned/operational holdings and owned/operated areas in the jth size class of holdings.
9. In this state, the value of Gini coefficient for operational holdings is higher than the same for ownership holdings. This is because the largest group of holdings (those operating more than 20.0 ha of land) captured 21.27 per cent of all operated areas although their share in total operational holdings was only 0.21 per cent.
10. Considering 17 major states, the rank correlation between the two Gini coefficients is worked out to be 0.72. Thus the areas experiencing less-restrictive tenancy situation have high inequalities in both the distributions—ownership and operational.
11. The researchers examining the determinants of extent (incidence) of tenancy in India have concluded that it is in fact related to a host of economic, demographic and technological factors (see Bardhan, 1984, Ch. 10; Sharma, 2010).
12. It is not to be supposed that all these lands belonged to the 'pure lessors'. It is very much possible that many lessors have non-farming activities as their dominant activity even though they have some land under personal cultivation part of which they lease out.
13. It is to be noted that the two states in this group present a fully contrasting picture. While no tenanted land in Jammu and Kashmir has been drawn

from lessors who are engaged in non-farming activities, in Kerala 58 per cent of tenanted land has been drawn from them.

14. This does not necessarily mean that all such lands had flown to the lessees from lower categories of holdings. In fact, in the land-lease market, there is a general tendency among the big lessors to choose big lessees while leasing-out land (Bhaumik, 1993).
15. Some scholars have argued that with growing non-farm sector in the economy, liberalisation of land-leasing would encourage many large landholders to lease out their lands and join the non-farm sector which would increase the supply of land in the land-lease market eventually benefiting the land-poor households. Often, the experience of China is cited in support of this contention. In this context, we would argue that while such a possibility could not ruled out completely, the actual incidence of such cases becomes a matter of wild guess. After all, the structures of non-farm economies in China and India are not comparable. While the rural non-farm economy in China is dominated more by its productive and dynamic segment, the non-farm activities in rural India are more of subsistence type. As a consequence, in most areas in India, the rate of participation in the rural non-farm sector has been much high among the workers from landless, marginal and small households (which they do owing to distress conditions) as compared to those from upper landholdings categories (Bhaumik, 2008).
16. It is interesting to note that in majority of the states in India, the incidence of tenancy declined between 1991-92 and 2003 (see NSSO, 2006b, p. 32). This happened even in those states which have imposed no/fewer restriction on land-leasing (all states of 3rd and 4th groups, except Gujarat). Had tenancy been related to degree of legal restriction on land-leasing, one would expect the incidence of tenancy to rise at least in those states where no worthwhile legal prohibition on land-tenancy exists (e.g. Tamil Nadu and Rajasthan).
17. There is another state, namely Andhra Pradesh, where lands available for tenancy cultivation have been shared almost equally between the 'landless+marginal+small' holdings and 'semi-medium+medium+large' holdings.
18. Several scholars in the past argued in favour of imposition of ceiling on operational holdings in order to prevent capture of tenanted lands by the higher landholding categories (see, for example, Pushpendra, 2000, p. 55). This has also been one of the important recommendations of the *Committee on State Agrarian Relations and the Unfinished Task in Land Reforms* which we think is very well thought out.
19. This is so under the assumption that the propensities to lease-in by different categories of holdings remain unchanged.
20. In West Bengal, fixed-rent tenancy has emerged as the most dominant form of tenancy in spite of this being disallowed by law.

21. An average of crop-yield levels for past three years in the area may be considered for this purpose.
22. In China, the country which is frequently referred to for liberally allowing the practice of agricultural land-leasing, the tenant retains two-thirds of output on the tenanted land and the landlord gets one-third only (Jin and Deininger, 2009, p. 630).
23. We noted that more than three-fourths of the lessors are drawn from the category of marginal and small holdings in a large number of states although their share in total leased out land has been low.
24. This, however, does not cease the right of any individual to seek justice in the court of law.

References

Bardhan, Pranab (1983): *Land, Labour, and Rural Poverty: Essays in Development Economics*, Oxford University Press, New Delhi.

Bhaumik, Sankar Kumar (1993): *Tenancy Relations and Agrarian Development: A Study of West Bengal*, Sage Publications, New Delhi.

— (2008), 'Rural Non-farm Employment, Poverty and Inequality in West Bengal' in Sankar Kumar Bhaumik (ed.), *Reforming Indian Agriculture: Towards Employment Generation and Poverty Reduction*, Sage Publications, New Delhi.

Chadha, G.K. and H.R. Sharma (1992): 'Agrarian Relations in India: Comments on Quality of Available Data' in G K Kadekodi and G.V.S.N. Murthy (eds.), *Poverty in India: Data Base Issues*, Vikas Publishing House, New Delhi.

De Janvry, Alian and Elisabeth Sadoulet (2008): 'Access to Land and Development', in *The New Palgrave Dictionary of Economics*, Palgrave Macmillan, 2nd Edition, Vol. I, pp. 7-9.

Deininger, Klaus (2003): *Land Policies for Growth and Poverty Reduction*, The World Bank, Washington, DC.

Deininger, Klaus and Songqing Jin (2005): 'The Potential of Land Rental Markets in the Process of Economic Development: Evidence from China', *Journal of Development Economics*, Vol. 78, Issue 1, October, pp. 241-270.

Deshpande, R.S. (2010): 'Land Policy Issues in the Development Context' in Asian Development Bank, *Agriculture, Food Security, and Rural Development*, Oxford University Press, New Delhi.

Government of India (2002): *Tenth Five Year Plan 2002-2007*, Vol. II, Planning Commission, New Delhi.

Government of India (2008): *Eleventh Five Year Plan 2007-12*, Vol. III, Planning Commission, New Delhi.

Government of India (2009): *Report of the Committee on State Agrarian Relations and the Unfinished Task in Land Reforms*, Ministry of Rural Development, New Delhi, December [Available at www.dolr.nic.in].

Hanstad, Tim, Robin Nielsen and Jennifer Brown (2004): *An Analysis of the Effectiveness of the Legal Framework Governing Rural Land Policy in India*, Rural Development Institute, USA, August.

Hanstad, Tim, T. Haque and Robin Nielsen (2008): 'Improving Land Access for India's Rural Poor', *Economic and Political Weekly*, Vol. 43, No. 10, March 8, pp. 49-56.

Haque, T. (2001): *Impact of Tenancy Reforms on Productivity Improvement and Socio-Economic Status of Poor Tenants*, Policy Paper # 13, National Centre for Agricultural Economics and Policy Research, New Delhi.

Jin, Songqing and Klaus Deininger (2009): 'Land Rental Markets in the Process of Rural Structural Transformation: Productivity and Equity Impacts from China', *Journal of Comparative Economics*, Vol. 37, Issue 4, December, pp. 629-646.

Lipton, Michael (2009): *Land Reform in Developing Countries: Property Rights and Property Wrongs*, Routledge, London.

NSSO (2006a): *Household Ownership Holdings in India, 2003*, Report No. 491 (59/18.1/4), Ministry of Statistics and Programme Implementation, Government of India, New Delhi.

NSSO (2006b): *Some Aspects of Operational Land Holdings in India, 2002-03*, Report No. 492 (59/18.1/3), Ministry of Statistics and Programme Implementation, Government of India, New Delhi.

Planning Commission (2010): *Mid-Term Review of Eleventh Five Year Plan*, Government of India, New Delhi, May.

Pushpendra (2000): 'Liberalisation and Agrarian Reforms: Some Recent Controversies', in B.K. Sinha and Pushpendra (eds.), *Land Reforms in India: An Unfinished Agenda*, Sage Publications, New Delhi.

Rawal, Vikas (2008): 'Ownership Holdings of Land in Rural India: Putting the Record Straight', *Economic and Political Weekly*, Vol. 43, No. 10, March 8, pp. 43-47.

Sharma, H.R. (2010): 'Magnitude, Structure and Determinants of Tenancy in Rural India: A State Level Analysis', *Indian Journal of Agricultural Economics*, Vol. 65, No. 1, January-March, pp. 80-100.

World Bank (2007): *India: Land Policies for Growth and Poverty Reduction*, Oxford University Press, New Delhi.

10

Equity and Efficiency Impacts of Land Leasing Restrictions: Evidence in India

Klaus Deininger, Songqing Jin and *Hari K. Nagarajan*

Introduction

Researchers and policy-makers alike now realize that unequal access to opportunities arising from a skewed distribution of assets can be harmful for sustained long-term growth and therefore of concern for researchers and policy-makers alike (Aghion *et al.*: 1999, World Bank: 2005). In India, given the inequality in the distribution of productive assets, in particular land, inherited from colonial times, land reform and land policy have occupied a central stage in the policy debates for a long time. Following successful abolition of intermediaries immediately after independence, award of property rights to sitting tenants through tenancy laws and expropriation with subsequent transfer of 'above-ceiling land' from large landowners to small farmers were the main mechanisms to improve operational and ownership distribution of land. Although initial progress was slow, the country has made significant progress in implementing such policies during the 1970s and 1980s when, under both interventions together, rights to almost 10 mn ha of land were transferred. This is an impressive quantitative achievement; in fact, the area affected amounts to more than three times what was involved in the well-known land reforms of Japan, Korea, and Taiwan together (King: 1977). Moreover, even though in most cases beneficiaries did not receive full ownership rights, studies suggest that land reform has helped to reduce poverty while being neutral in terms of efficiency (Besley and Burgess: 2000), thus helping to make a contribution towards greater equity. However, although the historical impact of land reform legislation is undisputed, scholars have increasingly questioned the need to continue maintaining restrictions that were put in place to make redistributive land reform feasible a few decades ago. The most obvious reason for doubts is that a combination of inertia and litigation, together with population pressure and associated land subdivision, has slowed progress in transferring land through land reform legislation to a trickle. In the decade since 1995, the average number of households who, according to official statistics, received land through land reform amounted to less than 3,000 per year. A second reason is that the view of land rental as a backward, exploitative and inefficient institution that tends to benefit the rich, the deficiencies of which can easily be rectified by a benevolent bureaucracy may no longer correspond to reality. In fact, with about 15 mn., the number of households participating in rental markets in 1999/2000 alone is larger than the total who benefited from land reforms since independence. If experience from other Asian countries such as China and Vietnam is any guide for India, increased availability of non-agricultural employment opportunities implies that land rental markets can be expected to positively affect productivity and equity. It also shows that it will often be more appropriate for government to reduce the transaction costs associated with land transfers between private parties rather than intervene directly.

If government's ability to enforce legislation that will help transfer land to the less fortunate is limited and markets are not *a priori* biased against the poor, many well-intentioned land market interventions of the past may have outlived their period of usefulness. Indeed, there is now evidence from India suggesting that rental restrictions may do more to undermine landlords' willingness to supply to the rental market and reduce the welfare of poor tenants by forcing them to enter informal arrangements in contravention of the rules (Haque: 2001). This is consistent with a significant body of international (largely urban) literature that documents and quantifies the undesirable impact of land rental restrictions in India and in other countries in detail. However, while all of this anecdotal evidence is suggestive, more detailed quantitative assessment of the functioning of rental markets, and in particular the impact of land reform policies on these markets, ideally representative at a national level, would be needed to guide the policy debate.

In this chapter we use nationally representative household-data from 1982 and 1999 to assess the ability of different groups of producers, in particular the more productive and the landless, to gain access to land through market mechanisms. In addition, we explore the impact of rural rental restrictions as one factor that increases the costs of market participation on land market functioning and outcomes. To do so, we make use of the variation in land reform policies, and the extent to which such policies are implemented, across states that arises from the fact that land policy is responsibility of the states rather than the Central Government. Joint analysis of these two issues with a large nation-wide sample allows us to answer broader questions concerning the role of land markets in India and derive some interesting conclusions.

The chapter is structured as follows. Section two describes the history of land reforms and tenancy regulations in rural India, reviews evidence on the impact of land rental market restrictions and key differences between rural and urban land markets, and uses this to lay out a conceptual model and empirical strategy. Section three presents the data used and reviews descriptive statistics for policy variables and household characteristics. Section four contains econometric results for the production function and different models of land rental market participation and discusses the results and their robustness. Section five concludes by putting results into context and draw out possible policy implications.

Background and Conceptual Framework

Land reform policy, through abolition of intermediaries, imposition of land ceilings, and regulation of tenancy contracts, played a key role in India from the moment it started its existence as an independent state. In this section we review main elements of land rental and ceiling legislation, the way it was implemented, and discuss the literature that explored links between such legislation and land market outcomes. We use this to develop a conceptual model for households' land market participation and derive hypotheses that can be tested with the data at hand.

Origins and Nature of Rural Tenancy Restrictions in India

Under colonial rule, the main goal of India's land administration system was to obtain government revenue. The *de facto* award of land rights to revenue collectors (*zamindars*) in large parts of the country has consequences that affect development up to this day (Banerjee and Iyer: 2004).[1] Given the severity of the burden on individual cultivators, agrarian reform was one of the top demands by the Indian National Congress. Similar to a number of other countries where the state lacked the power to expropriate landowners or the financial means to compensate them, land reform was adopted as part of a redistributive agenda that aimed to increase the property rights enjoyed by tenants.[2] Three elements of agrarian reform were emphasized, namely (i) abolition of intermediary revenue collectors; (ii) tenancy reform including the imposition of rent ceilings and the prohibition of evictions; and (iii) imposition of a ceiling on landholdings that could be owned by any individual or family with a forced transfer of any land beyond this limit to small cultivators (Mearns: 1999). Following independence and the transfer of legal competence on land issues to states, significant variation ensued in timing of legislation, its substantive content, and the efforts made to ensure implementation of reform across states (Appu: 1997). Abolition of intermediaries was tackled swiftly and successfully while other elements of the agenda, i.e. ceilings and tenancy reform, encountered considerable from landlord interests. In many states, passage of laws was held up to give landlords time to "prepare"—leading to spurious subdivisions, eviction of tenants and resumption of self-cultivation, or the transformation of tenants into wage workers.[3] Below we briefly review the context of and sequencing ceiling and tenancy legislation in different states. The *ceiling legislation* imposed after independence (in most cases only in the early 1960s) is widely judged to have been ineffective due to high ceilings (often defined in terms of individuals rather than a household), exemptions if land was "used productively", and limited implementation effort. National guidelines adopted in 1972 prompted most states to amend their legislation to establish greater consistency and was followed by the only period in Indian history when ceiling legislation led to large-scale land transfers.[4] While modalities of such transfers, in particular whether and how much beneficiaries had to pay, varied across states, one element common to most of them is that recipients' ability to transfer the land thus received to others remains restricted.[5] The main differences in *tenancy legislation* across states refer to definition of a tenant,[6] the conditions for permitting tenancy, the amount of rent than can be charged, and the type of property rights awarded to tenants, in particular relating to the circumstances under which landlords can "resume" rented land for 'personal cultivation' and the amount of compensation, if any, which they have to pay. Most states restrict the legality of land leasing.[7] Even where leasing is allowed, the rents that can be received by the landlord are limited[8] and tenants' rights are inheritable but contingent on personal cultivation, i.e. cannot be legally transferred to others through sub-lease arrangements by original tenants.

While the variations of legislation across states provides ample scope for analyzing the impact of policies on observed outcomes, capturing the fine differences in timing, applicability, modalities of implementation, and definitions inherent in such legislation, appears difficult if not impossible. Rather than trying to do so, we use the share of households who benefited from key policies as an indicator for policy-induced constraints to the operation of rental markets. Specifically, we construct for each state the share of households who were awarded tenancy rights and the share of ceiling surplus area that was actually transferred to beneficiaries.[9] As none of the Indian states permit sub-leasing of lands to which tenants had received permanent rights and most states also impose restrictions on transfers of land received in the course of implementing ceiling legislation, this is an indicator of direct restrictions on the operation of land rental markets. Both figures also provide a good proxy for a state government's level of implementation effort, a variable that is exogenous to households' decisions but that was shown to be of great importance in earlier studies (Banerjee *et al.*: 2002).

Table 10.1 provides summary statistics by state for total agricultural area and 1980 population (columns 1 and 2) as well as our measures of land reform implementation around 1980 (columns 3 and 4).[10] It illustrates

Table 10.1: Land Reform Implementation in Various Indian States

State	*Total agric. area (mn. ac)*	*1980 population (mn)*	*Share of households receiving tenancy rights*	*Share of area redistributed under ceiling legislation*	*Mean level of land tenancy laws*
Andhra Pradesh	17.11	75.73	0.75%	8.34%	0.53
Bihar	22.22	82.88	0.00%	4.42%	2.64
Gujarat	8.44	50.60	11.20%	1.95%	1.47
Haryana	10.60	21.08	0.01%	1.26%	0.00
Himachal Pradesh	2.08	6.08	3.19%	0.06%	n.a.
Karnataka	18.79	52.73	5.29%	1.71%	1.42
Kerala	0.94	31.84	12.49%	1.30%	2.42
Madhya Pradesh	43.75	60.39	0.61%	2.69%	0.94
Maharashtra	33.54	96.75	10.68%	7.74%	0.97
Orissa	13.54	36.71	1.43%	2.24%	1.94
Punjab	15.45	24.29	0.04%	1.50%	0.58
Rajasthan	27.04	56.47	0.16%	6.63%	0.00
Tamil Nadu	10.12	62.11	3.23%	2.47%	4.03
Uttar Pradesh	51.17	166.05	0.00%	5.81%	1.42
West Bengal	16.98	80.22	10.80%	14.91%	3.83
All India	**298.77**	**930.57**	**5.35%**	**4.41%**	**1.49**

Source: Kaushik and Haque (2005), based on annual reports by the Ministry of Rural Development for columns 1-4; Besley and Burgess (2000) for column 5.

that the relative emphasis on different types of intervention, and the extent to which such measures were implemented, varies across states. More than 10 per cent of households received tenancy rights in Kerala, Gujarat, West Bengal, and Maharashtra and more than 5 per cent of area was redistributed under ceiling surplus legislation in West Bengal, Andhra, Maharashtra, Rajasthan, and UP. Some states (e.g. West Bengal, Maharashtra) heavily relied on both measures, others (e.g. Andhra, Rajasthan and U.P.) focused exclusively on ceiling surplus, and some (e.g. Kerala and Gujarat) emphasized tenancy laws. Although the number of tenancy laws enacted is, in the absence of more detailed knowledge on content and implementation effort, at best an imperfect proxy for the number and severity of restrictions on land rental, it can be useful as a point of comparison. We thus include in the mean number of tenancy laws in each state (column 5) from Besley and Burgess (2000). One notes that the mean number of laws at any point amounts to 1.5, from no legislation in Haryana to more than 4 pieces in Tamil Nadu. However, the correlation between the number of laws and the number of tenants who received rights is low ($\rho = 0.28$), supporting the notion that legal provisions alone may have limited impact and that implementation effort makes a perceptible difference.

Although not differentiated in the Table, a more detailed look at the time dimension of these measures allows a number of conclusions (Kaushik and Haque: 2005). First, land reform has been a major effort; up to 2000, land reform laws resulted in the transfer of almost 10 mn ha, 2.5 mn ha under programmes to redistribute of ceiling surplus land, and 7.35 mn ha under tenancy legislation.[11] Second, after a spurt of land transfers in the 1970s and 1980s, progress has slowed down considerably; in fact between 1995-96 and 2003-04, i.e. for almost a decade, progress in awarding land rights to tenants had come to a complete standstill and the increment in ceiling surplus land transferred during the period amounted to only 10,800 ha. The latter represents about one-tenth of the land declared ceiling surplus, with the remainder being tied up in litigation. This suggests not only that further progress in achieving redistribution of ceiling land could be slow—it would take almost 90 years to dispose of remaining ceiling surplus cases if the current pace is maintained but also that, by clogging up the court system and preventing it from quickly dispensing justice in other urgent matters, the ceiling legislation may impose external effects beyond land rental markets (Moog: 1997).[12]

Evidence on Incidence and Impact of Tenancy Legislation and Rent Controls

Even though empirical evidence on the impact of rent ceilings and other forms of tenancy control in rural areas is limited, a number of studies have analyzed this issue in urban contexts. In fact, rent control has served as a textbook example for policies that can be effective to transfer resources in the short-term but will be associated with inefficiencies in the medium to long-run (Arnott: 2003). The key reason, which also formed the basis for analytical approaches, is that, by fixing rents below their equilibrium level, rent controls reduce the supply of new housing (or maintenance of existing units) by landlords who face an artificially reduced price (Gyourko and Linneman: 1990). Rental restrictions would indeed transfer resources from landlords to sitting tenants when they are imposed but also make access to rental property for those who were not renting when the controls were imposed more difficult (Basu and Emerson: 2000). With a constant or decreasing number of beneficiaries and an increasing number of new entrants who need to access to land through now distorted markets, the social cost of keeping land rental restrictions in place is expected to increase over time (Glaeser: 2002). As they are generally ill-targeted and also reduce tenant mobility (Munch and Svarer: 2002), rather crude first-generation rent controls have been replaced by a more flexible and sophisticated policy mix in many urban areas of the world (Arnott: 1995).[13] While initial studies were focused on developed countries, the adverse impact of rent restrictions may be as significant or even more severe in the developing world. Computing the cost imposed for cities in Ghana, Brazil, India, and Egypt, supports the notion that urban rent controls are generally effective in reducing rents, by between 4 per cent and 64 per cent.[14] However, benefits were generally ill-targeted and, in most cases, reduced significantly by the welfare losses resulting from reduced housing supply. While rent controls or similar restrictions may be useful to deal with emergencies in the short-term, other policies such as vouchers or supply-side incentives are generally believed to have greater long-run potential (Malpezzi and Ball: 1991).

Even though rigorous quantitative analysis of the impact of rent restrictions in rural areas is more limited than in urban ones, a general equilibrium model of land leasing supports the notion that, in the presence of restrictions on tenancy, landlords will provide less land to tenancy markets, implying less equalization of the farm size structure than without this intervention (Conning and Robinson: 2005). In addition, the impact of rental restrictions in rural areas may be more pronounced than in urban ones for four reasons. First, as owners of urban housing stock have less opportunities to revert to self-cultivation (or cultivation with wage labour) than rural landowners, the supply of housing to urban markets will be less elastic, and thus the negative supply response less pronounced, than in the case of rural land. Evidence of widespread tenant evictions as a response to actual or anticipated imposition of tenancy regulations (Appu: 1997) supports this argument. Second, to the extent that rural rents are defined in kind—often as a share of output-contract terms in rural areas will be less flexible than in urban ones, limiting the scope for circumventing such restrictions by adjusting rental rates (Basu and Emerson: 2000). Third, the rights given to tenants are often non-transferable and heritable but incomplete (i.e. still requiring them to pay rent to the landlord). This would reduce both parties' incentive for making land-related investments.[15] Limits on sub-leasing could also have a pronounced impact if, with generational change, original tenants are no longer able to farm the land they received in the most efficient way while their offspring may have taken up non-agricultural occupations. Finally, in the case of rural (but not urban) land, arrangements for land use or ownership will have a clear impact on productive efficiency due to the disincentive effects associated with wage-labour based cultivation (Binswanger *et al.*: 1995) and the fact that, if tenancy laws are combined with land ceiling legislation, landlords have an incentive to artificially subdivide land that is subsequently left idle. All this would imply that the impact of rental restrictions in rural areas will go far beyond the price effects on which the attention of the urban literature has focused. It also suggests that, while tenancy legislation and the associated restrictions on terms under which land can be transferred in rural areas may have been a good way of transferring residual ownership rights, the cost of maintaining such restrictions is likely to increase over time. Empirical analysis of their impact on equity, i.e. land access by landless producers as well as efficiency of land use at the household level over time will thus be of great interest.

Although the literature on land reforms in India and beyond is large (Warriner: 1969, Thorner: 1976, King: 1977, Haque and Singh: 1986), an often rather limited emphasis on impact constrains the extent to which it can be relied upon to assess the cost-effectiveness of specific interventions and to draw broader policy conclusions.[16] Quantitative studies that use a rigorous analytical framework and aim to construct a counterfactual have made little effort to explore the—direct or indirect—

costs of such reforms. For example, analysis of the large-scale registration of sharecroppers in West Bengal after 1978 (Banerjee *et al.*: 2002) finds that this policy led to large productivity gains, in line with earlier studies (Lieten: 1996, Rawal: 2001). Although these gains were contingent on massive tenant mobilization at the grassroots level, the immediate costs of implementation and the possible longer-term effects due to limited transferability of and overlapping property rights over land reform lands are difficult to assess. At an aggregate level, land reforms are found to have had a significant impact on poverty reduction but not on productivity (Besley and Burgess: 2000) and a follow-up study based on the same data tentatively concludes that the poverty impact is likely to be worth the cost in terms of productivity (Besley *et al.*: 2004).

Obviously, providing an assessment of the direct costs of implementing land reform will be difficult and mainly of historical interest. At the same time, exploring how land rental markets work and to what extent they are affected by tenancy restrictions will allow to put the policy debate on land markets in India on a more robust empirical footing. Doing so will allow us to draw conclusions as to whether the benefits from maintaining restrictions on the functioning of land rental markets are still sufficiently large to outweigh the costs of such measures, something that is of clear policy relevance.

Conceptual Framework

A key rationale for households to engage in land markets is the desire to adjust for differences in their existing endowments of land and family labour. Following similar models in the literature (Carter and Yao: 2002), let household i be endowed with fixed amounts of labour (L_i) and land (A_i), and a given level of agricultural ability (α_i). Relative land scarcity, together with the cost of supervising labour (Frisvold: 1994) makes wage-labour based cultivation undesirable in equilibrium (Binswanger *et al.*: 1995), implying that households allocate their labour endowment between farming their own land and off-farm employment at an exogenous wage (w_i). Renting of land incurs three main types of transaction costs (T^{in} or T^{out}), that include three key elements, namely (i) the need to acquire information on market conditions; (ii) the need to negotiate and enforce payments; and (iii) the transaction costs imposed by rental regulations that restrict transferability or completely outlaw certain contract types. While the first is more akin to a fixed cost, the other two elements, which make up the largest share, are likely to be proportional to the amount of land transacted. To keep matters tractable, we assume that households have the possibility of structuring rental contracts in a way that allows those lacking liquidity to enter into sharecropping or other arrangements that would allow them to defer rental payments until the harvest. With this, household i will choose l^{a*}, l^{0*} as well as A^* by solving the maximization problem:

Max

$$p\alpha_i f(l^a_i, A_i) + w\, l^0_i - I^{in}(A_i - A_i)(r + T^{in}) + I^{out}(A - A_i)(r - T^{out}) l^a, A \quad \ldots \text{(P)}$$

where p is the price of agricultural goods, l^0 is the amount of time allocated to off-farm labour ($l^0 = L_i - l^a_i$), I^{in} is a indicator variable for rent-in (=1 for rent-in, and 0 otherwise), similarly I^{out} is an indicator for rent out (=1 for rent-out, and 0 otherwise), T^{in} and T^{out} are the transactions costs respectively associated with rent-in and rent-out, all other variables in (a) are as defined above. The optimal choices of l_i^{a*}, l_i^{0*} and A_i^* will solve the first order conditions (FOC) of problem (P), i.e.

$$p\alpha_i f_l(l^a_i, A_i) = w \quad \ldots (1)$$

or for households who rent in,

$$p\alpha_i f_A(l^a_i, A_i) = r + T^{in} \quad \ldots (2a)$$

or for households who rent out,

$$p\alpha_i f_A(l^a_i, A_i) = r - T^{out} \quad \ldots (2b)$$

and for autarkic households,

$$(r - T^{out}) < p\alpha_i f_A(l^a i, A i) < r + T^{in} \quad \ldots (2c)$$

The first order conditions allow derivation of three empirically testable propositions [see Deininger and Jin (2005) for a more detailed derivation] as follows:

Proposition 1. The amount of land rented in is strictly increasing in ability, α, and strictly decreasing in the original land endowment A. Other things constant, rental markets will thus transfer land to "land-poor but efficient" producers, causing large landowners and those with low agricultural ability to rent out to smaller ones while still enjoying the credit-market and potential non-monetary (status) advantages associated with land ownership. This is indeed consistent with what is observed in the literature (Sadoulet *et al.*: 2001).

Proposition 2. Rental restrictions as well as other types of transaction costs drive a wedge between those renting in and those renting out with any increase in *T* decreasing α_l and increasing α_l[17] thereby expanding the range of producers who remain in autarky, reducing the number of households who are able to gain access to land by participating in rental markets, and decreasing the amount of land transacted through rental markets. Reduction in transaction cost will therefore increase social welfare.

Proposition 3. Increases of the wage for off-farm employment will increase the supply of land to the rental market and thus the amount of land transacted in rental markets. This will result in a reduction of the rental rate and, in a risk-free environment this will make everybody better off.

Estimation Strategy

According to equations 2a-2c, a household's decision to enter land rental markets depends on their marginal productivity in autarky compared to the effective land rental, either the rental payment inclusive of state specific transaction costs $r^{in}(T)$ for those renting in or the rental net of transaction costs $r^{out}t(T)$ for those renting out. In addition to agricultural ability (α), the productivity impact of rental market participation will depend on the household's endowment of land (*A*), family labour (*L*), assets (*K*), and the opportunity cost of labour which will be affected by the level of education (*E*) and the presence of opportunities in the local off-farm labour market (*O*). Defining a well-behaved net earning function *g(A, L, K, O)* with first derivative *g'()* allows us to express the market participation regimes as:

I. Rent-in regime ($A_i^* > A_i$) : $g'(\alpha, A, L, K, O) + \varepsilon_i > r^{in}(T)$
II. Autarky regime ($A_i^* = A_i$) : $r^{out}(T) < g'(a, A, L, K, O) + \varepsilon_i < r^{in}(T)$
III. Rent-out regime ($A_i^* < A_i$) : $g'(\alpha, A, L, K, O) + \varepsilon_i r^{out}(T)\ C$

... (3)

This switching regime model is estimated using ordered probit with upper and lower thresholds being a function of tenure security and market development variables (Sadoulet *et al.:* 1998). Transforming (8) into ordered probit model notation, we have:

$$pr\,(A_i^* > A_i) = pr\,\{\, e_i > r^{in}\,(T) - g'(\alpha, A, L, K, O)\}$$

$$pr\,(A_i^* = A_i) = pr\,\{\, e_i > r^{out}\,(T) < g'(\alpha, A, L, K, O) \cap \varepsilon_i < r^{in}\,(T) - f(a\, A, L, K, O)\}$$

$$pr\,(A_i^* < A_i) = pr\,\{\, \varepsilon_i < r^{out}\,(T) - g'(\alpha, A, L, K, O)\} \qquad \ldots (4)$$

We assume that the functions $g\,(\langle, A, L, K, O)$, $r\,(T)$ and $r\,(T)$ are linear, allowing us to express them as:

$$g'(\alpha, A, L, K, O) = \beta_0 + \beta_{1\alpha} + \beta_2 A + \beta_3 L + \beta_4 K + \beta_5 O$$

$$r^{in}\,(T) = \delta_0 + \delta_1 S$$

$$r^{out}\,(T) = \eta_0 + \eta_1 S \qquad \ldots (5)$$

respectively. The ordered probit model is then estimated by maximum likelihood.

A key element of the regression is households' agricultural ability α. As the data available are a panel of households and their offspring who were observed in 1982 and again in 1999, one way of recovering this parameter would be to estimate a panel production function and then use household (dynasty) fixed effects as a proxy for ability (Deininger and Jin: 2003). For a variety of reasons, including intergenerational transfers, the fact that a significant portion of cultivators had moved out of agriculture, and the expansion of the sample from about 4,000 to more than 7,000 households[18], such an approach would have involved the loss of a large number of observations. We, therefore, chose to estimate a stochastic frontier production function to determine producers' level of technical efficiency in each of the periods. The stochastic frontier model assumes that the disturbance term is composed of two additive components, one that is pure white noise (v_i), and a one-sided term [($ui \sim N+(0, \delta u^2)$] that follows a truncated normal with zero mean 0 that captures producers' level of technical inefficiency, *t* (Coelli: 1995). Assuming a Cobb-Douglas functional form allows to write the model as

$$qi = \phi_0 + \phi x_i + u_i - v_i \qquad \ldots (6)$$

where q_i is the logarithm of crop output obtained by household i, x_i is a input vector including sown area, land, material inputs (i.e., fertilizer, seed, etc.), fixed assets (all in logs), and household characteristics that may affect crop output, and) are the parameters to be estimated. The technical efficiency variable TE_i, defined as $TE_i = \exp(-ui)$, can be predicted after estimating equation (6).

Other household-level variables included in the ordered probit model (5) include a dummy for households who were landless at the time of the survey and the logarithm of the total land endowment to represent A, the number of members in the 14-60 and below 14-year age group to represent L, the total value of assets and the share of agricultural assets, defined to include livestock, implements, and structures directly useful for agricultural production to represent K, a dummy for whether the head has primary or above education and the head's age as a proxy of experience to represent human capital E, and the mean level of village income to represent wage labour opportunities in the local off-farm labour market O.

According to proposition 1, land rental markets would transfer land to more productive producers with lower land endowments and high levels of labour leading us to expect $\beta_1 > 0$, $\beta_2 < 0$, $\beta_3 > 0$. A positive sign on the element of β_4 that correspond to assets would point towards wealth bias in rental markets, possibly caused by credit market imperfections. Based on proposition three, we expect households whose past experience in off-farm labour markets reflects greater opportunities for such employment to be less likely to rent in and more likely to rent out and the level of rental market activity in general to increase with development of the off-farm economy, a variable that is represented empirically by the share of income derived from non-agricultural sources. We thus expect β_5 to be negative and the coefficient on the share of village income from agriculture to be positive and significant.

In addition to the above variables, a key focus of our analysis is on assessing the impact of exogenous restrictions on land market performance and households' land market participation (proposition 2). We do so by having policy variables that will have an impact on T — which are represented by implementation of restrictions and number of tenancy laws as explained earlier — affect the lower and upper thresholds δ and η in (5). Beyond these, we allow thresholds to vary over time to allow that overall economic development will systematically shift supply or demand of land in the rental market, and the household's caste status as a proxy for the ability to find out about land on offer in the local market and to enforce contracts.

Data Sources and Descriptive Evidence

Before discussing econometric results, we highlight some salient features of India's rural economy by presenting evidence on socio-economic characteristics as well as land market participation and changes in these variables over time. This allows us to explore changes of the economic structure, growth, and asset accumulation by individuals and at the village level. Concerning rental markets, our interest is to find out whether our hypotheses are supported by descriptive evidence, or whether, possibly as a result of "reverse tenancy" rental markets lead to transfers from small to larger producers rather than to land-poor ones.

Household Characteristics

The data used in the analysis come from two rounds of NCAER's ARIS/REDS survey that were conducted in 1982 and 1999, respectively. This survey, the first rounds of which were implemented in 1968-71 to evaluate the impact of an agricultural development programme, covers all of India's major states. Even though the first round sample, which is stratified by farm size and wealth class, was limited to project areas, the survey was significantly expanded in 1982 to make it more representative at the national level, covering slightly less than 5,000 households (Foster and Rosenzweig: 1996). The 1999 sample contains all of the households included in 1982 as well as replacements for those who were no longer present. If the original household had split, all of the households belonging to the same dynasty in the original village plus a subsample of successor households outside the village were interviewed, bringing the total to about 7,500 households (Foster and Rosenzweig: 2004).

Table 10.2 reports descriptive statistics for the whole country and its four main regions in both periods.[19] We note that there has been a marked increase in educational attainment, as illustrated by the fact that the share of household heads who had at least primary schooling completed increased from about 26 per cent in 1982 to 50 per cent in 1999. Also the gap that had earlier separated Northern and Southern states has narrowed considerably. At the same time, population growth has led to a decline in the average land endowment, from 3.3 ha in 1982 to 2.1 ha in 1999, and a small increase in landlessness, from 22 per cent to 24 per cent. The size

Table 10.2: Key Household Characteristics by Region and Time

	1982					*1999*				
	All	*North*	*West*	*East*	*South*	*All*	*North*	*West*	*East*	*South*
Basic characteristics										
Household size	6.86	7.20	7.28	6.97	6.01	6.02	6.64	6.10	6.30	5.21
Members aged below 14	2.36	2.58	2.61	2.36	1.87	1.86	2.15	2.02	1.93	1.36
Members aged 14 – 60	4.16	4.22	4.35	4.28	3.83	3.75	4.01	3.68	4.02	3.44
Members older than 60	0.34	0.41	0.32	0.33	0.32	0.41	0.48	0.40	0.36	0.41
Head's age	50.10	51.52	49.59	48.89	49.96	49.20	49.45	48.55	48.43	50.30
Female head dummy (%)	6.86	4.14	4.39	4.63	13.29	6.66	4.28	5.14	4.65	11.89
Head with primary or above.	0.26	0.22	0.17	0.26	0.40	0.50	0.56	0.42	0.54	0.51
Land endowment (ha)	3.29	2.65	4.78	2.31	2.52	2.12	1.99	2.82	1.29	1.86
Land endowment p.c.	0.52	0.42	0.72	0.36	0.42	0.39	0.35	0.52	0.24	0.37
Landless dummy (%)	22.28	22.43	18.78	20.30	27.02	23.93	27.16	21.10	23.41	25.03

(*Contd.*)

Consumption and asset ownership										
Per capita income	1538.62	1915.45	1470.32	1226.49	1424.24	2595.59	2994.16	2125.07	1979.44	3246.56
Per capita consumption exp.	1305	1506	1200	1058	1362	1561	1803	1526	1287	1579
Gini (per capita cons. exp.)	0.32	0.30	0.29	0.31	0.36	0.31	0.31	0.30	0.36	0.28
Value of all assets (Rs.)	17710	24654	17947	10170	14743	47749	66529	43810	29298	48688
Financial and off-farm (%)	26.69	25.73	29.56	28.30	23.62	23.21	18.56	19.87	25.14	31.66
Farming and livestock (%)	15.81	16.69	20.67	7.16	10.58	19.86	23.53	26.27	8.08	13.01
House & cons. durables (%)	57.50	57.58	49.79	64.54	65.79	56.93	57.91	53.86	66.78	55.33
Gini (all assets per capita)	0.60	0.54	0.56	0.55	0.64	0.56	0.55	0.53	0.59	0.54

(Contd.)

Participation in economic activities (%)	*1982*					*1999*				
	All	*North*	*West*	*East*	*South*	*All*	*North*	*West*	*East*	*South*
Crop production	70.24	68.57	75.64	71.49	65.07	62.89	67.45	68.41	61.29	53.15
Livestock production	78.11	81.31	87.35	66.27	70.36	62.85	72.90	66.88	55.16	54.26
Non-farm self-employment	11.33	10.63	11.16	13.13	11.22	10.96	7.86	10.77	20.73	7.41
Salaried employment	17.47	24.71	12.65	22.24	14.18	17.27	27.33	13.59	16.22	13.90
Off-farm wage employment	8.59	5.63	11.10	10.15	7.62	19.96	17.01	18.48	32.36	16.17
Wage employment	37.59	21.27	39.49	42.69	47.71	43.29	28.27	44.98	53.94	47.05
No. of observation	4980	1279	1613	670	1417	7476	1705	2479	1307	1985

Source: Own computation from 1982 and 1999 ARIS/REDS surveys.

All values are in 1982 Rs. with 1999 values having been deflated by state level deflators.

of the average household decreased, from 6.9 to 6.0, with about 4 in the 14 to 60 age category, 0.4 aged above 60, and 2.4 as compared to 1.8 below the age of 14. The share of female headed households is, with 6.8 per cent, almost the same in both periods, although female headship is more pronounced in the South than in other regions.

Survey results also point towards an annual increase of per capita income of 3.08 per cent during the period under concern. This aggregate masks pronounced differences across regions with the South having caught up and even replaced the North as the region with the highest income in the second period. Use of the information from the listing exercise allows us to compute the Gini coefficient for self-assessed income in a way that includes all households in the sampled villages. Doing so highlights that, with a Gini coefficient of 0.32 in the first and 0.31 in the second period, inequalities in income remain modest and have not appreciably increased between the two periods. The overall improvement in living standards as illustrated by increased income is mirrored by a significant rise in asset values of approximately 6 per cent per year.[20] It is worth noting that, with a Gini coefficient of 0.60 in 1982, inequality in assets is higher than income inequality as is found in other parts in the world. However, this coefficient has actually decreased to 0.56 in the second period. While overall asset endowments have increased, the broad composition of households' asset portfolio shows greater stability; the house and consumption durables make up the largest share in both periods (57.5% and 56.9%, respectively in 1982 and 1999), followed by financial and off-farm assets (26.7% and 23.2%), and farm assets including livestock (15.8% and 19.9%).

The bottom panel illustrates the diversification of livelihood sources achieved during the period by highlighting the share of households who participated in various activities, noting that households generally rely on more than just one income source. Three observations are of particular interest. First, despite a drop by about 7 and 15 points in the share of households engaging in crop and livestock production, respectively (from 70% to 63% for crop and from 78% to 63% for livestock), self-employment in agriculture continues to constitute the single most important source of employment in India's rural economy. The fact that this is followed by agricultural wage employment, the share of which has actually increased over the period (from 37.6% to 43.3%) illustrates the continued importance of agriculture. As wage employment is one of the few sources of livelihood open to the landless, agriculture is particularly relevant for the poor. A second finding of interest is that the share of households engaging in self-employment in rural India has been more or less constant, implying that growth of the rural non-farm sector has been just large enough to absorb population growth. This contrasts with other countries where the rural non-farm economy develops largely as a result of households taking up non-farm self employment (Lanjouw and Lanjouw: 2001). In fact, according to our data, the main source of income diversification was the growth of non-farm wage employment, participation in which increased from 8.6 per cent in 1982 to 20 per cent in 1999. This can be explained as a result of the government's emphasis on creating rural employment through a plethora of mechanisms that range from direct subsidies for firms setting up in remote areas to industrial and labour market regulation such as small scale reservation acts, as described and analyzed in recent contributions to the literature (Besley and Burgess: 2004).

Operation of Land Markets

To provide evidence on the extent to which our hypotheses on land markets are borne out descriptively, Table 10.3 presents the above variables for households disaggregated by their land rental market participation (rent-in, rent-out, or autarky). The Table points towards a large increase in the level of land market activity over the period; from 5.3 per cent and 2 per cent for renting out and renting-in in 1982, the share of market participants has increased to 10.7 per cent and 4.1 per cent, respectively, in 1999. While this is a large change, the level of rental market activity increased more rapidly, and in a shorter period, in other Asian countries such as China or Vietnam, despite the fact that the more egalitarian land ownership distribution in these countries would put greater limits on the potential of land markets to equalize operational holdings than in India.[21] Exploring econometrically the extent to which the impact of factors such as lower non-farm growth, landlessness, and land reform policies, affected the scope for land markets to improve land access and allocative efficiency will be of interest. Descriptive figures support the propositions from our model and point towards improved rental market functioning in the second, as compared to the first period. The first panel illustrates that in both periods rental provided an opportunity for relatively land-scarce and labour-abundant households to gain access to land as evident from comparing the per capita land endowment for landowners who are autarkic (0.51 ha and 0.36 ha in 1982 and 1999), rent-in (0.28 ha and 0.20 ha), or rent-out (0.68 ha and 0.64 ha) in the two periods. Land markets can be

Table 10.3: Key household characteristics by rental market participation status (1982 and 1999)

	1982			1999		
	Autarkic	*Rent-in*	*Rent-out*	*Autarkic*	*Rent-in*	*Rent-out*
Basic Characteristics						
Household size	6.92**	8.15**	5.34**	6.04	6.91**	5.54**
Members aged below 14	2.38**	2.75	1.83**	1.87	2.38**	1.53**
Members aged 14 – 60	4.20**	4.90**	3.10**	3.77	4.17**	3.45**
Members older than 60	0.34**	0.49*	0.41	0.40**	0.36	0.56**
Land endowment (ha)	3.34*	2.31*	2.93	2.02**	1.27**	2.87**
Land endowment p.c.	0.51	0.28**	0.68**	0.36**	0.20**	0.64**
Landless dummy (%)	23.76**	11.83*	0.00**	26.29**	37.34**	0.00**
Head's age	49.97*	51.85	51.71	48.98**	47.41*	51.65**
Female head dummy (%)	6.67*	2.15	12.03**	6.54	3.30*	8.90**
Head with primary or above (%)	25.34**	29.03	35.71**	48.51**	49.50	61.53**
Consumption and asset ownership						
Per capita consumption exp. (Rs.)	1280.42**	1426.98	1697.84**	1549.19**	1346.19**	2213.63**
Value of all assets (Rs.)	17215**	34783**	20333	46568**	33839**	62466**
Financial and off-farm (%)	26.47	19.48	34.20**	22.69*	19.23**	27.160**
Farming and livestock (%)	15.70**	32.12**	7.69**	20.91**	21.67	13.26**
House & cons. durables (%)	57.83	48.40	58.10	56.41	59.10*	59.58
Participation in activities (%)						
Crop production	72.60**	100.00**	19.17**	66.12**	100.00**	23.07**
Livestock production	78.66*	97.85**	61.65**	63.57**	81.82**	49.88**
Non-farm self-employment	11.3	5.38	13.91	9.9**	14.61*	17.96**
Salaried employment	16.84**	18.28	28.2**	15.98**	10.71**	30.05**
Wage employment	38.82**	26.88*	19.92**	44.93**	59.74**	23.94**
Number of observations	4621	93	266	6366	308	802

Source: Own computation from 1982 and 1999 ARIS/REDS surveys.
All values are in 1982 Rs. with 1999 values having been deflated by state level deflators.
*, ** and *** refer to statistical significance at 10 per cent, 5 per cent and 1 per cent respectively.

seen to have transferred land from households whose head is more educated and female headed to male headed ones with educational lower attainment. We also note that the share of landless who had gained access to land through rental markets increased from 12 per cent in the first to 37 per cent in the second period, suggesting a marked increase of outreach towards this group over time. Noting that our sample represents a total rural population in India of about 130 mn, this suggests that in 1999 about 15 mn households—a quarter of them landless—were able to use markets as a means to get access to land. It is worth noting not only that this figure is much larger than the number of households who got access to land through land reform but also that, given the magnitudes involved, even policies that have only a "modest" impact on the functioning of land rental markets, could have implications for a large number of households. Comparing levels of consumption and assets for households who differ in the nature of their land market participation reinforces the notion that rental provided opportunities for poor segments of the population to access productive resources and thereby improve their well-being, especially in the second period. The value of all assets owned by rent-in households in 1999 is, with Rs. 33,839, more than 25 per cent below the average, compared to asset ownership that is similar to the mean for autarkic households and about 33 per cent higher than the mean for those renting out, supporting the notion that it is the asset-poor who benefit from the access to land which rental markets provide. The narrowing of the gap between rent-in and average households with respect to per capita expenditure supports the hypothesis of land markets making a positive contribution to the livelihood of participants. Finding significant differences in the composition of the asset portfolio between rent-in and rent-out households, with the former having relatively more of their wealth in farming and livestock and the latter in off-farm and financial assets is not too surprising.

While higher rates of participation in crop and livestock production by rent-in households should not come as a surprise, the high share of rent-in households engaging in (agricultural) wage employment suggests that land rental provides opportunities for earning additional income to wage labourers, consistent with earlier evidence. At the same time, the fact that, contrary to what was found in 1982, non-farm self-employment is significantly higher among rent-in households than the mean (or autarkic ones) suggests that land rental is not an obstacle to participation in the rural non-farm economy. To the contrary, renting may provide an opportunity to accumulate experience and capital to replace the "agricultural ladder" (Spillman: 1919) with a general increase in occupational mobility including the non-farm sector (Alston and Ferrie: 2005).

Econometric Evidence

Consistent with descriptive evidence, results from production function and rental participation equations point towards the importance of land as a factor of production, suggesting that households with larger family labour endowments and less non-farm involvement will use rental markets as an avenue to access land. We also find that measures to effect land reform, whether proxied by the number of laws passed, the area transferred under ceiling legislation, or the number of beneficiaries from tenancy Acts, constrain rental market activity, thereby reducing both equity and efficiency. They make it more difficult for producers with high ability, including the landless, to gain access to land while at the same time constituting an obstacle for giving up farming by those with low ability who want to get out of agriculture.

Production Function

As discussed earlier, a panel approach to recovering to recover a measure of households' (or dynasties') agricultural ability would have required dropping households or dynasties for whom production is observed only in one of the periods, either because they entered the sample in 1999 as a replacement for an earlier household or because they moved out of agriculture during the 20-year period that separates the two surveys.[22] To avoid this, we rely on a frontier production function in both periods and, in cases where a household's production is observed only in one period, assume that the level of ability is time-invariant.[23]

Appendix Table 1 contains descriptive statistics for the variables included in the production function in both periods while Table 2 reports results from estimating the Cobb-Douglass production functions for output, based on value of production from the 40 most important commodities[24], using OLS and a frontier approach represented in Appendix Table 2. Production function estimation for each of the two time periods, High values of the R^2 for the OLS regressions points towards a very good fit, Coefficient estimates from the maximum likelihood estimation of the Cobb-Douglas stochastic production frontier are similar to those obtained by OLS. Concerning the individual variables, land is estimated to be by far the most important input to crop production; doubling cultivated land area alone would lead to an almost 50 per cent increase in total crop production. This is followed by labour with an estimated elasticity between 18.8 per cent and 20.8 per cent. Compared to these two variables, returns to labour are estimated to be more moderate with elasticities of about 5 per cent, 4 per cent, 2.5 per cent, and 2.3 per cent for fertilizer, assets, other, and pesticide expenditures. While neither education nor the gender of the household head are significant, having all of the farm area under irrigation is estimated to increase output by between 31.5 per cent and 34.2 per cent. The fact that ability varies considerably between households could imply that, even without a strong pull from non-agricultural employment opportunities as in the case of China, the scope for market-mediated transfers to bring about efficiency gains could be large.

Determinants of Land Market Participation

Results from ordered probit estimation of the rental market participation equation on the pooled sample for 1982 and 1999 as reported in Table 10.4 which provide insights on the factors affecting rental market participation as well as the impact of rental market restrictions. For the latter, we use three measures and, for each of them, present equations with and without producers' estimated ability to ensure that results do not depend on inclusion of this measure. While regressions without ability (columns 1, 3 and 5) include the whole sample, the fact that ability is defined only for households or dynasties who engaged in agricultural production in either of the periods reduces the sample by about 2,500 observations while not altering substantive results.[25]

Factor affecting participation on either the supply- or demand-side of the rental market as indicated from the main equation are consistent with our hypothesis (recall the coding convention of 1 for renting-out, 2 for autarky, and 3 for renting-in). In particular, as proposition 1 and descriptive statistics led us to expect, land rental provides an important opportunity for landless households to gain access to land. One indication for this is the significance and magnitude of the coefficient on the landless dummy. Furthermore, higher land endowments are estimated to increase households' propensity to supply land to the rental market while higher endowments with family labour, especially of members in the 14-60 year age group, increase the propensity to rent-in land. This supports the notion that, by transferring land to those with higher endowments of family labour,

Table 10.4: Determinants of participation in land rental markets (ordered probit regression)

	Policy measure in the upper/lower bound equations					
	Tenants recognized		*Ceiling land redistributed*		*No. of tenancy laws*	
Main equation						
Cultivation ability		0.248***		0.252***		0.247***
		(2.63)		(2.63)		(2.63)
Landless dummy	0.623***	0.746***	0.626***	0.786***	0.622***	0.762***
	(18.09)	(15.16)	(17.81)	(15.43)	(17.91)	(15.24)
Land endowment (ac)	-0.012***	-0.029***	-0.013***	-0.029***	-0.011***	-0.029***
	(4.63)	(8.56)	(5.14)	(8.63)	(4.61)	(8.35)
Members below 14 years	0.054***	0.054***	0.055***	0.056***	0.056***	0.058***
	(6.22)	(5.65)	(6.18)	(5.80)	(6.38)	(5.96)
Members aged 14-60 years	0.063***	0.056***	0.062***	0.057***	0.060***	0.055***
	(7.97)	(6.42)	(7.74)	(6.53)	(7.55)	(6.29)
Head's age	0.021***	0.026***	0.022***	0.026***	0.021***	0.026***
	(3.44)	(3.56)	(3.62)	(3.51)	(3.45)	(3.59)
Head's age squared/100	-0.025***	-0.028***	-0.025***	-0.028***	-0.025***	-0.029***
	(4.34)	(4.14)	(4.36)	(4.08)	(4.34)	(4.24)
Head has primary or above	-0.148***	-0.117***	-0.153***	-0.119***	-0.161***	-0.135***
	(4.59)	(3.23)	(4.77)	(3.30)	(4.99)	(3.67)
Mean village income (log)	-0.090***	-0.106***	-0.077***	-0.100***	-0.072***	-0.078***
	(3.42)	(3.56)	(2.91)	(3.32)	(2.77)	(2.64)
Total assets (log)	0.01	0.009	0.008	0.002	0.011	0.007
	(0.59)	(0.44)	(0.50)	(0.11)	(0.65)	(0.32)
Off-farm share in total assets	-1.194***	-1.617***	-1.180***	-1.588***	-1.216***	-1.685***
	(5.43)	(4.63)	(5.24)	(4.53)	(5.42)	(4.65)

(*Contd.*)

	Policy measure in the upper/lower bound equations					
	Tenants recognized		*Ceiling land redistributed*		*No. of tenancy laws*	
Lower bound (rent out to autarky)						
Policy variable	-12.300***	-4.294*	-1.502**	-2.154***	-0.110***	-0.121***
	(6.50)	(1.90)	(2.53)	(2.97)	(6.07)	(5.38)
ST/SC dummy	-0.200***	-0.138**	-0.178***	-0.127**	-0.187***	-0.136**
	(3.85)	(2.16)	(3.38)	(1.98)	(3.54)	(2.10)
OBC dummy	-0.105**	-0.04	-0.104**	-0.024	-0.093**	-0.03
	(2.49)	(0.81)	(2.42)	(0.48)	(2.23)	(0.60)
1999 dummy	0.527***	1.426***	0.454***	1.391***	0.451***	1.359***
	(8.73)	(13.09)	(7.49)	(12.79)	(7.53)	(12.51)
Upper bound (autarky to rent in)						
Policy variable	12.697***	8.816***	2.551***	4.256***	0.018	-0.023
	(4.18)	(2.64)	(2.71)	(4.06)	(0.90)	(1.00)
ST/SC dummy	0.166**	0.101	0.148**	0.09	0.165**	0.11
	(2.52)	(1.44)	(2.24)	(1.28)	(2.52)	(1.57)
OBC dummy	0.148**	0.166***	0.116*	0.139**	0.147**	0.170***
	(2.42)	(2.64)	(1.87)	(2.20)	(2.42)	(2.72)
1999 dummy	-0.239***	0.143*	-0.245***	0.156*	-0.258***	0.097
	(3.41)	(1.75)	(3.43)	(1.89)	(3.69)	(1.19)
Observations	11331	8820	11147	8820	11221	8711
Log likelihood	-4564.94	-3513.55	-4450.96	-3501.93	-4514.77	-3436.48

Robust z statistics in parentheses; * significant at 10%; ** significant at 5%; *** significant at 1%. Constants and regional dummies included throughout but not reported.

markets will improve the scope for gainful employment of labour in rural areas. To interpret the regression coefficients, note that the probability for a landless household to gain access to land through rental is 4.1 to 7.4 percentage points higher than of those who are endowed with land. At the same time, the propensity to supply land of the largest landowners in our sample is between 26.8 and 40.3 points higher than that of a landless household.

Contrary to widespread belief which assumes that markets will favour those with higher levels of wealth, the lack of significance of households' total asset ownership suggests that, consistent with descriptive statistics, land rental markets are not biased against the poor. However, the composition of households' asset portfolio does matter; the significant coefficient on the share of non-farm assets highlights that those with a higher share of their assets in agriculture will be more likely to rent-in land. Compared to those with no off-farm assets at all, those who initially had all the assets related to off-farm activities is more than 25 per cent more likely to rent-out their land. The results also suggest that higher levels of education, measured by whether the head has completed at least primary education, significantly increase the propensity to supply land to rental market due to higher opportunity cost of labour for more educated individuals (proposition 2 and 3). The probability to rent-out (rent-in) land for a household whose head attained primary education or above is 2.3 per cent higher (1% lower) than one without such education. Mean village income similarly increases the tendency to rent-out, implying that as, with overall development, the level of income increases, households will be more likely to move out of agriculture and supply their land to the rental market, thereby allowing those with higher levels of agricultural ability to increase their holdings and income levels, as also observed in China (Deininger *et al.*: 2005). Compared to villages in the lowest income docile, those in the highest are 4-5 percentage points more likely to supply land to the rental market. This would imply that the importance of land rental markets will increase with overall economic development and that imposing exogenous restrictions on the operation of such markets could have far-reaching implications and may inflict efficiency losses the magnitude of which increases over time.

The highly significant coefficient on ability also supports our hypothesis of rental markets transferring land from less to more efficient producers. To illustrate, the probability of the most efficient household in the sample to receive land is about 4 percentage points higher than that for the least efficient producer while the least efficient producer is about

7 percentage points more likely to rent-out than the most efficient one. This points towards the potential for land transfers through rental markets to significantly enhances efficiency.

Regarding the lower bound equation, the regressions suggest that the three different measures used to represent land market restrictions (labelled as "policy variable") are associated with a significant and quantitatively important reduction of supply of land to the rental market.[26] This is in line with our discussion suggesting that landlords will be less ready to rent out their land if regulations either imply that part or all of their property rights to land that is rented out may be lost or limit their ability to freely negotiate the amount of rent to be paid. This result is obtained irrespectively of the variable chosen to empirically represent policy although comparing the coefficients on policy measures across specifications suggests that the number of laws passed had the least impact, followed by the implementation of ceiling laws and the implementation of tenancy legislation. This is consistent with the notion that ceiling laws pose less of a threat than tenancy regulation, both because the latter apply to all market participants irrespectively of the size of their holding—in contrast to ceiling legislation which is applicable only to large owners—and because enforcement of tenancy regulation is less politically controversial and administratively complex than that of land ceilings.

Coefficients on other variables in the lower bound equation suggest that, even after adjusting for the factors discussed earlier, Scheduled Castes and Tribes as well as other Backward Castes are less likely to rent out land than the remainder of the population, a finding that may be related to lower levels of social capital and thus either less opportunity to find partners in rental markets or to enforce protection of property rights for land that had been rented out. At the same time, the highly significant coefficient on 1999 dummy illustrates that, over and above the activation of land rental markets that comes about as a result of increases in village income, the supply of land to rental markets has increased significantly over time.

Turning to the (upper) bound between autarky and renting-in, the fact that estimates for all of the policy variables are positive suggests that such measures not only reduced supply (as indicated by the lower bound equation) but also depressed demand and made it more difficult for households to obtain land through rental markets.

Across specifications, the level of significance is much higher for variables relating to the intensity of enforcement; in fact the number of laws, while of the expected sign, is not significantly different from zero at conventional levels, suggesting that such legislation affects market outcomes only if measures are taken to implement it, consistent with what has been observed in the literature (Banerjee *et al.*: 2002). The magnitude of the coefficients which can be interpreted as a rough measure of the difference in impact between tenancy and ceiling legislation is similar to what had been observed for the lower bound, suggesting that these variables have contributed to a roughly equal expansion of supply and demand in the rental market. Parallel to what emerged for the lower bound, we find that backward and Scheduled Castes and Tribes are more likely to remain in autarky. Over time the size of the autarky area has decreased, i.e. land rental markets have become more active, partly offsetting the negative impact of rental regulation. While this is an encouraging sign, one should note that, given the magnitude of the coefficients, almost a century will be required for the time trend to fully offset the impact of tenancy legislation. This, together with evidence that circumventing such legislation is normally easier for the rich than the poor (Thangaraj: 1995, Yugandhar: 1996) would imply that expecting the passage of time alone to eliminate the negative effect of tenancy regulation is unlikely to be a realistic policy option.

Equity and Efficiency Impact of Land Reform Policy

The above analysis provides strong support for the hypothesis that different forms of land rental legislation have reduced activity in land rental markets and constrained the ability of the poor and landless to acquire land that would help them to improve their income and social status. At the same time, many scholars argue that tenancy regulations can still be socially desirable because of an underlying trade-off between equity and efficiency. Under this interpretation, the equity and social gains they bring about would, at least for some welfare weights, more than outweigh the associated efficiency losses (Besley *et al.*: 2004). More importantly, the equity and efficiency-impact of land reform legislation can be obtained from the signs of appropriate coefficients in an expanded version of the ordered probit regression (6) where the policy coefficient T is interacted with a dummy for landlessness or the level of productive efficiency as obtained from the frontier production function discussed earlier.[27] Inclusion of producers' initial level of productivity requires little justification and the focus on landless households is motivated by equity concerns, in particular the apparent high production elasticity of land and the presumption that gaining access to land will be associated with a qualitative and discrete increase in economic opportunities and social status (Finan *et al.*: 2005).

Results from three specifications of the ordered probit equation that interact the level of tenancy restrictions with landlessness, efficiency, and both (in columns 1, 2, and 3, respectively), are reported in Table 10.5. While it is not surprising to find that the coefficients in the main equation are broadly consistent with those in Table 10.4, a number of additional insights are of interest. As to possible equity impacts, inspection of the upper bound equation suggests that the main impact of tenancy regulation on the demand side is to prevent access to land by the landless (column 1) and more efficient producers (column 2); in fact after accounting for both these effects, the coefficient on the policy variable by itself is negative and highly significant. One explanation consistent with this finding is that sitting tenants who already own land but are not necessarily the most efficient producers[28] benefit from tenancy regulation at the cost of landless and more productive producers are constrained by the transaction costs imposed through tenancy restrictions and unable to effectively express their demand in the market.

Concerning the impact of tenancy restrictions on efficiency, the upper bound equation suggests that, in addition to rationing out efficient producers on the demand side, and contrary to what is intended, tenancy regulations encourage more efficient producers to supply their land to the rental market, possibly because they are in a better position to sidestep the transaction costs imposed by this intervention. Instead, possibly because tenancy restrictions make it more difficult for them to gain access to additional land that would allow to make best use of their skills, producers with high levels of ability are estimated to supply land to the rental market. This suggests that tenancy restrictions can make the economy worse off and that the gains associated with their removal should be more than sufficient to compensate possible losers and ensure that reforms are Pareto-improving. Comparing the coefficient for the policy variable and its interaction with ability between columns 1 and 2 or 3, respectively, suggests that, in states where tenancy restrictions are prevalent, producers with low levels of ability are less likely to rent out their land. This would imply that, where such restrictions exist, possible efficiency gains from low-ability producers moving out of agriculture and making their land available to those with higher levels of ability are less likely to materialize.

Table 10.5: Ordered probit regression interacting policy with efficiency and landlessness

	Interaction considered	
	Landlessness	*Landlessness & Productivity*
Main equation		
Ability		0.384***
		(3.35)
Landless in 1982 dummy	0.768***	0.850***
	(20.07)	
Land endowment	-0.012***	-0.029***
	(4.68)	(8.50)
Members below 14 years	0.053***	0.054***
	(6.16)	(5.63)
Members aged 14-60 years	0.063***	0.056***
	(8.01)	(6.40)
Head's age	0.020***	0.025***
	(3.34)	(3.52)
Head's age squared/100	-0.025***	-0.028***
	(4.26)	(4.11)
Head has primary or above	-0.148***	-0.122***
	(4.59)	(3.36)
Mean village income (log)	-0.096***	-0.106***
	(3.62)	(3.58)
Total assets (log)	0.01	0.011
	(0.59)	(0.50)
Share of off-farm in total assets	-1.254***	-1.632***
	(5.69)	(4.70)
Lower bound (rent out to autarky)		
Share receiving tenancy rights	-13.250***	-26.524***
	(7.08)	(3.09)
Tenancy rights* ability		32.764***
		(2.60)
ST/SC dummy	-0.187***	-0.134**
	(3.56)	(2.09)
OBC dummy	-0.105**	-0.042
	(2.50)	(0.85)
1999 dummy	0.528***	1.427***
	(8.62)	(13.07)
Upper bound (autarky to rent-in)		
Share receiving tenancy rights	2.884	-0.109
	(1.07)	(0.01)
Tenancy rights* landlessness	55.098***	35.528***
	(4.56)	(2.82)
Tenancy rights* ability		7.689
		(0.51)
ST/SC dummy	0.147**	0.092
	(2.23)	(1.31)
OBC dummy	0.143**	0.170***
	(2.35)	(2.71)
1999 dummy	-0.180**	0.144*
	(2.54)	(1.75)
Observations	11331	8820
Log likelihood	-4528.49	-3502.61

Robust z statistics in parentheses; * significant at 10%; ** significant at 5%; *** significant at 1% Constants and regional dummies included throughout but not reported.

Exploring Variation Over Time

A key argument to justify adoption of rental regulations when they were first introduced was that presence of multiple market imperfections in rural areas, exacerbated by immobility of land, would allow landlords to extract large rents from tenants who lack alternative opportunities. Comparing tenants' returns from cultivating rented land per day of family labour spent, net of any cash or kind outlays for inputs and the rental payment, for tenants with the prevailing agricultural wage rate provides a simple way of assessing the extent to which participation in rental markets can help to improve overall welfare. Doing so with the data at hand highlights the potentially far-reaching and positive impact of rental markets; even after deducting the rental fee (in cash or kind) from the net return per day of Rs. 150 which is very close to the value marginal product of family labour from a Cobb-Douglas production function, the average tenant earns more than double the wage rate for agricultural labour (46.4 for men and 33.7 for women) for everyday spent on rented-in land. Although unobserved differences in ability could play a role, anecdotal evidence these are unlikely to be sufficient to explain such large differences. In view of the fact that wages for women are significantly lower than those for men, the potential for rental markets to contribute economically empower women is particularly noteworthy and in line with anecdotal evidence suggesting that women are among the most avid users of programmes aiming to make land available for rental, either on an individual or on a group basis.

An explanation that is consistent with the above evidence is that, with economic development and greater availability of non-agricultural employment options, the impact of other markets imperfections (e.g. barriers to credit market access) may have become less pronounced over time, a possibility that has indeed been discussed at a theoretical level in the literature (Mookherjee: 1997). While lack of wage rates in 1982 prevents us from directly examining how this gap has evolved over time, we use an econometric approach to empirically test for this. In particular, probit regressions for renting-in and renting-out where most of the coefficients included earlier are interacted with a time dummy allows us to test for the extent to which the structure of land access through rental markets has changed over time, and whether this would reduce or even obviate the need for tight regulation of such markets that may have been justified earlier.

Table 10.6: Determinants of rental market participation (probit results)

	Type of participation				
	Renting in			*Renting out*	
Cultivation ability		0.016*	0.017**		−0.022*
		(1.88)	(2.05)		(1.74)
Log area owned (α1)	−0.002***	−0.003***	−0.003***	−0.001	0.001
	(4.49)	(4.74)	(4.70)	(1.37)	(1.49)
Area owned * 1999 (α2)	0.008	0.012*	0.013*	0.069***	0.053***
	(1.46)	(1.78)	(1.86)	(5.56)	(3.41)
Head has primary or above (β1)	−0.004	−0.008	−0.008	−0.026*	−0.023
	(0.74)	(1.03)	(1.08)	(1.95)	(1.63)
Head primary or more * 1999 (β2)	−0.003**	−0.004**	−0.004**	0.009***	0.004***
	(2.28)	(2.10)	(2.07)	(5.25)	(3.49)
Log total assets (γ1)	0.009**	0.013**	0.012**	0.006	0.016**
	(2.39)	(2.33)	(2.25)	(1.07)	(2.01)
Total assets * 1999 (γ2)	−0.009**	−0.011**	−0.011*	−0.012*	−0.017*
	(2.13)	(1.99)	(1.86)	(1.80)	(1.92)
Share of off–farm in total assets (δ1)	−0.090*	−0.095*	−0.096*	0.111	0.047
	(1.81)	(1.87)	(1.92)	(1.60)	(0.79)
Share of off–farm assets * 1999 (δ2)	0.036	0.017	0.02	0.265**	0.166**
	(0.58)	(0.25)	(0.28)	(2.55)	(2.07)
Log village income (ζ1)	0.008	0.01	0.01	0.006	0.011
	(1.64)	(1.54)	(1.52)	(0.56)	(0.71)
Village income level * 1999 (ζ2)	−0.019***	−0.021***	−0.021***	0.007	−0.002
	(3.43)	(2.99)	(3.05)	(0.51)	(0.14)
Landless dummy (η1)	−0.012*	0.570***	0.573***		
	(1.92)	(5.83)	(5.83)		
Landless * 1999 (η2)	0.019*	−0.042***	−0.041***		
	(1.80)	(4.88)	(4.88)		
Members below 14 years	0.002***	0.002***	0.002***	−0.011***	−0.008***
	(2.86)	(2.83)	(2.70)	(5.55)	(5.15)
Members aged 14–60 years	0.002***	0.002***	0.002***	−0.015***	−0.008***
	(3.26)	(3.17)	(3.12)	(7.79)	(6.21)
Head's age	0.000	0.001	0.001	−0.004***	−0.003***
	(0.91)	(0.91)	(0.91)	(3.31)	(3.81)
Head's age squared	0.000	0.000	0.000	0.000***	0.000***
	(0.97)	(0.86)	(0.85)	(4.26)	(4.49)
Share receiving tenancy rights	−0.712***	−0.649***	−0.434**	−1.865***	−0.339
	(4.60)	(3.25)	(2.11)	(5.67)	(1.28)
Share receiving tenancy rights * landless		−1.652**	−1.626**		
		(2.39)	(2.45)		
Share receiving tenancy rights * ability			−1.051**		−1.379**
			(2.41)		(2.52)
1999 dummy	0.309***	0.250***	0.242***	0.106	0.260**
	(4.52)	(3.90)	(3.83)	(1.28)	(2.54)
Observations	11331	8820	8820	8660	7816
Log likelihood	−1542.8	−1363.31	−1360.98	−2695.67	−1956.37
Tests:					
α 1 + α2=0	16.70***	17.75***	17.83***	27.32***	23.05***
β 1 + β2 0	1.35	1.4	1.55	26.38***	23.45***
γ1 + γ2 = 0	0.08	0.42	0.71	2.06	0.01
δ 1 + δ2 = 0	2.07	2.37	2.39	23.42***	16.17***
ζ1 + ζ2 = 0	20.25***	15.06***	16.09***	4.51**	4.22**
η1 + η2 = 0	0.11	12.39***	12.67***	n.a.	n.a.

Robust z statistics in parentheses; * significant at 10%; ** significant at 5%; *** significant at 1%. Regional dummies included throughout but not reported. The reported test statistics for the joint tests in the bottom panel are ÷2 statistics.

Results from doing so (Table 10.6) suggest that there have indeed been structural changes in the almost 20 years separating our two surveys. Results from tests for the hypothesis that the value of one of the time-varying coefficients equalled zero in 1999 are reported in the bottom panel. Most importantly, a significant wealthbias in the ability to access land—which may have been the result of capital market imperfections—that was present in the first period is no longer discernible in 1999. While in 1982, households with higher levels of assets were more likely to access land through rental markets (see the positive sign of γ_1 in Table 10.6), this had been reversed in the second period as we can no longer reject the hypothesis of no wealth effect, i.e. $\gamma_1 + \gamma_2 = 0$ as indicated in the bottom of Table 10.6 and consistent with the findings from earlier regressions. While more detailed investigation of the factors underlying this results transcends the scope of this paper, one explanation consistent with this finding would be that supply of land as well as the ability of potential renters to access (formal or informal) credit markets have increased in the course of economic development. This interpretation is consistent with the elimination of a similar, albeit slightly less significant bias of land rental markets in favour of more educated individuals between the first and the second period (δ_1 and δ_2 in Table 10.6). It is also of interest to observe that, while the propensity of large land owners to rent in, which was already negative in the initial period (β_1) has further decreased and that in villages with higher levels of income, the demand for land rental is now significantly lower than it was earlier, both factors presumably due to better availability of non-agricultural opportunities. This suggests that, at least in the areas represented by our sample, concerns about a trend towards reverse tenancy that could undermine livelihood options for the poorest are not justified. At the same time, we note that the strong advantage which landless had enjoyed in the land rental market has decreased slightly, an issue that would be worth to be further explored. On the supply side, we note that the propensity of large land owners to supply land to the rental market also has increased significantly between the two periods, implying that, instead of encountering transactions between small landowners that have little impact on the agrarian structure, one will be more likely to find exchanges of land that modify the structure of operational landholdings.

Conclusion and Policy Implications

In rural India, there is an increasing recognition of the importance of land rental markets to bring land to more productive uses while at the same time providing a basis for development of the rural non-farm economy. Even though the presence of numerous restrictions on the operation of land rental markets in the case of India makes this of even greater interest, cross-state evidence on this issue has thus far been limited. By using a national sample, we are able to provide such evidence.

We find that, contrary to what is often assumed, land rental markets help to improve not only productivity but also equity. They do so by allowing landless and land-poor households to access land and improve their livelihood. More importantly, the pro-poor nature of land markets has significantly improved over time as wealth bias that had characterized such markets earlier has been eliminated. The fact that land markets are more active in locations with higher levels of non-farm activity supports the notion that such markets make an increasingly important contribution to diversification of livelihoods in rural areas. In fact, higher incidence of opportunities and options open through such diversification could be a key factor underlying the elimination of wealth-related barriers to land rental market participation. At the same time, the finding that presence of tenancy laws and ceiling legislation, as well as high levels of implementation of these, reduce rather than increase land access by the landless and possibly very efficient producers is cause for concern. It suggests that, while such laws may have led to significant social gains in the past, their maintenance does no longer have a positive impact on equity but instead is developing into an increasingly potent obstacle to land access and greater productivity. Our findings provide an empirical underpinning to support the desirability of eliminating land rental market restrictions that is articulated in the Government's most recent 5-year plan (Government of India: 2002).[29] In fact, by demonstrating the significance and magnitude of these impacts quantitatively, our results reinforce the importance of moving from policy principles towards action on this issue.

While our sample is too small (and the survey too general) to provide specific guidance on how to move the debate from the "whether" to the "how" of reform, our results, together with the literature, provide two pointers. The literature suggests that sitting tenants are the main beneficiaries from tenancy regulation or other rental restrictions—and thus stand to lose most from reform. In fact, elimination of rent controls on rural lands without protection of sitting tenants in Egypt seems to have been associated with undesirable consequences (Bush: 2000). Ways to compensate tenants who have been on the same plot for a long time will increase the likelihood that reforms be feasible politically. This could include innovative mechanisms to clarify overlapping property rights to land reform land as experimented within some Indian states (Hanstad and Nielsen: 2004). On the other hand, many Indian scholars contend that liberalization of rental markets should be complemented by strong further regulation that would, for example, limit the amount of land that can be rented-in by a producer (Saxena: 2002). The evidence that land rental markets increasingly transfer land towards the land-poor that emerges from our analysis suggests that there may be no need for such regulation. To the contrary, experience in other countries suggests that such regulation is difficult to enforce and, by creating incentives for discretionary interference by bureaucrats, impose considerable transaction costs (Zepeda: 2000) and the need for it should be evaluated very carefully. Opening up of land leasing markets in India has the potential to generate significant benefits for the rural poor and it would be a pity if these gains were jeopardized by a tendency of low-level bureaucrats to use well-intended regulation—a pretext for discretionary use of their power.

Appendix Table 1: Characteristics of crop production of rural India in 1982

	All India	*Regions*			
		Northern	*Western*	*Eastern*	*Southern*
1982					
Value of crop production	22820.86	30739.11	21492.66	16301.58	20407.8
Crop area (ha)	2.12	2.41	2.74	1.35	1.41
Share of crop area irrigated	0.27	0.43	0.19	0.18	0.26
Total labour in crop production (days)	472.19	419.4	449.48	437.17	571.43
Seed expenditure (Rs.)	749.88	865.04	821.73	346.59	752.32
Fertilizer expenditure (Rs.)	2161.45	3501.84	1506.81	1058.67	2330.39
Pesticide expenditure (Rs.)	268.09	266.27	184.24	84.65	477.77
Other expenditure (Rs.)	626.32	1163.98	462.11	258.02	523.41
Total farming assets (Rs.)	3504.22	5404.92	4520.87	649.34	1798.99
Head's age	50.91	52.77	50.12	49.87	50.71
Head attained primary education or higher	0.26	0.21	0.16	0.28	0.43
Share of household with female head	0.05	0.02	0.03	0.03	0.1
No. of observations	3,328	834	1,171	452	871
1999					
Value of crop production	44434.29	61075.79	40082.64	25425.6	47803.93
Crop area (ha)	2.54	2.78	3.11	1.61	1.99
Share of crop area irrigated	0.63	0.92	0.55	0.56	0.51
Total labour in crop production (days)	166.62	136.57	170.34	143.97	213.16
Seed expenditure (Rs.)	2171.38	2086.84	2621.99	958.04	2455.94
Fertilizer expenditure (Rs.)	3589.31	5359.63	3403.45	1806.13	3276.48
Pesticide expenditure (Rs.)	730.25	1027.46	863.09	174.33	597.83
Other expenditure (Rs.)	2335.45	3341.76	2751.46	950.65	1548.77
Total farming assets (Rs.)	45240.54	69732.29	50568	8318.05	36984.85
Head's age	50.04	50.71	49.59	49.33	50.6
Head attained primary education or higher	0.51	0.56	0.41	0.57	0.56
Share of household with female head	0.05	0.04	0.04	0.04	0.08
No. of observations	4,208	1,037	1,557	713	901

Appendix Table 2: Estimation of Cobb-Douglass crop production function (1982 data)

	1982		1999	
	OLS	Frontier	OLS	Frontier
Total crop area sown (log)	0.488***	0.494***	0.584***	0.607***
	(22.95)	(31.74)	(35.13)	(37.54)
All labour in crop production (log)	0.208***	0.188***	0.154***	0.138***
	(11.15)	(14.15)	(14.47)	(13.56)
Seed expenditure (log)	0.093***	0.078***	0.290***	0.283***
	(8.49)	(10.02)	(32.43)	(33.47)
Fertilizer expenditure (log)	0.054***	0.045***	0.058***	0.051***
	(10.73)	(10.91)	(12.02)	(11.24)
Pesticide expenditure (log)	0.023***	0.023***	0.029***	0.028***
	(5.11)	(5.36)	(10.17)	(10.18)
Other expenditures (log)	0.025***	0.026***	0.014***	0.011***
	(4.11)	(5.26)	(4.27)	(3.67)
Value of farming assets (log)	0.040***	0.039***	0.009**	0.011***
	(6.25)	(8.28)	(2.30)	(2.89)
Share of crop area irrigated	0.342***	0.315***	0.110***	0.109***
	(5.03)	(5.51)	(6.07)	(6.15)
Head's age	0.001	0.002**	-0.001	-0.001
	(1.63)	(2.24)	(1.18)	(1.12)
Head attained primary or above education	0.009	0.034	-0.037***	-0.029**
	(0.33)	(1.48)	(2.60)	(2.05)
Female headed	-0.032	-0.024	-0.017	-0.016
	(0.67)	(0.54)	(0.53)	(0.52)
Constant	5.847***	6.595***	5.818***	6.302***
	(61.98)	(88.92)	(86.64)	(90.60)
Observations	3328	3328	4208	4208
R^2 or log-likelihood	0.77	-2721.08	0.87	-2284.66

Absolute value of t statistics in parentheses: * significant at 10%; ** significant at 5%; *** significant at 1%

Notes

11. The replacement of a system of fixed land rent with revenue collectors, called *zamindars*, who were empowered to collect as much land revenue as they saw fit in 1757, is commonly linked to a deep agrarian crisis that manifested itself in a series of devastating famines, most notably the Bengal famine of 1769/70 in which 10 mn. out of a population of 30 mn. were estimated to have died (Maharatna: 1994). To reduce the incentives for short-term overexploitation, the "permanent settlement" of land revenue in 1793 established three types of revenue system, namely (i) the *zamindari* system where revenue administrators were given permanent rights to land and its

revenue; (ii) the *ryotwari* system where arrangements were made between individual cultivators and the administration; and (iii) the *mahalwari* system where revenue arrangements were made with the village.

2. In countries as diverse as Uganda, Thailand, Egypt, Mexico, and India, rent ceilings and/or award of non-transferable occupancy rights have been used to redistribute land resources to occupants in an environment where more far-reaching measures of land reform were politically infeasible.
3. Using census figures, Appu (1997) estimates that, in order to avoid having to give rights to tenants, landlords evicted about 30 mn. tenants or about one-third of the total agriculturally active population. This is paralleled by evidence in other countries where landlords tried—and often succeeded to evict tenants in anticipation of legislation to protect tenants against eviction or limit the rents they would have had to pay (Binswanger *et al.*: 1995).
4. Almost 95 per cent of the total 7.34 mn. acres that had been declared surplus had been identified by 1980 with coverage increasing to 98 per cent by 1985. Of these, 5.4 mn. acres were actually distributed to beneficiaries.
5. The states of Punjab, Madhya Pradesh, Andhra Prdaesh, Himachal Pradesh, Haryana, Gujarat, Maharashtra and Karnataka stipulate that beneficiaries will gain ownership rights only once they have reimbursed the government for the compensation that had been paid to the original landowner, plus in some cases administrative expenses. In Uttar Pradesh, beneficiaries did not receive ownership rights but were instead made permanent tenants of the government and any transfer, including sub-leasing, was prohibited. Transfer of land received under ceiling legislation is also prohibited in West Bengal, Bihar, Maharashtra and Gujarat, while being subject to a moratorium (of 10, 15, and 20 years after award of full ownership rights) in Orissa, Karnataka, and Himachal Pradesh, respectively.
6. In a number of states including Andhra Pradesh, Bihar, Karnataka, Madhya Pradesh, Uttar Pradesh, Haryana, Punjab, Rajasthan and Tamil Nadu, the definition of 'tenant' does not include sharecroppers, implying that the extent to which they will benefit from protection will depend on local discretion.
7. This includes a complete prohibition of leasing for all except disabled landowners in parts of Andhra Pradesh and in Bihar, Karnataka, Madhya Pradesh, Orissa, Uttar Pradesh, Himachal Pradesh, Kerala and Jammu and Kashmir. Other states (Gujarat, Haryana, Maharashtra, Punjab, Orissa, Rajasthan, West Bengal, and the Andhra Part of Andhra Pradesh) permits leasing but make leases heritable and non-transferable with some (Assam, Gujarat, Haryana, Maharashtra and Punjab) establishing a right for the renter to purchase the property after sometime.
8. The rent ceiling is defined either in terms of multiples of the land revenue or as a share of output, such as a sixth of output in Gujarat (Kutch area) Maharashtra, and Rajasthan (in case the landlord does not provide any inputs), a fifth or fourth of output in AP, Bihar, Rajasthan (if the landlord provides inputs), Orissa, Tamil Nadu, and West Bengal (unless the landlord

provides capital, manure, and seed in which case the ceiling is half of the produce), and a third of the crop in Punjab and Haryana.

9. We use area rather than beneficiaries because in some cases ceiling surplus land was distributed to a collective entity such as a cooperative so that the number of beneficiaries would be misleading. Also, the existence of large discrepancies between the amount of land expropriated and actually distributed—which is due to the fact that in some cases land that had been distributed could not occupied by beneficiaries or was taken back after sometime—led us to focus on land actually distributed.
10. Wherever available, the level of the respective variables in 1980 and 1998, respectively, is used as a right hand side variable in the regressions.
11. The amount of land involved is much larger than what was redistributed in other Asian land reforms such as Japan (2 mn ha), Korea (0.58 mn ha) and Taiwan (0.24 mn ha). In terms of total area distributed, this puts India on par with Mexico which, in a much more land-abundant setting, and starting in 1917, managed to distribute slightly more than 13 mn ha (Deininger *et al.*: 2002).
12. In addition to cases related to contestation by landlords, there is an unknown number of instances where beneficiaries were allocated land but were either unable to establish effective possession or were subsequently evicted. A field survey to explore this issue in Andhra Pradesh pointed to at least 20 per cent of beneficiaries who were not able to access the property they had received although the number of those who are able to file court cases calling for their (re)instatement is much more limited.
13. Recognition of this fact has led policy-makers to implement a modified set of interventions most commonly referred to as "tenancy rent controls" that would allow unlimited rent increases between tenancies and may include some other provisions to increase returns to landlords, in the hope of avoiding some of the negative impacts associated with earlier forms of rent control (Arnott: 1997). As even such measures would be subject to the deadweight losses associated with rent controls. (Basu and Emerson: 2003), the main policy issue boils down to deciding whether potential gains in tenure security that can be achieved through such controls will outweigh the deadweight losses of such a policy instrument as well as the costs of implementing it (Glaeser: 2002, Arnott: 2003), an issue that needs to be resolved empirically.
14. In the cases studied, rent to income ratios decreased from about 20 per cent in uncontrolled regimes to about 10 per cent in controlled ones and in the case studies, rent control reduced the price paid by the typical tenant by between 4 per cent and 64 per cent. At the same time, the fact that house price to income ratios increase from about 4 in uncontrolled to between 7 and 8 in controlled regimes suggests that rent control does reduce the supply of housing.
15. The tenant will be unlikely to invest as doing so will result in an immediate

increase of the rent whereas investment by the landlord is unlikely as part of the benefits will go to the tenant.

16. Successive generations of IAS officers in the Lal Bahadur Shastri National Academy of Administration in Mussoorie have made a significant contribution to the area by conducting case studies of land reform implementation in almost all of India's states that have thus far resulted in the publication of 10 volumes on the subject that are replete with empirical evidence.
17. We denote by αl () and αlu () the level of ability where households with given other characteristics switch from rent out land to autarky or from autarky to renting-in land, respectively. In other words, households with $\alpha i < \alpha l$ will rent out land, those with $\alpha i > \alpha_u$ rent-in, and those with $\alpha_u > \alpha i < \alpha l$ stay autarky. We can solve for α_u, and αl jointly with $l^{\alpha}i$ from the FOCs:

$$\alpha_1 = \frac{r-T}{pfA\left(l^a i, Ai\right)}, \quad \alpha_u = \frac{r+T}{pfA\left(l^a i, Ai\right)}$$

18. Of 4981 households who were included in the 1982 sample, 3880 could be retrieved in the 1999 round; 2697 who continued under the same head and 1183 who had split up. Replacements for households who had dropped out together with multiple ones who had emerged from the split ones resulted in a total sample size of 7474 households in 1999.
19. We group states into four regions as follows: The North includes the states of Haryana, Himachal Pradesh, Punjab, and Uttar Pradesh; the West includes Gujarat, Maharashtra, Madhya Pradesh, and Rajasthan; the East includes Assam, Bihar, Orissa, and West Bengal; and the South includes Andhra Pradesh, Karnataka, Kerala, and Tamil Nadu.
20. While part of this increase may be due to improvements in the survey instrument that resulted in better measurement of assets in the second period, it is impossible to test this hypothesis as disaggregated data for 1982 were not available.
21. In Vietnam, the shrae of households renting in increased form 3.8 per cent to 15.8 per cent in the 5-year period between 1993 and 1998 (Deininger and Jin: 2003). In China, the same figure increased from 2.3 per cent in 1996 to 9.4 in 2001 (Deininger and Jin: 2005).
22. In fact, our sample suggests that it was households with higher productivity who left the agricultural sector and that the difference in the level of productivity between this group and those who remained in agriculture is statistically significant.
23. For the subset of households that were observed in both periods, we estimated 1999 ability from estimating production frontier using 1999 data and 1982 ability using 1982 data. The pairwise correlation between the 1999 and 1982 abilities is 0.70, significant at 1 per cent level.
24. These commodities include paddy, wheat, barley, maize, jowar, bajra, ragi, pulses, small millets, groundnut and other oilseeds, cotton, jute and other fibre crops, sugarcane and other sugar and starch crops.

25. There are 4,346 observations in the 1982 and 7,476 in the 1999 survey, 491 of which lack information on village income, yielding 11,331 ones that can be used in the regression. Columns 2, 4 and 6 include 2,889 producer households from 1982 who split into 4,605 in 1999, in addition to 1,419 ones who were added in the second period, excluding 93 for whom no village level income was available.
26. To interpret the coefficients in the lower bound equation marking the transition form renting out to autarky, note that a positive coefficient implies expansion of the rent-out regime and better functioning of rental markets while a negative coefficient suggests the opposite.
27. As landless households will obviously not be able to supply land to the rental market, the interaction between landlessness and the policy variable is included only in the upper bound equation.
28. Especially when they become old and their children do not want to continue in farming as was found to be the case in the Philippines (Deininger *et al.*: 2002).
29. "...freedom in leasing of land, both 'leasing in' and 'leasing out' will help generate income for both lessee and lessor/contractor. A legislation needs to be enacted to facilitate the land utilisation by making land transactions easier and facilitating leasing and contract farming." (Government of India: 2002, p. 528).

References

Aghion, P., E. Caroli and C. Garcia-Penalosa. (1999): "Inequality and Economic Growth: The Perspective of the New Growth Theories." *Journal of Economic Literature 37* (4): 1615-60.

Alston, L.J. and J.P. Ferrie. (2005): "Time on the Ladder: Career Mobility in Agriculture, 1890-1938." *NBER Working Paper 11231*. Cambridge MA: National Bureau of Economic Research.

Appu, P.S. (1997): *Land Reforms in India: A Survey of Policy, Legislation and Implementation*. New Delhi: Vikas Publishing House.

Arnott, R. (1995): "Time for revisionism or rent control?" *Journal of Economic Perspectives 9* (1): 99-120.

Arnott, R. (1997): "Rent control." *The new Palgrave Dictionary of Economics and the Law*.

Arnott, R. (2003): "Tenancy rent control." *Swedish Economic Policy Review 10* (1): 89-121.

Banerjee, A.V., P.J. Gertler and M. Ghatak. (2002): "Empowerment and Efficiency: Tenancy Reform in West Bengal." *Journal of Political Economy 110* (2): 239-80.

Banerjee, A.V. and L. Iyer. (2004): "History, Institutions, and Economic Performance: The Legacy of Colonial Land Tenure Systems in India." *MIT Working Paper 02-27,* Cambridge, MA.

Basu, K. and P.M. Emerson. (2000): "The Economics of Tenancy Rent Control." *Economic Journal 110* (466): 939-62.

Basu, K. and P.M. Emerson. (2003): "Efficiency Pricing, Tenancy Rent Control and Monopolistic Landlords." *Economica 70* (278): 223-32.

Besley, T. and R. Burgess. (2000): "Land Reform, Poverty Reduction, and Growth: Evidence from India." *Quarterly Journal of Economics 115* (2): 389-430.

Besley, T. and R. Burgess. (2004): "Can Labour Regulation Hinder Economic Performance? Evidence from India." *Quarterly Journal of Economics 119* (1): 91-134.

Besley, T., R. Burgess, and B. Esteve-Volart. (2004): "Operationalising pro-poor growth: India case study." London: DFID, BMZ, GTZ, KfW, and the World Bank.

Binswanger, H.P., K. Deininger and G. Feder. (1995): "Power, Distortions, Revolt and Reform in Agricultural Land Relations." *Handbook of Development Economics 3B,* 2659-772.

Bush, R. (2000): "An Agricultural Strategy without Farmers: Egypt's Countryside in the New Millennium." *Review of African Political Economy 27* (84): 235-49.

Carter, M.R. and Y. Yao. (2002): "Local versus Global Separability in Agricultural Household Models: The Factor Price Equalization Effect of Land Transfer Rights." *American Journal of Agricultural Economics 84* (3): 702-15.

Coelli, T.J. (1995): "Recent Developments in Frontier Modelling and Efficiency Measurement." *Australian Journal of Agricultural Economics 39* (3): 219-45.

Conning, J. and J.A. Robinson. (2005): "Property Rights and the Political Organization of Agriculture." *Working Paper 405*. New York: Hunter College.

Deininger, K., M. Maertens, P. Olinto, and F. Lara. (2002): "Redistribution, Investments and Human Capital Accumulation: The Case of Agrarian Reform in the Philippines." *World Bank Discussion Paper*. Washington DC.

Deininger, K. and S. Jin. (2003): "Land Sales and Rental Markets in Transition: Evidence from Rural Vietnam." *World Bank Policy Research Working Paper 3013*. Washington, DC: World Bank.

Deininger, K. and S. Jin. (2005): "The potential of land markets in the process of economic development: Evidence from China." *Journal of Development Economics 78* (1): 241-70.

Deininger, K., S. Jin, B. Adenew, S. Gebre-Selassie and B. Nega. (2005): "Tenure Security and Land-Related Investment: Evidence from Ethiopia." *European Economic Review forthcoming.*

Finan, F., E. Sadoulet and A. de Janvry. (2005): "Measuring the Poverty Reduction Potential of Land in Rural Mexico." *Journal of Development Economics 77* (1): 27-51.

Foster, A.D. and M.R. Rosenzweig. (1996): "Comparative advantage, information and the allocation of workers to tasks: Evidence from an agricultural labour market." University of Pennsylvania.

Foster, A.D. and M.R. Rosenzweig. (2004): "Agricultural Productivity Growth, Rural Economic Diversity, and Economic Reforms: India, 1970-2000." *Economic Development and Cultural Change 52* (3): 509-42.

Frisvold, G.B. (1994): "Does Supervision Matter? Some Hypothesis Tests using Indian Farm-Level Data." *Journal of Development Economics 43* (2): 217-38.

Glaeser, E.L. (2002): "Does rent control reduce segregation?" *HIER Discussion Paper 1985*. Cambridge MA: Harverd University.

Government of India. (2002): "10th Five Year Plan (2002-2007).". New Delhi: Planning Commission.

Gyourko, J. and P. Linneman. (1990): "Rent Controls and Rental Housing Quality: A Note on the Effects of New York City's Old Controls." *Journal of Urban Economics 27* (3): 398-409.

Hanstad, T. and R. Nielsen. (2004): "West Bengal's Bargadars and Landownership." *Economic and Political Weekly 39* (8): 821-9.

Haque, T. (2001): "Impact of tenancy reforms on productivity improvement and socio-economic status of poor tenants." *Policy Paper 13*. New Delhi: National Centre for Agricultural Economics and Policy Research.

Haque, T. and A. Singh. (1986): *Agrarian Reforms and Institutional Changes in India*. New Delhi: Concept Publishing Company.

Kaushik, A. and T. Haque. (2005): "Nature of land market interventions in India." Institute for Sustainable Development, Noida.

King, R. (1977): *Land Reform: A World Survey*. London: G. Bell and Sons.

Lanjouw, J.O. and P. Lanjouw. (2001): "The Rural Non-farm Sector: Issues and Evidence from Developing Countries." *Agricultural Economics 26* (1): 1-23.

Libecap, G.D. (2002): "Indians, markets, and rainforests: Theory, methods, analysis (book review)." *Journal of Economic Literature 40* (2): 564-6.

Lieten, G.K. (1996): "Land Reforms at Centre Stage: The Evidence on West Bengal." *Development and Change 27* (1): 111-30.

Maharatna, A. (1994): "The Demography of the Bengal Famine of 1943-44: A Detailed Study." *Indian Economic and Social History Review 31* (2): 169-215.

Malpezzi, S. and G. Ball. (1991): *Rent control in developing countries*. Washington DC: World Bank.

Mearns, R. (1999): "Access to Land in Rural India." *Policy Research Working Paper 2123*. Washington DC: World Bank.

Moog, R.S. (1997): *Whose interests are supreme? Organizational politics in the civil courts in India*. Ann Arbor, Michigan: Association for Asian Studies.

Mookherjee, D. (1997): "Informational Rents and Property Rights in Land." *Property relations, incentives and welfare: Proceedings of a conference held in Barcelona, Spain, by the International Economic Association* 3.

Munch, J.R. and M. Svarer. (2002): "Rent Control and Tenancy Duration." *Journal of Urban Economics 52* (3): 542-60.

Rawal, V. (2001): "Agrarian Reform and Land Markets: A Study of Land Transactions in Two Villages of West Bengal, 1977-1995." *Economic Development and Cultural Change 49* (3): 611-29.

Sadoulet, E., A. de Janvry and C. Benjamin. (1998): "Household Behaviour with Imperfect Labour Markets." *Industrial Relations 37* (1): 85-108.

Sadoulet, E., R. Murgai and A. de Janvry. (2001): "Access to Land via Land Rental Markets." In de Janvry, A., G.

Gordillo, J.P. Platteau and E. Sadoulet, eds., *Access to Land Rural Poverty and Public Action*. Oxford: Oxford University Press.

Saxena, N.C. (2002): "Tenancy Reforms vs Open Market Leasing—What would Serve the Poor Better?" New Delhi: Planning Commission, Government of India.

Spillman, W.J. (1919): "The agricultural ladder." *American Economic Review 9* (1): 170-9.

Thangaraj, M. (1995): "Trends in fragmentation of operational holdings in India: An exploratory analysis." *Indian Journal of Agricultural Economics* (India).

Thorner, D. (1976): *Agrarian Prospect in India*. New Delhi: Allied Publishers.

Warriner, D. (1969): *Land Reform in Principle and Practice*. Oxford: Clarendon Press.

World Bank. (2005): "World Development Report 2006: Equity and Development." Washington, DC: The World Bank and Oxford University Press.

Yugandhar, B.N. (1996): *Land reforms in India. Volume 3. Andhra Pradesh—People's pressure and administrative innovations* New Delhi; Thousand Oaks, Calif. and London: Sage Publications.

Zepeda, G. (2000): "Transformación Agraria. Los Derechos de Propriedad en el Campo Mexicano bajo el Nuevo Marco Institucional." *CIOAC*, Mexico.

11

Changing Tenancy Relations in Rural India: A Case for Legalizing Tenancy

H.R. Sharma

Introduction

The broad features of the land lease policy that was followed in the post-independent India got evolved through the recommendations of the Congress Agrarian Reforms Committee, report on Panel on Land Reforms and successive Five Year Plans. Based on the recommendations of different committees, the following national guidelines were laid down to enact tenancy legislations to ensure 'land to the tiller'. These are to provide security of tenure; fixation of fair rent ranging between one-fourth to one-fifth; landowners to be allowed to resume the land for self-cultivation up to a limited area; the surrender of tenancy rights with mutual consent; disabled persons, defense personnel, and such exemptions be allowed to lease out land; the term personal cultivation should be clearly defined if landlords are allowed to evict the tenants to resume land for self-cultivation; the tenancy records should be corrected and oral tenancies should be abolished and on non-resumable land, the landlord-tenant nexus to be ended and tenants in these areas to be brought into direct contact with the state. The tenancy legislations enacted and implemented in different states following these guidelines aimed at providing security of tenure, fixing fair land rent ranging from one-fourth to one-fifth of the gross produce, allowing landowners to resume land for self-cultivation up to a limited area and ending the landlord-tenant nexus on non-resumable land to bring tenants in these areas into direct contact with state.

Insofar as the current status of provisions in tenancy legislations in different states are concerned, all states can be divided into five broad categories. First, Kerala and Jammu and Kashmir have legally banned leasing-out of agricultural land without any exceptions Orissa have legally prohibited leasing-out of agricultural land excepting by certain disabled categories like widows, minors, defense personnel, etc. Second, Telengana region of Andhra Pradesh, Bihar, Jharkhand, Karnataka, Himachal Pradesh, Madhya Pradesh, Chattisgarh, Orissa, Uttar Pradesh, Uttarakhand and Orissa have legally prohibited leasing-out of agricultural land excepting by certain disabled categories like widows, minors, defense personnel, etc. Third, the states of Punjab, Haryana, Gujarat, Maharashtra and Assam have not specifically banned leasing. While in Punjab and Haryana tenants acquire rights to purchase land after six years of continuous possession, in Maharashtra they are entitled to purchase leased-in land within one year of the commencement of tenancy. In Assam, an ordinary tenant acquires right of occupancy after three years of continuous possession and an occupancy tenant has a right to purchase leased-in land after three years. Fourth, in Andhra region of Andhra Pradesh, Rajasthan, Tamil Nadu and West Bengal there are no restrictions on land leasing although in West Bengal only share cropping leases are legally permitted. Fifth, in Scheduled Tribe areas of Andhra Pradesh, Orissa, Madhya Pradesh and Maharashtra transfer of tribal land to non-tribals even on lease basis, can be permitted by competent authority.

The actual implementation of the tenancy legislations has, however, fallen far short of the provisions of tenancy legislations. The tardy implementation of tenancy legislations across states has been discussed and commented upon extensively in the literature.[1] A review of the literature on the implementation of tenancy legislations in the sixties and the seventies bring out a number of deficiencies. First, although rent has been fixed yet these were fixed much above the recommended levels in various plan documents in some states. Second, the definition of tenants often excluded the share croppers who constituted a fairly large number of proportions. Third, ejectment of tenants from their holdings was permissible on a number of pretexts. Fourth, a provision of voluntary surrender provided a handy excuse to evict tenants under pressure, coercion and sometimes even physical intimidation. The tenants were allowed to surrender land voluntarily and the definition of personal cultivation did not include physical labour as was recommended by Panel on Land Reforms in 1956. Fifth, the provision regarding the resumption of land proved to be most pernicious in spite of the fact that many states granted only a limited right for the resumption of land for personal cultivation. Sixth, the provision for the conferment of ownership rights on the tenants of non-resumable land was far from real. Seventh, the amount of compensation to be paid by the tenants to acquire ownership rights was generally high and beyond the capacity of small and marginal tenants. In brief, the tenancy legislations in different states were riddled with numerous deficiencies and loopholes. The remarks made by the Task Force on Agrarian Relations way back in 1973 that 'several states in the matter of tenancy reform legislations fall far short of the accepted policy and what is worse, implementation of enacted laws has been half-hearted, halting and unsatisfactory in large parts of the country' reflected the sordid state of affairs in the implementation of tenancy laws. Naturally, therefore, the actual implementation of the tenancy legislations fell far short of the provisions of tenancy legislations. At the national level, 125.86 lakh tenants have been cumulatively conferred the ownership rights till March, 2006 on an area amounting to 167.14 lakh hectares.

I. Changing Tenancy Relations: Macro Evidence

The tenancy relations over the years have undergone significant changes thanks to the combined effect of institutional, technological and demographic factors. The important changes in different aspects of tenancy relations such as magnitude of tenancy, terms of tenancy, reverse tenancy and duration of tenancy contracts in seventeen major states of the country emanating from NSS data on landholdings between 37th round (1982) and 59th round (2003) have been summarised below.[2] It is important to mention here that though NSS data are gross underestimates of the extent of tenancy; these do serve an important purpose in providing information about different aspects of tenancy relations at the state level inasmuch as the findings of micro studies cannot be generalized because of their limited coverage.[3]

(i) Magnitude of Tenancy

The changes in the incidence of tenancy across seventeen major states brought out by the NSS reports on landholdings of various rounds since 1981-82 (37th Round) have been summarised here (Appendix Table A-11.1). First, the proportion of entirely owned self-cultivated holdings increased continuously since 1981-82 in as many as ten major states, namely, Bihar, Haryana, Jammu and Kashmir, Kerala, Maharashtra, Punjab, Rajasthan, Tamil Nadu, Uttar Pradesh and West Bengal while in other seven states like Andhra Pradesh, Assam, Gujarat, Himachal Pradesh, Karnataka, Madhya Pradesh and Orissa, the proportion of such holdings has increased by varying degrees between 1991-92 and 2002-03. Second, the proportion of mixed holdings decreased by varying degree continuously both during the eighties and the nineties in as many as twelve major states, namely, Andhra Pradesh, Assam, Haryana, Jammu and Kashmir, Karnataka, Maharashtra, Orissa, Punjab, Rajashtan, Tamil Nadu, Uttar Pradesh and West Bengal. In other five states like Bihar, Gujarat, Himachal Pradesh, Kerala and Madhya Pradesh while there was a significant decrease in the proportion of such holdings during the eighties, the proportion of such holdings increased during the nineties. Third, the proportion of entirely leased-in holdings increased during the eighties but decreased during the nineties in Assam, Bihar, Gujarat, Haryana, Himachal Pradesh, Karnataka, Kerala, Madhya Pradesh, Rajasthan, Uttar Pradesh, while in other states like Andhra Pradesh, Jammu and Kashmir and Punjab the proportion of such holdings decreased during the eighties but increased during the nineties. In Maharashtra and Tamil Nadu, the proportion of entirely leased-in holdings declined continuously. Fourth, incidence of tenancy in terms of proportion of operated area leased-in (LIA) declined continuously both during the eighties and the nineties only in six states, namely, Assam, Haryana, Maharashtra, Rajasthan, Tamil Nadu and West Bengal. In other states, the pattern was mixed. For example, while the proportion of operated area leased-in declined during the eighties but increased during the nineties in Bihar, Gujarat, Kerala and Orissa, a reverse pattern was discernible in states like Andhra Pradesh, Assam, Haryana, Himachal Pradesh, Jammu and Kashmir, Karnataka, Madhya Pradesh and Maharashtra.

(ii) Households not Operating Land

The changes in the proportion of households not operating land, who comprise households owning but not operating land and those neither owning nor operating land, also provide some insights about the changing incidence of tenancy (Appendix Table A-11.2). For example, households who own but do not operate land comprise, among others, lessors of various categories which, *inter alia*, is an indicator of the magnitude of tenancy in a particular state. Likewise, those who neither own nor operate any land are landless households many of whom have no access to lease market may be because of onerous terms of tenancy and the participation of medium and large farmers in lease market. The data in this regard show that during the nineties, the proportion of households not operating land increased hugely in almost all the states. In fact, in some states, most notably, Andhra Pradesh, Bihar, Gujarat, Haryana, Karnataka, Kerala, Maharashtra, Tamil Nadu, Uttar Pradesh and West Bengal, the increase in the proportion of such households was truly mind boggling. The data further reveal that in as many as eleven states (Assam, Bihar, Gujarat, Jammu and Kashmir, Kerala, Madhya Pradesh, Orissa, Punjab, Rajasthan, Uttar Pradesh and West Bengal) the increase in the proportion of households not operating land was largely contributed by those who owned but did not operate land. In the remaining six states (Andhra Pradesh, Haryana, Himachal Pradesh, Karnataka, Maharashtra and Tamil Nadu), the proportion of households who neither owned nor operated any land also recorded significant increase.

(iii) Terms of Tenancy

The terms of tenancy have been clubbed into four main types, namely, fixed money (FM), fixed produce (FP), share tenancy (ST) and other terms (OT) to ensure temporal comparability.[4] The changes in these terms of tenancy over the period have been discussed in terms of per cent distribution of leased-in holdings and leased-in area under them. The data since the early eighties throw up the following broad trends. First, the proportion of holdings and area leased-in accounted for by fixed rent tenancy, including fixed money and fixed produce, increased continuously both during the eighties and the nineties in as many as thirteen major states of the country (Andhra Pradesh, Bihar, Gujarat, Haryana, Himachal Pradesh, Karnataka, Kerala, Madhya Pradesh, Punjab, Rajasthan, Tamil Nadu, Uttar Pradesh and West Bengal). In the remaining states, while the proportion of holdings and area leased-in accounted for by fixed rent tenancy decreased during both the periods in Assam and Jammu and Kashmir, no neat pattern was discernible in Maharashtra and Orissa. Second, the holdings and operated area leased-in accounted for by share tenancy increased continuously in Andhra Pradesh, Gujarat, Karnataka and Rajasthan and decreased in Haryana and West Bengal. In seven other states (Assam, Bihar, Himachal Pradesh, Jammu and Kashmir, Kerala, Maharashtra and Uttar Pradesh), the per cent share of holdings and area accounted for by share tenancy decreased during the eighties but increased during the nineties. No neat pattern was evident in the remaining states. Third, the importance of 'other terms' in terms of per cent of holdings and area leased-in accounted for by these terms decreased by varying degree in a majority of the states both during the eighties and the nineties.

(iv) Reverse Tenancy

The changes in the extent of reverse tenancy have been studied in terms of changes in the amount of land leased-in by landless, marginal and small households and leased-out by the marginal and small households and their per cent share in the total land transacted in the lease market. The salient features in this regard are summarised here (Appendix Table A-11.3). First, the proportion of leased-in land accounted for by landless, marginal and small households increased between 1971-72 and 1991-92 in as many as ten states (Andhra Pradesh, Assam, Gujarat, Jammu and Kashmir, Madhya Pradesh, Maharashtra, Rajasthan, Tamil Nadu, Uttar Pradesh and West Bengal) but decreased during the subsequent period between 1991-92 and 2002-03. More importantly, however, in all these states except Madhya Pradesh, Maharashtra and West Bengal, the extent of decrease in the per cent share of these households in total leased-in land was truly significant. There was a mixed pattern in the remaining states. For example, while in Bihar and Orissa, the proportion of leased-in land accounted for by landless, marginal and small households increased during both the periods, in four states (Haryana, Karnataka, Kerala and Punjab) it decreased between 1971-72 and 1991-92 but increased thereafter between 1991-92 and 2002-03. Second, in so far as changes in the share of small and marginal households in the total land supplied in the lease market are concerned, the data show that in nine states (Andhra Pradesh, Gujarat, Karnataka, Madhya Pradesh, Maharashtra, Rajasthan, Tamil Nadu, Uttar Pradesh and West Bengal), per cent contribution of small and marginal households towards the total land supplied in the lease market increased continuously in both the periods. In four other states (Assam, Haryana, Orissa and Punjab), the per cent share of small and marginal households in the total leased-out land increased between1971-72 and 1991-92 but decreased thereafter between 1991-92 and 2002-03.

(v) Duration of Tenancy Contracts

In NSS reports on landholdings, the data on duration of tenancy contracts have been provided in terms of the distribution of leased-in area by duration of tenancy contracts. The tenancy contracts have been classified into six main categories viz. contracts for less than one agricultural season, one agricultural season but less than one agricultural year, one to two agricultural years, two to less than five years, five to less than twelve years and contracts for more than twelve years (Appendix Table A-11.4). The data show that in 2002-03 more than 50 per cent of the leased-in land in ten major states (Assam, Bihar, Haryana, Karnataka, Madhya Pradesh, Maharashtra, Orissa, Punjab, Rajasthan and Uttar Pradesh) was under tenancy contracts of less than two years duration. In the remaining states (Andhra Pradesh, Gujarat, Himachal Pradesh, Jammu and Kashmir, Kerala, Tamil Nadu and West Bengal), more than half of the leased-in land was under those contracts which extended beyond two years. Further, the duration of tenancy contracts between 1991-92 and 2002-03 decreased in a majority of the states in that per cent of leased-in land accounted for by contracts for less than two years increased in as many as twelve major states like Assam, Bihar, Himachal Pradesh, Jammu and Kashmir, Karnataka, Kerala, Madhya Pradesh, Maharashtra, Punjab, Rajasthan, Uttar Pradesh and West Bengal.

II. Changing Tenancy Relations: Micro Evidence

The estimates of tenancy available in NSS reports on landholdings are not only gross underestimates, but they also do not throw information on many other important aspects of tenancy relations like inputs use and productivity levels associated with different levels and types of tenancy, land rent, factors affecting incidence and type of tenancy, tenancy arrangements for different types of crops, interlocking of land lease market with other factor markets, and so on. Therefore, studies based on NSS data do not capture full dimensions of changes in tenancy relations. The changes in the above-mentioned important aspects of tenancy relations are available in a plethora of micro studies from different regions of the country. A fairly exhaustive review of micro studies from different regions of rural India since the early eighties throws up a number of interesting features about the actual functioning of lease market.[5] Some of these changes are summarised below. First, the proportion of leased-in land is significantly higher than shown by NSS data; in some areas, it is as high as 20-25 per cent while in some others more than 50 per cent of the gross cultivated area is reported to be leased-in. Again, the incidence of tenancy is significantly higher in agriculturally developed/irrigated regions compared to agriculturally backward regions and all classes of households participate in the lease market either as lessors or lessees or even both. Among different crops, the proportion of leased-in land is higher in case of non-foodgrain crops compared to cereal crops. Tenancy contracts are oral, and most of them are for a short period.[6] Second, while in agriculturally backward regions, the traditional pattern of tenancy wherein the small and marginal farmers dominate lease market as lessees and large and medium farmers as lessors is more common, in agriculturally developed regions the lease market is in a state of transition where trend towards reverse tenancy has become more pronounced. Third, more recent studies show that small and marginal farmers have started leasing-out agricultural land consequent to increase in the cost of production, growing scarcity of water, falling returns and increasing uncertainty due to erratic weather conditions and also because of increasing opportunities of rural non-farm employment. Fourth, there is an anecdotal evidence to suggest that many farmers, including small and marginal ones, who have abandoned farming, are leaving their land fallow in view of restrictive tenancy laws. This tendency is particularly more pronounced in those states where tenancy is legally prohibited. Fifth, share tenancy with input cost sharing continues to be an important mode of leasing-in land, particularly for small and marginal farmers. Though output sharing ratios vary from region to region, most of the studies report 50:50 sharing. In agriculturally backward regions, share tenancy with input cost sharing is more common compared to agriculturally developed regions where fixed cash rent tenancy is more popular. The micro studies also seem to suggest that small and marginal farmers prefer to lease-in land under share tenancy with input cost sharing, because of two reasons; one, lack of resources to pay cash rent in advance under fixed money and second, their inability to bear entire risk of crop failure which has increased in recent times. Further, almost all studies show that the area leased-in for growing non-foodgrain crops is under fixed money. Sixth, studies examining the effect of tenancy on the inputs use and agricultural productivity have thrown up mixed results. While some studies find use of low amount of inputs and low level of yields on the leased-in plots compared to owned plots, the findings of others are just contrary to these. An exhaustive survey of literature, however, suggests that there is no conclusive evidence to support the hypothesis that yields under share tenancy are lower than under owner farming or fixed rent leasehold tenancy.[7] Likewise, there is also no conclusive evidence to suggest that yield levels for households involved in interlocking of factor markets are lower than their counterparts not involved in such arrangements. Seventh, findings of micro studies broadly support the hypothesis that households of different size categories participate in the lease market to utilise their indivisible and non-tradable inputs and capital resources like family labour, bullock labour and machinery more optimally.[8] However, more recent studies have also reported a variety of other important factors like absenteeism of landowners who are engaged in non-farm activities, inferior quality of land, land not suitably located, escalation in the cost of production, growing uncertainty in crop production, etc. that oblige landowners to lease-out land. Eighth, micro studies from the states/regions like Bihar which were notorious for exploitative semi-feudal agrarian relations have reported a significant transformation in these relations. In this context, a study conducted in six villages of north Bihar in 1997 concludes 'the study challenges the prevailing notion of North Bihar stagnant agriculture strapped by feudal production relations. This notion was perhaps valid a decade ago but seems no longer true. The agrarian scene in the region seems in the throes of massive transformation' (Shah and Ballabh: 1997, p. A. 189). The qualitative changes in some aspects of agrarian relations in Bihar were also reported by a more recent study (Jha: 2004). Ninth, there is also enough evidence to indicate that today's lessors in large parts of the country are different from their predecessors in the fifties, the sixties and the seventies when many amongst them did not take any interest in agricultural operations and were leasing-out land to keep control over tenants through semi-feudal exploitation in a variety of ways. Today, a majority of the lessors take keen interest in agriculture in terms of input cost sharing and assume greater risk and uncertainty.

III. Legalization of Tenancy

A comprehensive review of the evidence thrown up by the NSS data emanating from different rounds of landholding survey including the more recent round in 2003 and an exhaustive survey of micro studies available for different regions for the eighties, the nineties and for more recent years show that tenancy relations in different parts of country have undergone significant qualitative changes. These studies show that tenancy relations in their diverse manifestations today are qualitatively different as compared to those prevailing during the fifties, the sixties and the seventies when the existing tenancy legislations were enacted. Against this background, there is a strong case for amending tenancy legislations to legalize lease market in the interest of different categories of farmers including landless, marginal and small households, agricultural production and more efficient use of scarce land resources.

In the ongoing debate on regulating tenancy relations, some scholars suggest the recording of tenants on the pattern of "Operation Barga" (OB) in West Bengal as a viable option to provide tenurial security and protect the interests of the poor tenants. However, even though recording of share tenants under OB in West Bengal conferred tenurial security on the tenants, there is no unanimity on its effect on agricultural production. There are studies to show that OB did have a depressing effect on productivity because *bargardars* lacked entrepreneurial skills and supporting facilities and that security of tenure that follows from OB does not necessarily work towards increasing investment on land. Likewise, it has been pointed out that the notion that recording of tenancy rights will enable tenants to draw most of the credit from the institutional sources also seems to have been belied (Haque: 2001). Further, studies have shown that OB has affected the creation of future tenancies and that the creation of share tenancies in the aftermath of the implementation of OB has declined significantly. The tenancies are now created for a short period that is for boro (spring summer crop) paddy (Pal: 2001; Place and Otsuka: 2001). In view of this, OB does not offer a permanent solution to the problem which lies in conferring the ownership rights on the land cultivated by the share tenants. It has also been suggested that entry to lease market should be restricted to petty peasants and medium and large farmers should be banned from leasing-in land which also does not appear to be practical solution.

We would, however, argue that even if it is possible to ban tenancy completely by implementing existing legislations, this may not be a desirable option in the existing socio-economic milieu. In a growing agrarian economy like India factors like increase in population coupled with lack of alternative employment opportunities will always create strong demand for land, especially from the land-poor households. In such an economy, there will always be a continuous increase in the proportion of population who would switch over to non-farm activities and migrate to urban areas. Such households would be willing to lease-out land because of their inability to cultivate it themselves. Therefore, a complete ban on tenancy would push it underground and households in such situations would either prefer to keep their land fallow or lease-out to their kith and kin. This would hardly serve the land-poor households and would also have deleterious effect on agricultural productivity. Likewise, in agriculturally developed regions, there would always be a demand for leased-in land from medium and large farmers to expand their scale of production and utilize their lumpy capital inputs optimally.

In view of above, the leasing-in and leasing-out land should be legalized with adequate safeguards to protect the interests of the small and marginal farmers. The legalization of tenancy does not mean abrogation of the existing tenancy legislations. In today's context, it means suitably amending these laws allowing leasing-in and leasing-out land incorporating some provisions like fixing the tenure of lease, recording of lease and allowing landowners to resume land for self-cultivation after the expiry of lease. In fact, the need of the hour is to separate the ownership rights from the use rights of land. While ownership rights should be protected and made secure and non-alienable, the use rights should be allowed to be traded freely in the lease market. Further adequate safeguard should be provided to protect the lessees and lessors of lower categories. For example, under no circumstances, the lessees, particularly medium and large ones, should be allowed to buy the leased-in land from lessors belonging to lower categories. This will remove the fear of small and marginal farmers being wiped out and their land increasingly getting alienated and transferred to medium and large farmers. The legalization of tenancy shall allow all sections of rural population to participate in the lease market depending upon their resource endowment. For instance, in agriculturally developed states like Punjab and Haryana, the small and marginal farmers may be tempted to lease-out and medium and large farmers to lease-in land. The studies have shown that large and medium farmers who had leased-in land from small and marginal farmers had been able to invest in modern inputs, reap scale economies and raise their level of productivity. In a similar vein, the marginal and small farmers who had leased out their land also gain in terms of their occupational mobility and realizing higher annual income. There is an evidence to suggest it has proved to be win-win situation for both the parties. In other states like Bihar and Orissa, where wages are low and adequate employment opportunities are not around, the small and marginal farmers shall be in a position to enlarge their holding size and thus afford a reasonable level of living with all attendant benefits of tenancy like borrowing from financial institutional agencies. The medium and large farmers in these states are likely to migrate to urban areas to take non-farm employment opportunities without any risk of losing their land. Such households over the years when their livelihoods become more secure in the non-farm sector are expected to sell their land. The legalization of tenancy shall also give rise to long-term tenancy contracts which have important positive implications towards improving agricultural productivity.

In conclusion, a well-functioning lease market can contribute towards broad-based rural development in several ways like facilitating land transfer from less productive to more productive uses, encouraging rural households to take up non-farm jobs without the risk of losing their land and increasing investment incentives in that those who make such investment can enjoy the benefits even if they are not in a position to use the land personally. In today's context, legalization of tenancy is a better option not only to the Market-Friendly Land Reforms promoted by the World Bank (Deininger: 2003) but also to Confiscatory Land Reforms advocated by the so-called neo-populist school (Griffin, Rahman and Ickowitz: 2002).

Notes

The present paper extensively draws on the earlier papers published by the author on different aspects of tenancy relations:

1. See Appu, 1973, 1996 and Planning Commission, 1973.
2. The changes in different aspects of tenancy relations from the early fifties till the early eighties have been discussed in Sharma, 1992a, 1992b, 1994, 1995 and 1997. The discussion on more recent changes is available in Sharma, 2000a, 2000b, 2006b, 2008, 2009 and 2010 and Chadha, Sen and Sharma, 2004.
3. The quality of data on tenancy thrown up by two main sources viz. NSS reports on landholdings and agricultural census has been widely discussed and commented upon in the literature on tenancy studies. There is a near unanimity among scholars that the data from the former source is more reliable compared to the latter source. Likewise, the design and concepts underlying data on tenancy emanating from NSS reports on landholdings of various rounds and their temporal comparability have also been discussed and commented upon. For details, see Chadha and Sharma, 1991; Sharma, 1992, 1995 and Chadha, Sen and Sharma, 2004, Chapter 3.
4. The clubbing of different tenancy contracts into four broad groups and their implications have been explained in Chadha and Sharma, 1991; Sharma, 1995 and Chadha, Sen and Sharma, 2004, Chapter 3.
5. The findings of a number of more recent micro studies on different aspects of tenancy relations from different regions of the country have been summarised in Sharma, 2005 and 2006.
6. A more recent study of three villages in Andhra Pradesh has reported almost all of these characteristics of tenancy. The incidence of tenancy in the village

selected from agriculturally developed region was as high as 67 per cent of total area of operational holdings cultivated by sample households. The proportion of leased-in land was also higher in case of cash crops as compared to cereal crops. For details, see V.K. Ramachandran, *et al.* 2010.

7. Some of the earlier studies that have examined the effect of tenancy on the use of inputs and output include Vyas, 1970; Rao, 1971; Rudra and Chakravarty, 1973; Shaban, 1987. The more recent ones include Bhaumik, 1993; Sharma, *et al.*, 1995; Chattopadhyay and Sengupta, 2001 and Awasthi, 2005. For an exhaustive review of the earlier studies for India and a number of other countries, see Otsuka, *et al.*, 1992.
8. A number of micro and macro studies in the seventies and the eighties and also more recent ones that have examined the effect of different factors on the incidence of tenancy have reported indivisible and non-tradable inputs like family labour, bullock labour, machinery (tractors, bore-well, etc.) and new agricultural technology as important determinants of the incidence of tenancy. For details see, Bardhan, 1976; Laxminaryan Tyagi, 1977a, 1977b; Bilss and Stern, 1982; Singh, 1989; Kalim, 1999 and Sharma, 2000a, 2000b and 2010.

References

Awasthi, Maya Kant (2005): 'Interspatial Total Factor Productivity Analysis of Alternative Land Lease Arrangements in Bundelkhand Region of Uttar Pradesh', *Indian Journal of Agricultural Economics*, Vol. 60, No. 3.

Appu, P.S. (1975): 'Tenancy Reforms in India', *Economic and Political Weekly*, Vol. X, Nos. 3, 4 and 5.

Appu, P.S. (1996): *Land Reforms in India: Survey of Policy, Legislation and Implementation*, Vikas Publishing House Pvt. Ltd., New Delhi.

Bardhan Pranab, K. (1976): 'Variations in Extent and Forms of Agricultural Tenancy: Analysis of Indian Data Across Regions and Overtime', *Economic and Political Weekly*, Vol. XI, Nos. 37-38.

Bhaumik, S.K. (1993): Tenancy Relations and Agrarian Development: A Study of West Bengal, Sage Publications, New Delhi.

Bliss, C.T. and N.H. Stern (1982): *Palanpur: The Economy of an Indian Village*, Clarendon Press, London.

Chadha, G.K. and H.R. Sharma (1991): 'Understanding Agrarian Relations in India: Some Comments on the Quality of Data', in *Poverty in India: Data Base Issues*, G.V.S.N. Murty and G.K. Kadekodi (Eds.) Vikas, New Delhi.

Chadha, G.K., Sucharita Sen and H.R. Sharma (2004): *State of the Indian Farmer—A Millennium Study: Land Resources*, Academic Foundation, New Delhi.

Deininger, Klaus (2003): *Land Policies for Growth and Poverty Reduction*, Oxford University Press, New Delhi.

Griffin, Keith, Azizur Rahaman Khan and Amy Ickowitz (2002): 'Poverty and Distribution of Land', *Journal of Agrarian Change*, Vol. 2, No. 3.

Haque, T. (2001): *Impact of Tenancy Reforms on Productivity Improvement and Socio-Economic Status of Tenants*, National Centre for Agricultural Economics and Policy Research, New Delhi.

Jha, Parveen (2004): 'Continuity and Change: Some Observations on the Landscape of Agricultural Labourers in North Bihar, India', *Journal of Agrarian Change*, Vol. 4, No. 4.

Laxminarayan, L. and S.S. Tyagi (1977a), 'Tenancy: Extent and Interstate Variations', *Economic and Political Weekly*, Vol. XII, No. 22.

Laxminarayan, L. and S.S. Tyagi (1977b), 'Interstate Variations in Types of Tenancy', *Economic and Political Weekly*, Vol. XII, No. 39.

Otsuka, Keijiro, Hiroyuki Chuma and Yujiro Hayami (1992): 'Land and Labour Contracts in Agrarian Economies', *Journal of Economic Literature*, Vol. XXX, December.

Pal, S.S. (2001): *Impact of Tenancy Reform on Production and Income Distribution: A case study of Operation Barga in West Bengal*, Policy Paper No. 1, National Centre for Policy Research in Agricultural Economics, New Delhi.

Place, F. and Otsuka, K. (2001): "Tenure Agricultural Investment and Productivity in the Customary Tenure Sector of Malawi", *Economic Development and Cultural Change*, Vol. 50, No. 1.

Planning Commission (1973): *Report of the Task Force on Agrarian Relations*, Government of India, New Delhi.

Ramachandran, V.K., Vikas Rawal and Madhura Swaminathan Eds. (2010): *Socio-Economic Surveys of Three Villages in Andhra Pradesh: A Study of Agrarian Relations*, Tulika Books, New Delhi.

Shah, Tushar and Vishwa Ballabh (1997) 'Water Markets in North Bihar: Six Village Studies in Muzzaffarpur District', *Economic and Political Weekly*, Vol. XXX, No. 52.

Sharma, H.R. (2010): 'Magnitude, Structure and Determinants of Tenancy in Rural India: A State Level Analysis', *Indian Journal of Agricultural Economics*, Vol. 65, No. 1, pp. 80-100.

Sharma, H.R. (2009): 'Changing Tenancy Relations in Rural India', in *Agrarian Reforms, Land Markets and Rural Poor,* D. Narasimha Reddy (ed), Concept Publishing Company, New Delhi.

Sharma, H.R. (2008): 'Tenancy Relations in India: Legislation, Grass-roots Realties and Policy Options', in *Contract Farming and Tenancy Reforms*, R.S. Deshpande (ed), Concept Publishing Company, New Delhi.

Sharma, H.R. (2007): 'Land Distribution and Tenancy among Different Social Groups', *Economic and Political Weekly*, Vol. 42, No. 41, pp. 4183-4185.

Sharma, H.R. (2006a), 'Liberalising the Lease Market', *Economic and Political Weekly*, Vol. XLI, No. 8, pp. 696-698.

Sharma, H.R. (2006b), Summary of Group Discussion on "Changing Agrarian

Relationships in Rural Areas", *Indian Journal of Agricultural Economics*, Vol. 61, No. 1, pp. 81-84.

Sharma, H.R. (2005): Rapporteur's Report on "Changing Agrarian Relationships in Rural Areas", *Indian Journal of Agricultural Economics*, Vol. 60, No. 3, pp. 561-570.

Sharma, H.R. (2000a), 'Tenancy Relations in Rural India: A Temporal and Cross-Sectional Analysis', *Indian Journal of Agricultural Economics*, Vol. 55, No. 3, pp. 210-222.

Sharma, H.R. (2000b), 'Agrarian Structure and Agricultural Development in Rural India: Emerging Trends and Patterns', *Man and Development*, Vol. XXII, No. 2, pp. 22-38.

Sharma, H.R. (1997): 'Land Reforms: Status and Opportunities', in *Agricultural Paradigm for the Ninth Five Year Plan Under New Economic Environment,* Bhupat M. Desai (ed.), Oxford and IBH, New Delhi, pp. 712-739.

Sharma, H.R., Parkash Mehta and Khetrmohan Mohapatra (1995): 'Tenancy Relations and Productive/Allocative Efficiency: A Study in Orissa Agriculture', *Indian Journal of Agricultural Economics,* Vol. L, No. 4, pp. 597-608.

Sharma, H.R. (1994): 'Distribution of Landholdings in Rural India, 1953-54 to 1981-82: Implications for Land Reforms', *Economic and Political Weekly,* Vol. XXIV, No. 13, pp. A12-A25.

Sharma, H.R. (1992): 'Evolution of Agrarian Relations in India', *Journal of Indian School of Political Economy,* Vol. IV, No. 1, pp. 80-105.

Sharma, H.R. (1992): 'Agrarian Relations in India Since Independence', *Journal of Indian School of Political Economy*, Vol. IV, No. 2, pp. 201-262.

Siddiqui, Kalim (1999): 'New Technology and Process of Differentiation: Two Sugarcane Cultivating Villages in UP', *Economic and Political Weekly*, Vol. XXXIV, No. 52.

Singh, Iqbal (1989): 'Reverse Tenancy in Punjab Agriculture: Impact of Technological Change'. *Economic and Political Weekly*, Vol. XXIV, No. 25.

Appendix

Table A-11.1: Incidence of Tenancy in Rural India, 1981-82 to 2002-03: Major States (Per cent)

State	Year	EOH	MH	ELIH	NONLIH	OW A	LIA	NONLIA
Andhra Pradesh	1981–82	83.45	15.27	1.04	0.23	92.98	6.23	0.79
	1991–92	79.69	13.32	2.43	4.56	88.49	9.57	1.94
	2002–03	85.10	7.85	5.18	1.85	89.17	9.97	0.85
Assam	1981–82	83.93	10.77	3.50	1.80	90.99	6.35	2.66
	1991–92	79.88	5.07	7.35	7.70	85.84	8.87	5.29
	2002–03	87.84	5.05	3.05	4.05	93.14	5.06	1.80
Bihar	1981–82	78.03	20.04	1.01	0.56	88.68	10.27	1.05
	1991–92	86.02	1.90	5.56	6.52	92.81	3.91	3.28
	2002–03	86.88	8.86	3.93	0.32	90.97	8.90	0.13
Gujarat	1981–82	94.61	3.80	1.34	0.25	97.8	1.95	0.25
	1991–92	92.83	2.36	2.35	2.46	92.7	3.34	3.96
	2002–03	94.29	3.16	1.81	0.75	94.39	5.08	0.53
Haryana	1981–82	72.29	25.51	2.20	–	80.31	18.22	1.65
	1991–92	78.96	12.41	4.83	3.80	66.07	33.74	0.19
	2002–03	88.15	9.95	0.95	0	84.97	14.38	0.65
Himachal Pradesh	1981–82	87.24	12.27	0.49	–	94.6	3.20	2.20
	1991–92	84.08	3.80	5.01	0.71	92.18	4.83	2.99
	2002–03	93.24	4.05	2.30	0.40	96.2	2.87	0.92
Jammu & Kashmir	1981–82	88.16	11.73	0.11	–	95.19	2.37	2.44
	1991–92	91.2	5.35	0.52	2.93	94.44	3.73	1.83
	2002–03	97.88	1.35	0.60	0.17	99.27	0.32	0.41
Karnataka	1981–82	83.14	14.57	1.78	0.51	92.12	6.04	1.84
	1991–92	80.35	6.14	3.73	9.78	84.53	7.43	8.04
	2002–03	94.33	3.01	2.20	0.46	95.61	3.68	0.71
Kerala	1981–82	90.89	4.87	2.92	1.32	95.52	2.05	2.43
	1991–92	90.99	2.54	3.21	3.26	95.37	2.88	1.75
	2002–03	93.96	3.35	1.24	1.45	94.95	4.18	0.87
Madhya Pradesh	1981–82	81.91	14.19	1.13	2.77	92.65	3.56	3.79
	1991–92	80.21	5.06	6.48	8.25	86.53	6.30	7.17
	2002–03	90.91	5.71	1.54	1.84	95.05	3.42	1.53

Maharashtra	1981–82	87.15	9.29	2.69	0.87	94.21	5.20	0.59
	1991–92	89.22	5.53	2.40	2.85	90.50	5.48	4.02
	2002–03	90.71	4.71	2.30	2.28	94.10	4.59	1.31
Orissa	1981–82	74.27	24.06	1.49	0.18	83.97	9.92	6.12
	1991–92	70.46	17.69	1.70	10.15	84.41	9.48	6.11
	2002–03	77.60	12.00	8.40	2.00	83.78	13.15	3.07
Punjab	1981–82	76.19	21.71	1.64	0.46	82.21	16.07	1.72
	1991–92	80.27	15.6	1.43	2.70	80.82	18.83	0.35
	2002–03	86.50	10.30	2.40	0.80	82.12	17.84	0.04
Rajasthan	1981–82	89.53	8.51	1.52	0.44	94.47	4.31	1.22
	1991–92	90.32	6.22	1.55	1.91	92.41	5.19	2.40
	2002–03	96.75	2.11	0.81	0.35	96.97	2.81	0.21
Tamil Nadu	1981–82	70.45	19.70	8.56	1.29	87.37	10.92	1.71
	1991–92	74.56	11.02	5.82	8.60	84.91	10.89	4.20
	2002–03	89.41	5.78	3.71	1.10	93.26	6.10	0.63
Uttar Pradesh	1981–82	78.53	20.61	0.82	0.04	88.05	10.24	1.70
	1991–92	80.84	14.47	2.64	2.05	88.45	10.49	1.06
	2002–03	87.51	9.42	2.58	0.48	89.95	9.60	0.44
West Bengal	1981–82	71.58	19.37	7.02	2.03	86.18	12.34	1.48
	1991–92	75.40	12.00	5.27	7.33	85.74	10.40	3.86
	2002–03	83.59	9.00	5.10	2.31	89.78	9.42	0.79
All-India	1981–82	80.58	16.24	2.37	0.81	91.08	7.18	1.74
	1991–92	81.98	8.87	3.85	5.30	87.91	8.28	3.81
	2002–03	88.6	6.99	3.06	1.35	92.57	6.60	0.84

Note: EOH = Entirely Owned Holdings; MH = Mixed Holdings; ELIH = Entirely Leased-in Holdings; NONLIH = Neither Owned Nor Leased-in Holdings; OWA = Owned Area; LIA= Leased-in Area; NONLIA = Neither Owned Nor Leased-in Area.

Source:
i. Report on Landholdings (2); 37th Round 1982, NSS Report No. 331.
ii. Report on Some Aspects of Operational Holdings; 48th Round 1991-92, Report No. 407.
iii. Report on Some Aspects of Operational Holdings, 2002-03, 59th Round, NSS Report No. 492.

Table A-11.2: Changing Proportions of Households not Operating Land, Owning but not Operating and Neither Owning nor Operating Land in Rural India, 1981-82 to 2002-03, Major States

State	Households not operating land			Owning but not operating			Neither owning nor operating		
	1981-82	1991-92	2002-03	1981-82	1991-92	2002-03	1981-82	1991-92	2002-03
Andhra Pradesh	40.47	37.49	57.06	37.22	31.84	43.93	3.25	5.65	13.13
Assam	13.27	14.01	18.08	11.15	7.89	14.97	2.12	6.12	3.11
Bihar	18.17	18.62	28.74	15.38	10.99	22.57	2.79	7.63	6.17
Gujarat	36.14	25.11	38.22	25.22	15.31	26.06	10.92	9.80	12.16
Haryana	34.57	19.05	32.34	29.64	16.18	23.10	4.53	2.87	9.24
Himachal Pradesh	17.61	12.22	20.90	16.53	7.83	7.14	1.08	4.39	13.76
Jammu & Kashmir	11.77	6.39	7.65	5.03	5.24	4.82	6.74	1.15	2.83
Karnataka	27.02	19.04	38.56	14.30	13.71	25.06	12.72	5.33	13.50
Kerala	12.99	6.55	37.28	9.09	3.70	34.10	3.90	2.85	3.18
Madhya Pradesh	24.06	21.88	27.26	14.80	11.46	16.58	9.26	10.42	10.68
Maharashtra	41.25	32.00	43.68	29.86	22.51	27.78	11.39	9.49	15.90
Orissa	27.78	27.20	33.69	24.60	21.97	26.75	3.18	5.23	6.94
Punjab	22.56	25.62	31.42	19.48	21.73	27.94	3.08	3.89	3.48
Rajasthan	12.29	12.54	17.74	7.21	9.98	12.57	5.08	2.56	5.17
Tamil Nadu	37.26	36.66	59.23	33.80	27.92	44.13	3.46	8.74	15.10
Uttar Pradesh	20.00	11.75	17.42	17.67	9.27	14.13	2.33	2.48	3.29
West Bengal	22.14	15.31	32.04	17.71	8.68	27.86	4.43	6.63	4.18
All-India	26.06	21.84	33.97	21.18	15.68	25.30	4.88	6.16	8.67

Sources:

i. Report on Landholdings (1); 37th Round 1982, NSS Report No. 330.
ii. Report on Landholdings (2); 37th Round 1982, NSS Report No. 331.
iii. Report on Landholdings; 37th Round 1982, NSS Report No. 338.
iv. Report on Some Aspects of Ownership Holdings; 48th Round 1991-92, Report No. 399.
v. Report on Livestock and Implements in Household Operational Holdings; 48th Round 1991-92, Report No. 408.
vi. Report on Some Aspects of Operational Holdings, 2002-03, 59th Round, NSS Report No. 492.

Table A-11.3: Reverse Tenancy: The Extent of Participation of Landless, Marginal and Small Households in the Lease Market as Lessees and Lessors in Rural India, Major States, 1971-72 to 2002-03

State	*Year*	*Land leased-in (00 ha)*	*Land leased-out (00 ha)*	*Per cent share in leased-in land*	*Per cent share in leased-out land*
Andhra Pradesh	1971–72	5831	2857	42.93	30.92
	1991–92	6969	2396	82.52	64.17
	2002–03	6128	1607	71.74	70.57
Assam	1971–72	6172	750	95.26	44.67
	1991–92	1804	402	95.04	58.69
	2002–03	1094	35	82.97	53.50
Bihar	1971–72	10393	2190	88.06	44.23
	1991–92	3118	981	95.03	41.50
	2002–03	5967	771	97.38	65.09
Gujarat	1971–72	2286	507	64.94	26.23
	1991–92	2571	940	90.31	38.68
	2002–03	2426	833	70.74	61.93
Haryana	1971–72	3418	230	63.21	14.79
	1991–92	4074	4234	28.76	64.06
	2002–03	1845	692	40.50	37.58
Himachal Pradesh	1991–92	–	–	97.57	65.15
	2002–03	194	126	94.78	91.58
Jammu & Kashmir	1971–72	396	129	66.00	75.88
	1991–92	166	–	96.51	70.71
	2002–03	23	1.52	73.08	93.55
Karnataka	1971–72	10998	2130	65.74	30.33
	1991–92	4187	1459	58.47	19.49
	2002–03	1588	743	66.53	22.53
Kerala	1971–72	884	212	90.11	65.63
	1991–92	262	–	82.13	–
	2002–03	428	57	98.74	63.97
Madhya Pradesh	1971–72	8227	1004	58.81	19.20
	1991–92	9084	1783	75.86	28.94
	2002–03	4350	1055	70.60	36.85
Maharashtra	1971–72	5563	1067	28.55	16.10
	1991–92	4734	1638	59.56	24.49
	2002–03	3301	1050	55.74	37.62
Orissa	1971–72	4731	1845	82.29	59.07
	1991–92	4484	1736	91.77	75.02
	2002–03	4560	1383	97.62	62.89
Punjab	1971–72	5228	693	64.41	16.56
	1991–92	1767	810	42.41	32.36
	2002–03	2337	604	48.02	25.89

(*Contd.*)

(Contd.)

State	Year	Land leased-in (00 ha)	Land leased-out (00 ha)	Per cent share in leased-in land	Per cent share in leased-out land
Rajasthan	1971-72	4324	428	59.16	7.81
	1991-92	5147	709	57:37	10.99
	2002-03	1609	471	32.33	23.16
Tamil Nadu	1971-72	5494	3032	91.31	25.15
	1991-92	3900	1745	86.91	64.57
	2002-03	1819	773	66.94	76.15
Uttar Pradesh	1971-72	19351	4358	88.34	46.83
	1991-92	16767	4158	91.37	49.30
	2002-03	11890	4583	81.24	58.12
West Bengal	1971-72	11691	1787	94.58	51.57
	1991-92	5029	1145	97.20	60.45
	2002-03	3517	643	94.40	63.97
All-India	1971-72	9874	2521	67.03	34.55
	1991-92	74772	19497	71.27	32.44
	2002-03	53638	15637	71.00	47.73

Sources:

(i) Report on Some Aspects of Landholdings: 26th Round, 1971-72, NSS Report No. 215.

(ii) Report on Some Aspects of Household Ownership Holdings (1): 48th Round, 1991-92, NSS Report No. 399.

(iii) Report on Household Ownership Holdings in India: 59th Round, 2003, NSS Report No. 491.

Table A-11.4: Distribution of Leased-in Area by Period of Lease in Rural India, 1991-92 to 2002-03, Major States (Per cent)

State	Year	Less than one agricul-tural season	One agricultural season but less than one agricultural year	One to less than two agricultural years	Two to less than five years	Five to less than twelve years	Twelve and more than twelve years	Not recorded	Total
Andhra Pradesh	1991-92	2.31	25.50	17.27	28.8	8.42	10.04	7.66	100
	2002-03	6.01	13.05	23.76	22.27	29.81	5.04	0.05	100
Assam	1991-92	0.84	29.05	15.26	14.49	1.93	6.33	32.01	100
	2002-03	5.97	19.79	23.24	15.93	9.70	5.44	19.93	100
Bihar	1991-92	3.64	9.30	34.26	17.82	7.76	7.55	19.40	100
	2002-03	2.13	16.51	45.78	19.78	9.27	5.09	1.44	100
Gujarat	1991-92	3.54	20.33	20.26	15.29	1.17	15.23	24.17	100
	2002-03	1.25	9.11	10.56	30.75	0.01	48.32	0.00	100
Haryana	1991-92	10.24	15.43	59.51	0.77	0.04	0.72	13.29	100
	2002-03	6.38	16.42	49.24	24.27	3.27	0.42	0.00	100
Himachal	1991-92	1.18	5.60	6.47	12.25	13.14	48.32	13.03	100
Pradesh	2002-03	2.02	10.71	8.38	22.22	24.28	32.38	0.00	100
Jammu &	1991-92	7.69	0.00	9.16	13.86	4.98	48.58	15.72	100
Kashmir	2002-03	3.31	25.32	16.17	50.00	0.24	4.98	0.00	100
Karnataka	1991-92	9.72	25.88	8.92	16.11	5.22	4.45	29.69	100
	2002-03	7.21	13.32	35.47	32.94	10.60	0.47	0.00	100
Kerala	1991-92	6.34	10.75	2.90	16.50	24.26	20.84	18.41	100
	2002-03	1.85	9.70	15.84	34.14	24.25	14.22	0.00	100
Madhya Pradesh	1991-92	0.95	14.81	25.09	8.77	7.03	20.56	22.78	100
	2002-03	10.04	23.85	29.48	14.45	11.52	9.01	1.65	100

(Contd.)

(Contd.)

State	Year	Less than one agricult-ural season	One agricultural season but less than one agricultural year	One to less than two agricultural years	Two to less than five years	Five to less than twelve years	Twelve and more than twelve years	Not recorded	Total
Maharashtra	1991-92	9.75	11.12	25.17	19.63	10.90	10.34	13.10	100
	2002-03	4.35	29.61	21.49	13.54	19.21	11.44	0.35	100
Orissa	1991-92	1.49	30.88	26.31	14.37	11.32	4.79	10.33	100
	2002-03	1.58	19.69	33.68	27.39	8.24	8.61	0.82	100
Punjab	1991-92	6.96	12.12	41.53	11.25	12.92	4.62	10.60	100
	2002-03	4.60	13.65	48.84	9.52	19.80	3.59	0.00	100
Rajasthan	1991-92	3.16	15.00	29.25	26.94	10.22	9.29	6.15	100
	2002-03	2.82	13.63	40.01	37.59	3.44	2.50	0.00	100
Tamil Nadu	1991-92	3.45	6.76	17.30	23.33	9.34	21.56	18.36	100
	2002-03	3.50	3.47	18.43	22.38	25.74	26.48	0.00	100
Uttar Pradesh	1991-92	2.64	15.08	38.83	11.81	10.48	9.43	11.80	100
	2002-03	3.52	23.65	36.76	17.57	7.55	7.66	3.29	100
West Bengal	1991-92	3.22	8.45	10.67	18.58	12.67	35.81	10.60	100
	2002-03	6.85	15.58	24.14	16.12	13.53	20.58	3.21	100
All-India	1991-92	4.82	16.44	30.67	14.64	7.81	10.87	14.72	100
	2002-03	4.62	17.96	32.91	20.67	12.82	9.52	1.49	100

Sources:

i. Report on Some Aspects of Operational Holdings; 48th Round 1991-92, Report No. 407.

ii. Report on Some Aspects of Operational Holdings, 2002-03, 59th Round, NSS Report No. 492.

12

Land Leasing, Poverty and Inequality

Naresh K. Sharma

Introduction

There has been a lovely debate on the issue of agricultural tenancy. Two prominent issues have been (1) the relationship between agricultural tenancy and productive efficiency; and (2) the distributive effects of tenancy. Another relevant issue concerns the relationship between the institution of agricultural tenancy and poverty. This issue has much practical relevance, since land reforms are often aimed at reducing rural poverty.

It is useful to distinguish between two broad types of land reforms: (1) Some reforms are aimed at changing ownership patterns, based on land redistribution (such reforms may be called *redistribution reforms*). In India, these have normally involved putting a ceiling on the permissible landholdings for households of different sizes (subsequent to abolition of intermediaries such as *zamindars* in earlier period). (2) Some reforms impose restrictions on leasing activities (these may be called *tenancy reforms*). These restrictions can be in the nature of controls or limits on the level of rents, the length of leases, the transferability and inheritability of tenancies, and the resumption of land by lessors. An extreme form of the last type of restriction is to prevent the resumption of lease altogether: Table 12.1 summarises the legal provisions regarding leasing-out of agricultural land for 15 major states of India. The relevant section from the U.P. Zamindari Abolition and Land Reforms Act, 1950 is reproduced in the Appendix.

Tenancy reform can also take the form of not allowing tenancies *at all.* This is often what is understood as the basis of "land to the tiller" programmes. As soon as land is leased-out, it is supposed to pass from

Table 12.1: Legal Restrictions on Leases

State		Restriction
I.	**In this category leasing is totally prohibited**	
1.	Kerala	Leasing prohibited
2.		
II.	**In this category a landholder is prohibited from leasing-out land [in his/her holding] except when he/she is:**	
3.	A.P.	Disabled, Armed Forces personnel, and those landowners who own not more than three times a "family holding" may lease out.
4.	Bihar	Disabled, or in armed Forces, or Govt. servants earning up to Rs. 250 per month.
5.	Karnataka	A soldier or a seaman.
6.	M.P.	Disabled, Armed Forces personnel, or those imprisoned. Others may lease out for one year in any three years.
7.	U.P.	Disabled, Armed Forces personnel, those imprisoned or *bona fide* students.
8.		
III.	**In this category leasing is permitted but the tenant acquires a right to purchase land as provided below:**	
9.	Assam	An ordinary tenant acquires right of occupancy after three years continuous possession and an occupancy tenant has a right to purchase leased land.
10.	Gujarat	Every tenant has a right to purchase leased land within one year of tenancy.
11.	Haryana	Tenant acquires right to purchase leased land after six years of continuous occupation.
12.	Maharashtra	Every tenant has a right to purchase leased land within one year of tenancy.
13.	Punjab	Tenant acquires right to purchase leased land after six years of continuous occupation.
14.		
IV.	**In this category no restrictions as who can lease out**	
15.	A.P. (AA)	No restrictions.
16.	Orissa	No restrictions.
17.	Rajasthan	No restrictions.
18.	Tamil Nadu	No restrictions.
19.	West Bengal	No restrictions but only sharecropping (bargadari) leases are permitted.

Source: Sharma (1992)

Notes:

1. Abbreviations used in this table are: A.P.=Andhra Pradesh; TA=Telangana Area; AA=Andhra Area; M.P.=Madhya Pradesh; U.P.=Uttar Pradesh.

Table 12.1 (*Contd.*)

2. Only those restrictions on leases are included here that concern rights to lease-in or lease-out and any restrictions on quantities of land that may be leased. Legal provisions about regulation of leases (e.g. rent regulation, period of leases etc.) are not considered. There is only one type of restriction on amount that a person may be permitted to lease-in, viz. that the operational holding of the lessee should not exceed the ceiling limits. All states except Madhya Pradesh, Uttar Pradesh and West Bengal make such explicit provision. In west Bengal, a *bargardar* is not allowed to cultivate more than 4 hectares of land including her/his own land.
3. Most of the States make a provision that a lease is not to be entered into if it results in a 'fragment". A "fragment" is defined to be smallest piece of land parcelling of land below which is considered counter-productive. The States that do not make explicit provision on this include: Andhra Area of Andhra Pradesh, Bihar, Madhya Pradesh, Tamil Nadu, Uttar Pradesh and West Bengal.

the hands of the lessor to the hand of the lessee – the actual tiller of the land.

This chapter focuses on the tenancy reforms of a particular type, i.e. those which prohibit leasing of land. The impact of such tenancy reforms on poverty is analysed with reference to Palanpur, a village in Western Uttar Pradesh. A more detailed analysis of tenancy in this village has been presented elsewhere (Sharma: 1994). The discussion will be concentrated on sharecropping, which is reported widely among all land lease contracts.

A brief discussion of tenancy in Palanpur is presented in section 2 below. The focus is on productive efficiency related issues. In section 3, we discuss how tenancy affects poverty in the village. In section 4, we take up the issue of land reforms. In particular, the popular slogan "land to the tiller" in its literal sense is scrutinised, with particular attention to its implications for inequality.[2] Consequences of restriction of leasing of land on poverty and inequality are analysed in section 5. Some concluding comments are offered in section 6. But to begin with, the next section provides a short introduction to the study village.

Background of the Village

Much has been written about Palanpur.[3] The village is situated in Moradabad district of western Uttar Pradesh. The village comprised of 143 households and 960 persons.[4] Educationally the village is backward.

The village is a multi-caste village and exhibits a diversified occupational structure – including a number of adult men engaged in various occupations outside the village, ranging from secure employment to daily wage labour in nearby town grain market. But agriculture remains the most important occupation and also the most important source of income. The population of the different castes and their main occupations are given in Table 12.2, which also gives landownership details. Table 12.3 provides distribution of ownership and operational holdings in the study village.

Palanpur belongs to the "green revolution" belt of Uttar Pradesh. *Zamindari* abolition has been effectively carried out in this village in early fifties. The erstwhile *zamindars* (three of them owned land in his village) now own a very small portion of the village land. Thus almost all of land is owned by the erstwhile tenants of those three *zamindars* or by those who have acquired land from them since. The part of U.P. Zamindari Abolition and Land Reforms Act, 1950 concerning leasing of

Table 12.2: Population, Land Owned and Occupation Structure of Village Palanpur

Caste	*No. of Households*	*Population*	*Land Owned (bighas)*	*Caste Occupation*	*Current Occupation*
1. Thakur	30	217	762	Landlords, Warriors	Cultivation Outside jobs
2. Murao	27	217	1080	Gardeners	Cultivation
3. Dhimar	13	74	65	Water carriers	Cultivation Outside jobs
4. Gadaria	12	83	195	Shepherds	Cultivation Outside jobs
5. Dhobi	4	27	21	Washermen	Cultivation Wage labour
6. Teli	16	92	98	Oil-pressers	Cultivation Wage labour
7. Passi	15	85	141	Mat-weavers	Outside jobs (immigrated)
8. Jatab	19	118	218	Leather workers	Cultivation Wage labour
9. Others	7	47	17	Miscellaneous	Miscellaneous
TOTAL	143	960	2597		

Note: This list follows a rough hierarchical caste order except for the last category of 'others'. Lower the caste code in this list is, the higher is caste status.

Table 12.3: Distribution of Ownership and Operational Holdings by Size-Class

Size-Class in bighas	*Number of Ownership Holdings*	*Number of Operational Holdings*
0	27	46
0.1 – 5	19	13
5.1 – 15	35	18
15.1 – 30	36	27
30.1 – 50	15	30
Above 50	11	9
Total	143	143
Maximum Holding	99	122
Minimum Holding	1	1
Average Holding	18	19
Gini Coefficient	0.59	0.60

land is observed most in violation – namely, the institution of tenancy is very much alive even though forbidden by law barring few exceptional cases.[5]

Of 143 households in the village, 27 households are landless[6] and the rest own between themselves 2600 bighas.[7] The distribution of landholdings by size class is given in Table 12.2. Distribution of landholdings slows inequality both in ownership holding pattern as well as operational holding pattern (Gini coefficient for either is around 0.6). But *there are no very big landowners.* The largest landowning household owns 99 bighas of land (equal to 7 hectares which is within ceiling limits as per law) and this is also a very big household consisting of 21 persons. Nearly all land is irrigated and much of it yields to crops in a year.

Incidence of tenancy in the village is widespread. Leased area consists of 28 per cent of all cultivated land.[8] Incidence of tenancy and its importance are seen to be even greater in terms of households involved in tenancy contracts. Nearly 75 per cent of all households and 87 per cent of landowning households were so involved. All this leasing activity is informal or oral tenancy.

Tenancy and Productive Efficiency

Importance of agrarian structure for production, growth, inequality of assets and income and poverty can scarcely be overemphasised. In this section we briefly discuss some issues related to tenancy.

First a few words on the problem of rent may not be out of place. Concept of ownership of land is inextricably mixed with this question. It has been held that the concept of private ownership of land in modern sense did not prevail in India till it was made so after onset of the British rule.[9]

Moreland (1929) says of Hindu system of land tenure that there was no concept of land rights. On the other hand, there was concept of duties. The cultivators had a duty to pay a share of produce to the king for the protection provided. Similarly, the king had certain duties and could not claim share of produce as a right from the subjects. At most it can be said that the king had a right of tax on crop in modern sense and there was no case for land rent. The sovereign can certainly not be thought to be the ultimate owner of the land.[10]

It is also widely accepted now that the *zamindars*, *talukdars*, *mansabdars* etc. were at best revenue collectors and not the owners of the land. In some cases they did not have even inheritable rights. However in due course these intermediaries came to acquire rights in land, which were akin to ownership rights. This land tenure system was highly iniquitous and also a barrier to unleashing of productive forces in agriculture. The actual tillers of land, who were generally tenants at will, had neither resources nor incentives to invest in improving the productive powers of the land.

The classical writers of eighteenth and nineteenth centuries held rent to be an unjust appropriation by the landlords. They devoted considerable time to talk about effect of tenancies on improvements in state of agriculture. Competitive theory of rent begins with Jevons and subsequently develops as a full-fledged theory with Marshall. Marshall also raised the issue of current resource use efficiency in a particular type of land tenure viz. sharecropping. Sharecropping was seen as an inferior form of tenancy not only for long-term investment but also as leading to inefficient use of productive resources. This issue (efficiency of resource use in share tenancies) became a source of vigorous debate from late sixties onwards after Cheung's important work. Scholars have been divided on this issue till today. The empirical tests have not been able to settle the issue of productive efficiency in tenancy either See Sharma (1994) for a detailed review of the theoretical debate and empirical evidence.

The evidence so far has not been able to clinch the issue of sharecropping being inferior to other forms of tenancy or owner cultivation in its use of resources and yields obtained. There has been much variations in the results reported from region to region. Many studies including our own have found no significant (statistically) difference either in input use or in productivity on share operated land

and owner cultivated land. Otsuka and Hayami (1988) after weighing the available evidence from a survey of studies, conclude that:

> "on the balance, evidence is stronger to support the hypothesis of equal allocative efficiency between share tenancy and other land tenure forms."

From the above it is ambiguous to claim that tenancy leads to inefficient outcomes. Restrictions on leasing of land cannot be justified on grounds of encouraging productive efficiency.

Tenancy and the Poor

An important question in content of a village economy dominated by agriculture is how the institution of tenancy influences the poor of the society. The first task is to identify the poor households. Anjou and Stern (1998a) have studied problem of identification of poor in Paramour by using all available information including not only quantitative figures like incomes and assets but also qualitative information gathered from discussion with the villagers.

They find that current income is not always the best variable to capture poverty indeed they suggest that the current income is more 'noisy' and less reliable to capture poverty – the current income overestimates poverty for those classes of households that are concentrated below the poverty line (e.g. the upper caste households) and underestimate poverty for those classes of households that are concentrated below poverty line (e.g. the landless). Thus they also used a *measure of long term income, and "Observed Means"* criteria for identifying the poor. "*Observed Means*" criteria basically involved ranking of all households in terms of their prosperity by two researcher – Jean Druze and Marsh Sharma – who had stayed in the village for over a year and had opportunity to closely observe all the village households. On the basis of these studies, they find that (a) the landless agricultural labour are exposed to the highest risk of falling into poverty, followed by (b) those belonging to caste *jatabs* (one of the most deprived social groups), (c) all landless households; and finally (d) those without any fit adult male in the family. We shall consider these groups of households to understand how the operation of the institution of tenancy affects the poor and what, for these vulnerable groups, are the implications of restricting lease of land by law.

Table 12.4: Land Ownership, Income and Leasing of Land

Classifying Criteria	*Area leased out (bighas)*	*Area leased in (bighas)*	*Net Area leased in (bighas)*
Land Ownership Class:			
Landless	0 (0)	71(10)	+71
0.1–5.0	28(4)	136(18)	+108
5.1–15.0	152(22)	198(27)	+46
15.1–30.0	205(30)	233(31)	+28
30.1–50.0	107 (16)	31(4)	–76
Above 50	195 (28)	73(10)	–122
Income Class:			
Up to 500	85 (12.4)	177 (23.8)	+92
501 to 800	174 (25.3)	93 (12.5)	–81
801 to 1200	276 (40.2)	151 (20.4)	–125
1201 to 1600	92 (13.4)	163 (22.0)	+71
1601 to 1800	36 (5.2)	50 (6.7)	+14
Above 1800	24 (3.5)	108 (14.6)	+84
Total (village)	687 (100)	742 (100)	+55

Note: Figures in brackets are column percentages. Land ownership classes are in bighas. Income classes are in per capita income in rupees.

How does tenancy touch these poor households? Tables 12.4, 12.5, 12.6 and 12.7 capture leasing activities for households belonging to different size classes of landholdings, income classes or prosperity classes according to the observed means criteria.

The first striking feature that emerges is that the distribution of these different classes among the lessors is very similar to their distribution among the lessees (Table 12.5 and 12.6). Thus the lowest income class (with per capita income below Rest. 500) contribute 21 per cent lessor households and 22 per cent lessee households or the marginal landowners (owning up to 5 bighas i.e. below 0.4 hectares) constitute 15 per cent of lessors and 13 per cent of the lessees (Table 12.5).

This shows that the land poor, the income poor or the "obviously" poor[11] find opportunities in the land lease market *both in* leasing-in land and also in leasing-out land. The poorest income class leases-in significantly large amounts of land than they lease-out (Table 12.4). On the other hand, the next two classes on the income scale are net lessors – indeed the largest suppliers of land in the lease market. Thus, *leasing-out also is a significant economic opportunity for the poor.*

Why do the poor lease out land? There are number of reasons including the following: temporary sickness of the 'bread winner';

Table 12.5: Land Leased as a Fraction of Land Owned and Land Operated by the Leasing Households

Classifying Criteria	*Number of households*	*Number of households leasing out*	*Leased-out area as % of area owned by lessor households*	*Number of households leasing-in*	*Leased-in area as % of area operated by lessee households*
By Land ownership Class:					
0	27	0 (0)	0	5 (9)	100
0.1–5	19	10 (15)	95	7 (13)	91
5.1–5	35	21 (31)	79	15 (28)	62
15.1–30	36	18 (26)	53	19 (35)	36
30.1–50	15	11 (16)	25	4 (7)	17
Above 50	11	8 (12)	36	4 (7)	19
By Per Capita Income Class:					
Up to 500	33	14 (21)	47	12 (22)	67
501–800	33	16 (23)	47	11 (20)	31
801–1200	35	17 (25)	50	12 (22)	35
1201–1600	15	9 (13)	47	7 (13)	42
1601–1800	14	8 (12)	23	5 (9)	38
Above 1800	13	4 (6)	19	7 (13)	48
Total	**143**	**68 (100)**	**43**	**54 (100)**	**43**

Notes: In column 4 and 6 of this table, area leased-in (out) is expressed a percentage of area operated (owned) by *only those* households in a given category that are leasing-in (out) some land. For example, in the lowest per capita income class there are 33 households. Of these, 14 households leased-out 47% of land owned by these 14 households and 12 households lease-in an area constituting 67% of area cultivated by these 12 households.

In column 3 and 4 the figures in parentheses are column percentages

uneconomic holding; need for money (sometimes to repay past debts); and the households lacking in the labour. For these households the opportunities offered by the market for tenancy arise only from the possibilities of being able to lease out their small plots of land to someone more resourceful. This rent income is generally rather small for the small piece of land that he or she can possibly lease out. Yet for a poor destitute, it's a great support in absence of any other means. Thus the lease contracts are mean some opportunities for the destitute in the form of rent. Many of these households are families without fit adult males. Banning or severely restricting leasing of land will adversely impact all of them,

Table 12.6: Distribution of Tenants and Landlords by Level of Apparent Prosperity by "Observed Means" Criteria

Level of apparent Prosperity	*Number of households belonging to the specified group*		
	"Landlord" households	*"Tenant" households*	*All households*
Very rich	6(9)	7(13)	11(7)
Rich	4(6)	3(6)	6(4)
Prosperous	13(19)	12(220	29(20)
Secure	19(28)	13(24)	38(27)
Modest	18(26)	19(35)	43(30)
Poor	4(6)	0(0)	8(0)
Very Poor	4(6)	0(0)	8(0)
Total	**68(100)**	**54(100)**	**143(100)**

Note: In brackets, percentage of all households in the relevant group.

(1) For further details regarding the notion of "apparent prosperity" and the identification of the respective classes by Jean Dreze and Naresh Sharma, see Dreze (1988) and Dreze and Stern (1989) or Lanjouw and Stern (1998).

since the rental income provides a cushion to such households and provides them with some security.

These poor households also significantly lease-in land. The three poorest classes lease-in more than half of all leased land. (Table 12.4). Restricting tenancies does not provide them security but will result in resumption of their leased land by the owners.

The second striking feature emerges concerning the landless. It is observed that very few landless households are able to lease-in land. Only 19 per cent of the landless were able to lease-in land while in contrast among the marginal farmers (owning only small pieces of land below 5 bighas) 37 per cent were leasing-in land. It appears that ownership of land, *however small*, is crucial in being able to lease-in land. The main reason for this is that the perspective lessor must be convinced of a households' capability for farming, including ownership of productive assets, investible resources, and the knowledge.

This has important implications for land reform policy to protect the rural poor *Redistributive land reforms that put even small pieces of one acre in the hands of the landless are far more important in providing economic security to them than providing tenurial security or restricting leasing-out of land.*

As far as, the landless agricultural labour households are concerned, the impact of prohibiting tenancy on them is ambiguous. Due to prohibition of leasing, the demand for agricultural labour will increase. But then supply of labour by the non-labour households will also increase. Suppose the tenancy has been banned by the law (which is actually the substance of the law in Uttar Pradesh).[12] And that this law is effectively enforced. In this scenario, the landowners will feel compelled to undertake cultivation by themselves employing family labour and supplementing it by hired labour, or lessee will leave parts of their lands uncultivated, or will have to give up part of their land, if they try to lease out. It is reasonable to assume that land under 'personal cultivation' will increase and *consequently there will be an increase in demand for hired labour.* At the same time, a number of former tenants will now enter the labour market. Augmentations of labour supply from the cultivating households will have adverse impact on labour employment opportunities for the landless agricultural labour households.[13]

It has been seen that the institution of tenancy can help some of the poor by providing opportunity to lease-out for widow headed households, for example. It may help for poor marginal landholders to lease-out and when they seek better job elsewhere. It also provides the only opportunity for other land-poor households to augment their operated area through leasing-in of land.

Since the land lease market does not sort the lessors and the lessees into two coherent classes in terms of their socio-economic conditions, common interests etc. and since the tenant is not necessarily the weaker partner in every lease or even in most of leases, "protecting" the tenant can be counter-productive. Some of these aspects are highlighted in the simulation undertaken in the next section.

"Land-to-the Tiller": Some Simulation Exercises in Land Reforms

Land reforms has been a subject of big debate for a long time now and has involved academicians, politicians, social workers and other activists.[14] Philosophy of land reform is based on two main planks: (a) existing land relation may be production inhibiting and it may be necessary to change them before any improvement in agriculture can be expected; and (b) existing land relations are exploitive and it is necessary to change them for social justice, for removing excessive concentration of power on the one hand and extensive poverty on the other. As indicated in section 2, the land tenure system involving the intermediaries was beset with both these problem and hence, the Zamindari Abolition in India was inspired by both of these. At the same time there has been a near consensus that implementation of these reforms left much to be desired.[15]

'Land to the Tiller' has been a popular slogan and theme and continues to echo[16] even in the face of widely observed fact of 'reversed tenancy'.[17] In fact a large number of States in India have made legal provisions that virtually prohibit tenancy and also provide, under certain specified circumstances, from transfer of land to the tenant when land is leased-in violation of law. Table 12.1 summarize these legal restrictions on leasing activity in 15 Indian States.

This section studies the impact of various 'land-to-the-tiller' policies on inequality of income distribution through some simulation exercises.

The Methodology and Assumption for Simulation of the Land Reforms

The idea of simulation is to assess the likely outcome – in terms of distribution of income – of a land reform programme that assigns full land rights to the actual tiller of the soil. Thus, the simulation exercise presumes that the tenant has been given ownership rights on his/her leased land. There remains no distinction between operational holdings and ownership holdings. Indeed, the pre-reform operation, holding becomes the post-reform ownership as well as operational holdings.

What changes are likely to take place the productive activity – for example in cultivation practices – and in distribution of income in the village? If the existing/original (i.e. pre-reform) land system is an intermediary land tenure system (such as *zamindari*), then this land reform is likely to have the following three main implications: (1) all the produce now belongs to the erstwhile tenant; (2) land and fertility improving efforts are put in place by the new owners and there is new investment, and (3) greater application of current inputs. But if the existing/original (i.e. pre-reform) land system is that of permanent ownership rights in land among a large number of households, then the cultivation practices are likely to remain as in the pre-reform period, as discussed below. In this second scenario, which closer to present day reality, the only change due to land-to-the-tiller reform would be in the claims to the produce. What was earlier given to the lessor, is now fully retained by the tenant.

This simple simulation requires detailed data on the terms of lease, agreed rent, the lease partners, complete matching of lease arrangements as obtained from the tenant and the lessors. The detailed investigation and matching ensured this completeness and quality of data. Thus the basic idea is to reallocate the whole of the net produce (including rent on land) to the tenant himself. This when combined with all other income, gives the post-reform income distribution. All simulations were on the real data and were performed with the help of Fortran programming. *The basic assumptions behind the simulation exercises*: First, cultivation practices remain the same whether a cultivator tills it as its owner or as a tenant. This is a reasonable assumption when land to the tiller programmes is not accompanied by other effective programmes to augment resources of the tillers. This assumption also derives strength from a study of cultivation practices in Palanpur.[18]

Second, all other incomes for a household remain same as obtained before the reform. Thus for example, the households continue to receive the same non-farm incomes as they were receiving before the reform.

Four Schemes are Devised for Simulation of Land to the Tiller Type of Land Reforms: The first scheme forbids any leasing activity. In all other schemes leasing-out allowed for certain categories of landowners. These are described below. It must be noted that the simulations exercises do not create any 'fictional' data (on incomes) of their own. Each exercise re-allocates land assets and accordingly re-distributes (actual) incomes generated.

> REFORM 1: Land ownership is transferred to the actual cultivators of the leased plots. No exception is made irrespective of initial ownership holdings of the respective tenant and the landlord. No leasing is allowed subsequently. Thus no rents are paid or received. The actual law in many Indian States e.g. Kerala, M.P., Uttar Pradesh, Bihar, Karnataka, etc. is very close to the land-to-the-tiller policy. (See Appendix)

All the subsequent REFORMS are modification of *REFORM 1* to take into account the vulnerability of the poor or small landholding lessors and exempt them from adverse consequences of land-to-the-tiller policy.

> REFORM 2: Same as REFORM 1, except that if a lessor owns not above 15 bighas, and then his/her land is not transferred to the tenant. In other words, only small landowners are permitted to lease-out land. The logic is that small landholders lease-out due to some genuine difficulties. The lessors who own above 15 bighas, loose their leased land to their tenants. In this scheme,

81 households owning below 15 bighas are exempted from the reform.

REFORM 3: The scheme is same as REFORM 2, except that the exemption criterion has been changed to exclude 46 poor households according to apparent prosperity by "Observed Means" criteria from the purview of tenancy and land reform (in place of the owners of landholdings below 15 bighas).

REFORM 4: In this scheme, 76 households, that are below poverty line, earning a per capita income of Rs. 960 or less in 1983-84, are exempted from the purview of the land reform.

The Results

The results are presented in Table 12.8. Two measures of inequality – Gini coefficient and coefficient of variation – have been estimated. Post-reform income distribution for all four reforms do slightly worse than the actual income distribution in both the measures. The corresponding Lorenz curves and the distribution function plots are very close to each other. These results are quite surprising. Given wisdom has been that the tenants are poorer and hence land to the tiller should result in reduced income inequality. On the contrary, we find that the income inequality has slightly increased.

We have earlier witnessed (section 3) the phenomenon of the leasing-out of land by some poor and very poor lessors. Most of them leased out to the tenants much richer than themselves.[19] Thus it is possible that this tenancy reform makes the poor lessors worse off while enriching his/her more prosperous tenant. This could to some extent explain the worsening of income inequality from 0.31 to 0.34 in terms of Gini Coefficient, as seen after REFORM 1 (Table 12.8).

Even more surprising are the results for other reforms where the poor lessors were sought to be protected from dispossession of land.

Thus one needs to explain the worsening of income distribution after transferring land to the tiller when the poor and land poor are protected from losing land by such reforms (REFORM 2 to REFORM 4).

For this, we look at the structure of leasing. Some interesting details of leasing emerge from a study of Table 12.9. This table provides extent of leasing within a class, from higher class to the lower class, and from lower to the higher class. Classification criteria include ownership holding, operational holding and current per capita income. It is confirmed that there is significant amount of leasing from lower classes to higher

Table 12.7: Participation in the Lease Market by Ownership Holding

Ownership per cent	*Number of Households*	*Number of households leasing out*	*Number of households leasing in*	*TO FILL*	*TO FILL*
0	27	0	5	0	18
0.1 – 5	19	10	7	11	37
5.1 – 15	35	21	15	40	43
15.1 – 30	36	18	19	50	53
30.1 – 50	15	11	4	73	27
Above 50	11	8	4	73	36
Total	143	68	54	48	38

classes. Even more significant are the following observations. In a much larger number of cases, the tenants belonged to the higher operational holding classes than *their* lessors (68 cases accounting for 75 per cent of leased area) where as only in 23 cases of lease contracts did the tenants belong to lower operational holding classes than *their* lessor counterparts (accounting for only 19 per cent of leased area). On the other hand, the lessors were more often bigger landowners than their tenants (55 cases involving 54 per cent of leased area) whereas the only 24 tenants were bigger landowners than their respective lessors involving 21 per cent of leased area. (all figures pertain to Rabi season. The picture is very similar for the Kharif season leases). *This indicates that a large number of lessors who own more land than their tenants, come to operate small land area after leasing has taken place.* Thus transferring of land from the lessors to tenants may make the inequality worse by increasing post-reform inequality of landholding. Indeed, operational holding is distributed slightly less equally than ownership holding for the village as a whole.

The land-reform exercises carried out here caution one against uncritically accepting any land reform programme. In particular, the laws that seek to restrict or ban tenancy and that renders the leased land liable to be taken away from the lessors and given to (their) lessees, are of doubtful value. 'Land to the tiller' may not be suitable slogan or policy in all types of poor economies. Suitability of a particular set of land reform measures can only be judged on the basis of ground realities.

Poverty, Inequality and Restrictions on Leasing of Land

It was observed from Table 12.4 to 12.7 that the lessors and lessees come from very similar caste, landownership class, income class or prosperity.

Table 12.8: Performance of "Land-to-the-tiller" Reforms Comparing Inequality Indices for per Capita Income for Actual Income vs. Income under Specified Land regime (Post-Reform)

Inequality Indices	*Actual Income*	*Under Reform 1*	*Under Reform 2*	*Under Reform 3*	*Under Reform 4*
Coef.of var.	0.546	0.618	0.602	0.605	0.576
(std.error)	(0.012)	(0.023)	(0.023)	(0.023)	(0.024)
Gini Coeff	0.307	0.337	0.327	0.329	0.312
(std.error)	(0.007)	(0.008)	(0.007)	(0.008)	(0.007)

Note: Definition and details of REFORM 1 to REFORM 4 are given in the text.

Table 12.9: Intra-Class leasing by Different Classification Criteria

Classifying Total Criteria	*% area leased within the same classes*	*% area leased out to higher classes*	*% area leased out to lower classes*
RABI			
Caste 100 (102)	35 (26)	39 (45)	27 (31)
Ownership Holding 100 (102)	25 (23)	21 (24)	54 (55)
Operational Holding 100(102)	6 (11)	75 (68)	19 (23)
KHARIF			
Caste 100 (98)	42 (33)	34 (42)	24 (23)
Ownership Holding 100 (98)	19 (20)	21 (18)	60 (60)
Operational Holding 100(98)	9 (14)	72 (56)	19 (28)

Note: Figures in brackets are number of lease contracts for a given category. Only those contracts are considered here for which both the leasing partners belong to Palanpur.

There are a lot of variations in ownership or operational holding across households (Gini coefficient being 0.60). However, there are no landowners that dominate all others. The largest landowner owns not more than seven hectares of land. In such scenario, *tenancy performs the role of facilitating adjustment of productive resources in cultivation* and hence not restricting leasing of land commends itself from both productive and distributive reasons.

The complexity of land leasing activity in such village is brought out in Table 12.10. First, leasing was widely prevailing across households from all possible backgrounds. Out of a total of 143 households, 106 were either leasing-in land or leasing-out land or both. Only 37

Table 12.10: Classification of households by Tenancy Status

	Landless	*Land Owning*	
		Leasing-out	Not leasing-out
Leasing in	Pure tenants (5)	Tenant-landlords (16)	Owner-tenants (33)
Not leasing in Cultivating		Cultivating landlords (28)	Pure owner cultivators (15)
Not cultivating	Landless non-Cultivators (22)	Pure landlords (24)	

Note: Figures in parentheses indicate number of households showing the specified tenancy status.

households did not participate in leasing activity and 22 of them were landless households. Second, the landless are the only class of households that are virtually out of the lease market. Conventionally, they were seen as typical tenants. This is disturbing because they are economically weaker and in need of economic, productive resources. Third, there were 16 households that were both leasing-in as well as leasing-out land. This clearly demonstrates that the leasing of land is being used as an adjustment device. Fourth, there are 24 households that lease-out all their land. Most of these households are small holders owning an average less than one hectare of land. Fifth, there are more lessors (68) than lessees (54) in the village.

Table 12.11 presents details about land and labour endowments according to these tenancy classes. There is clear indication that land operated per worker among the cultivators is greatly equalised compared to land owned per worker. Another important facet of the tenancy market is revealed when we look at turnover of tenancies from one year to another. There was change of tenure status of 45 households from Rabi season to Kharif season.[20]

Such rapid changes in tenure status underscore two main points. First there are only a few households in the village that can be sensibly classified as "tenants" or "landlords". For most households, changes in tenure status are too frequent and it shall be misleading to classify them under these terms on the basis of current leasing. Second, leasing-in and leasing-out in a season are outcome of adjustments that households make or are able to make depending on their current resource endowments. This means that even with little or no change in landownership, there are large changes in available family labour, drought power, cash resources and debt from

Table 12.11: Tenancy and Labour Endowments

Tenure Class	*Number of house-holds*	*Land owned per capita*	*Land Cultivated per capita*	*Land owned per worker*	*Land cultivated per worker*
Pure owner cultivator	15	2.7	2.7	15.3	15.3
Owner tenant	33	2.6	4.2	12.2	19.9
Tenant landlord	16	3.0	3.9	14.4	19.1
Cultivating landlord	28	4.8	3.1	35.3	22.8
Pure landlord	24	2.1	0.0	37.1	0.0
Pure tenant	5	0.0	2.5	0.0	17.7
Landless non-cultivator	22	0.0	0.0	0.0	0.0
Total	143	2.7	2.76	17.0	17.3

one year to another, that significantly influence decision to lease-in or lease-out or do neither.

The reason behind the change in tenure status (i.e. for the decision to lease or not) include: (a) illness in the household leading to labour and cash scarcity, (b) unusual expenditure (e.g. on birth, death or marriage ceremonies or remittances outside or on education) or a crisis (e.g. being a victim of robbery) leading to cash constraints, (c) acquisition of new assets (e.g. tractor or irrigation pump), (d) availability or non-availability of own drought power, (e) in or out migration of a working member, and (f) availability of land to lease-in. Any of these factors may change from year to year without seriously altering their more enduring household characteristics. Of course, frequent recurrence of some of the problems facing a household (mentioned above) in a short period can change their permanent characteristics, too.

Thus given the pattern of permanent or enduring household characteristics – land owned, family labour, outside regular jobs, level of prosperity etc. – there are very few permanent lessors (landlords) or lessees (tenants). Much of tenancy is explained by transient economic situation of a household. The tenancy, in a large number of cases, is a response to these transient changes and is an adjustment mechanism.

Two far-reaching changes in the past five decades impacting on agrarian relations have been (a) abolition of intermediaries and (b) technological change. There may be much regional variation in landownership patterns, crop specificities, local institutions, and natural resources and accordingly, lessons learnt from specific village studies may be more or less applicable to these diverse situations. However, with greater technological inputs leading to increasing mechanisation,

Table 12.12: Labour and Capital in Agriculture: 1957-1984

Item	*1957-8 (in %)*	*1983-4 (in %)*
1. Total inputs/outputs	14.4	37.0
2. Non-labour inputs/output	6.2	33.0
3. Inputs including labour & seed/output	1.4	25.0
4. Hired labour costs/Total input costs	56.8	11.0
5. Seed costs/Total input costs	33.2	21.0
6. Other input costs/total input costs	10.0	68.0

Note: All inputs and outputs are valued at current prices for respective years for computing the ratios. Input costs exclude land tax, interest on capital, depreciation etc. All numbers (i.e. input/output ratios and composition of inputs) are expressed in per cent.

and capital intensive agriculture, the tendencies analysed here are either already manifest or going to manifest sooner or later. One such tendency is the emergence of the large farmer as tenant and the pushing aside of the landless from the lease market. Table 12.12 throws light on increasing input intensity in agriculture. This process has only accelerated over past decade as we discovered on our re-visit to Palanpur in the new millennium.

With new technology, the economic size of holding has gone up, as the owner of lumpy machines wants to operate larger area. Partly, this problem is lightened with emergence of markets for hiring of services of most of these machines. On the other hand, higher cost outlays put small landholders and prospective landless tenants at a distinct disadvantage. Land reform policy with respect to institution of tenancy must respond to this situation.

Let us summarise the impact of restriction of lease market on the poor. If leasing of land is prohibited, then

(1) Income inequality is likely to increase.

(2) Poor lessees cannot lease-in land any more. Thus the landless and land poor households are denied access to land through lease market. Their economic opportunities are reduced.

(3) Poor lessors cannot lease-out land any more. This particularly hurts the poor households without fit adult males.

(4) Farmers are unable to adjust their cultivation holding in response to transfer exigencies. This results in misallocation of resources and reduction in production.

Further, the landless have much poorer access to the lease market. They are among the most poor. Land reforms must aim to facilitate them to have access to productive resources. A redistributive reform, providing a minimum landholding to every rural labour household, and a more open lease market should be the basis of pro-poor land reforms policy. To guard against concentration of operational holdings in a few hands, land ceiling must be applicable to both own as well as leased land.

Some Concluding Remarks

This paper addressed the problem of influence of the institution of tenancy on the rural poor. Our data permitted some land reforms simulation exercises to be carried out. The results of these simulation exercises are significant in that they are actual income figures and not some constructed data for the sake of simulation exercises. If conditions of simulation hold, then the results of the said land reform will be as shown by these exercises.

The lesson from the investigation suggest that prohibiting or severely restricting the land lease market is counter-productive both by equity and productivity criteria.

At the end of this discussion, it may seem intriguing why land to the tiller slogan with strong anti-tenancy sentiment has persisted for so long. To understand this it must be remembered that land reforms in India in the present form had their genesis in pre-independence *zamindari* days. Land inequalities were very great and the actual tiller was subjected to not only economic exploitation but to physical violence and various human indignities.[21] The present case study was about a village bearing no resemblance to the above conditions. Wherever *zamindari* abolition and extreme unequal land distribution has been undone, applying the same old remedy will not work. *This also underlines the need for undertaking a re-distributive land reform where old inequalities in and institution of landownership still persist.*

Land redistribution type of reform also will have to be judiciously applied taking into account regional factors. Its effectiveness and even appropriateness will vary from region to region. In particular initial inequality of landownership is important both in motivation for redistributive land reforms and also for the hurdles that will have to be faced in its implications.

Appendix

A significant part of section 157 as also section 156 of the U.P. Land Reforms and Zamindari Abolition Act, 1950 are given below:

S. 156. Letting of land forbidden except as in S 157 or to a recognised educational institution for agriculture etc.

S.157. Lease by a disabled person – (1) A *Bhumidhar* or an *Asami* holding the land in lieu of maintenance allowance under 11 who is—

(a) An unmarried woman, or if married deemed divorced or separated from her husband or whose husband suffers from any of the disqualifications mentioned in clauses (c) or (d) or a widow;
(b) A minor whose father suffers from any of the disqualifications mentioned in clauses (c) or (d) or has died;
(c) A lunatic or an idiot;
(d) A person incapable of cultivating by reason of blindness or other physical infirmity;
(e) Prosecuting studies in a recognised educational institution and does not exceed 25 years in age and whose father suffers from any of the disqualifications mentioned in clause (c) or (d) or has died;
(f) In the Military, Naval or Air Services of the Indian Dominion; or
(g) Under detention or imprisonment;

May let the whole or part of his holding (i-A) (Bank can let out land)

References

Ansari, N. (1964): Continuous Village Surveys, 1958-59, Number 41, Palanpur: Moradabad: UP: A study of its Economic Resources and Economic Activities, (Agricultural Economics Research Centre, University of Delhi).

Appu, P.S. (1975): "Tenancy Reform in India", *Economic and Political Weekly*, 10, 33-35, special number (August), pp. 1339-1375.

Berck, P. and Levy, A. (1986): "The cost of equal land distribution: The case of

Israeli Moshavim", *American Journal of Agricultural Economics*, 68, 3, pp. 605-14.

Bliss, C.J. and Stern, N.H. (1978): "Productivity, wages and nutrition: Part 1 and Part II", *Journal of Development Economics*, Vol. 5, pp. 331-362 and 363-398.

Bliss, C.J. and Stern, N.H. (1982): *Palanpur: The Economy of an Indian Village* (Delhi and Oxford: Oxford University Press).

Byres, C.J. (1988): "Charan Singh (1902-87): An assessment", *Journal of Peasant Studies*, Vol. 15, No. 2, pp. 139-189.

Cheung, S.N.S. (1968): "The private property rights and sharecropping". *Journal of Political Economy*, November-December 76, pp. 1107-1122.

Cheung, S.N.S. (1969): *The Theory of Sharecropping* (Chicago: University of Chicago Press).

Damle, C.B. (1989): "Land reform legislation in Karnataka: Myth of success", *Economic and Political Weekly*, Vol.24, No.33, August 19, pp. 1896-1906.

Dantwala, M.L and Shah, C.H. (1971): *Evaluation of Land Reforms*, Vol.1, (Bombay: University of Bombay).

Dreze, J. (1988): "Social insecurity in India", paper presented at a workshop on Social Security in Developing Countries held at the London School of Economics, July, 1988.

Dreze, J., and Sharma, N. (1998): "Palanpur: Population, Society, Economy", in Lanjouw, P. and Stern, N. eds (1998c)

Dreze, J., Lanjouw, P. and Sharma, N. (1998): "Economic Development in Palanpur, 1957-93", in Lanjouw, P. and Stern, N. eds (1998c)

Dreze, J., Lanjouw, P. and Stern, N. (1989): "Identifying and reaching the poor: Principles and some examples from north Indian village", London School of Economics.

Dreze, J. and Mukherjea, A. (1987): "Labour contracts in rural India: Theories and evidence", Discussion Paper No.7, STICERD, Development Economics Research Programme, London School of Economics.

Gill, S.S. (1989): "Changing land relations in Punjab and implications for land reforms". *Economic and Political Weekly*, Vol. 24, No.25, June 24, pp. A79-A85.

Government of India (1959): Reports of the Committees of the Panel on and Reforms, Government of India, New Delhi.

Government of India (1963): *Progress of Land Reforms*, New Delhi.

Government of India (1988): Occasional papers (First series), New Delhi: Department of Rural Development).

Gupta, M.R. (1989): "Nutrition, dependents and the mode of wage payments", *Oxford Economic Papers*, Vol. 41, No. 4, October, pp. 737-748.

Habib, I., Raychaudhari, T., Kumar, D. and Desai, M. (1982 and 1983): *Cambridge Economic History of India*, (2 Vols), Cambridge: Cambridge University Press)

Haque, T. and Sirohi, A.S. (1986): Agrarian Reforms and Institutional Changes in India, (New Delhi: Concept Publishing Company).

Jha, S.C. (1971): *A Critical Analysis of Indian Land Reform Studies*, (Bombay: Asian Studies Press).

Joshi, P.C. (1975): *Land Reforms in India* (New Delhi: Allied Publishers Ltd.).

Krishna, R. (1977): "Towards a theory of optimum land reform for a dualistic agriculture", *Journal of Indian Society of Agricultural Statistics*.

Lanjouw, P. and Stern, N. (1989): "Agricultural changes and inequality in Palanpur 1957-1984", DEP no.24, *STICERD*, London School of Economics.

Lanjouw, P. and Stern, N. (1989a): "Poverty in Palanpur", in Lanjouw, P. and Stern, N. eds (1998c).

Lanjouw, P. and Stern, N. (1989b): "Inequality", in Lanjouw, P. and Stern, N. eds. (1998c).

Lanjouw, P. and Stern, N. eds (1989): *Economic Development in Palanpur over Five Decades* (Oxford: Clarendon Press).

Lehmann, D. (ed) (1974): *Peasants, Landlords and Governments* (New York: Holmes and Meier Publishers Inc.) [Also published as Agrarian Reform and Agrarian Reformism, (London: Faber and Faber)].

Lipton, M. (1974): "Towards a theory of land reform", in Lehmann, D. (ed), agrarian Reform and Agrarian Reformism (London: Faber and Faber), pp. 269-315.

Lipton, M (1985): "Land Assets and Rural Poverty", *World Bank Staff Working Paper No.744*.

Kumarappa, J.C. (1951): *Report of the Congress Agrarian Reforms Committee*, (New Delhi: All India Congress Committee).

Malviya, H.D. (1955): *Land Reforms in India*, (New Delhi: All India Congress Committee).

Misra, K.C. (1990): *Land Systems and Land Reforms*, (New Delhi: Himalaya Publishing House).

Moreland, W.H. (1929): *The Agrarian Systems of Moslem India* (Allahabad: Central Book Depot).

Murty, C.S. (2004): "Large farmers in lease market: Are marginal farmers affected?", *Economic and Political Weekly*, Vol. XXXIX, No.29, July 17-23, pp. 3270-79.

Neale, W.C. (1962): *Economic Change in Rural India: Land Tenure and Reforms in Uttar Pradesh 1800-1955* (New Haven: Yale University Press).

Otsuka, K. and Hayami, Y. (1988): "Theories of share tenancy: A critical survey", *Economic Development and Cultural Change*, Vol.37, No.1, October, pp. 31-68.

Pant, Govind Ballabh, Kidwai, Rafi Ahmed and Tiwary, V.N. (1982): *Agrarian Distress in United Provinces*, (Delhi: Prabhu Publications).

Radhakrishnan, P. (1990): "Land reforms: rhetoric and reality", *Economic and Political Weekly*, Vol.25, No.47, November 24, pp. 2617-2621.

Sampath, R.K. (1984): "On the theory of optimum land reform", *Indian Journal of Economics*, Vol.64, No.254, pp. 257-264.

Saxena, K.B. (1990): "Access to land as instrument of poverty alleviation",

Journal of Rural Development, Vol.9, No.1, January, (special issue on poverty alleviation), pp. 115-143.

Sen, A. (1981): "Market failure and the control of labour power", *Cambridge Journal of Economics*, part 1: 5, 3, September, pp. 201-228, part 2: 5, 4, December, pp. 327-350.

Sharma, N.K. (1992): *Tenancy Legislation in India: A Comparative study of major Indian states*, Project Report, NIRD, Hyderabad.

Sharma, N. and Dreze, J. (1996): "Share-cropping in Palanpur", *Journal of Development Studies*, Vol.33, No.1 (October).

Sharma, N.K. (1994): *Agricultural Tenancy in Palanpur*, Ph.D. thesis to be submitted to Indian Statistical Institute, Calcutta. Singh, B. (1961): *Next Step in Village India: A Study of Land Reforms and Group Dynamics* (Bombay: Asia Publishing House).

Singh, B. and Misra, S. (1964): *Land Reforms in Uttar Pradesh* (New Delhi Oxford Book Company).

Srivastava, S.P. (1984): *Laws of Zamindari Abolition in Uttar Pradesh* (Allahabad: Alia Law Agency, second edition, 3 vols).

United Provinces Zamindari Abolition Committee (1948): Report of U.P. Zamindari Abolition Committee, (Chairman G.B. Pant), Allahabad: U.P. Government).

Vyas, V.S. (1970): "Tenancy in a dynamic setting", *Economic and Political Weekly*, 5, June, pp. A73-A80.

Walinsky, L.Y. (ed.) (1977): *Agrarian Reform as Unfinished Business: Selected Papers of Wolf Ladejinsky* (New York: Oxford University Press).

Notes

1. Workshop on "Land Market and Rural Poverty" (10-11 August, 2004) Centre of Rural Studies, Lal Bahadur Shastri Academy of Administration Mussoorie-248179.
2. No serious attempt is made here to analyze in detail the notion of rent, how it arises and its legitimacy. This is major issue of its own which cannot be dealt with adequately in this chapter.
3. See particularly Bliss and Stern (1982); and Dreze and Sharma (1998).
4. Throughout this chapter, the reference year is 1983-4. In addition to very detailed data on agricultural leases for this particular year, including, terms of contract, complete matching of details from lessors and lessees and relevant cultivation details, which was crucial for the simulation exercises undertaken in section 4.
5. See the Appendix. Tenancies are governed by the Uttar Pradesh Zamindari Abolition and Land Reform Act, 1950. Section 156 and 157 of the Act clearly place severe restrictions on leasing of land.
6. Some of these "landless' households are only temporarily so, having

separated from their land owning fathers. Sooner or later they inherit their share of land. About 15 households were genuinely landless.

7. Bigha is a local unit of land area. One acre =6.4 bighas and 1 hectare=15 bighas in this region.
8. Incidence of tenancy has remained about 25 per cent of total land area owned for most survey years as well including the one undertaken in 1993. Most village studies reveal far greater land under private leases than what is reported by large scale survey such as NSS or Agricultural Census. See Sharma and Dreze (1996) for evidence on this.
9. See *Cambridge Economic History of India* (1982-83), Neale (1962), Misra (1990) among others.
10. See Misra (1990: p.81): "In India individual ownership of land was an ancient system and this system remained sacrosanct till the beginning of foreign rule. It may be added that in this country property right over land was never an invisible concept. Three parties such as, the peasantry, the community and the sovereign, had been enjoying the privilege of property right over land within their jurisdiction all through ages without any clash of interest".
11. This is what the observed means criteria of apparent prosperity of the households basically provides. It is said that economic standing of the people is rather a common knowledge among the villagers. The long and close interaction of the researchers with the villagers offered the same opportunities to them the researchers).
12. See the Appendix for the prohibition of leasing of land according to the UP Zamindari Abolition and Land Reforms Act. The law allows only those landowners to lease out land who may suffer from disability, or are serving in armed forces of India or prosecuting studies. Violation of the law attracts penalty of loosing the land. With such serious consequences, there is no wonder that there are no formal lease contracts in Palanpur or indeed in most of India. For instance, though tenancy is completely prohibited in Kerala, yet there is quite a large scale of leasing of land in Kerala. See Murty (2004) for evidence from the NSS data.
13. This in conformity with the efficiency wage hypothesis. See Bliss and Stern (1978), Dreze and Mukherjee (1987) and Gupta (1989) on this.
14. Literature on land reform is vast. On theory and ideology of land reforms see Lehman (1974). Some important studies include: Appu (1975), Berck and Levy (1986), Byres (1988), Dantwala and Shah (1971), Gill (1989), Jha (1971), Joshi (1975), Krishna (1977), Kumarappa (1951), Lipton (1985), Malviya (1955), Singh (1961), Singh and Mishra (1964), UPZAC (1949) and Walinsky (1977). Important reviews of land reforms can be found in Haque and Sirohi (1986), Jha (1971), Joshi (1975) and Neale (1962).
15. Noted even in the Government reports. See Government of India (1988), which admits that, "It is widely recognised that the chief reason for poor implementation of the land reforms has been the lack of political will." (p.64).

16. "Going by the five year plans and the agenda Papers, the ultimate objective of national policy on tenancy is making the 'tiller of the soil' the owner". (Radhakrishnan (1990: p. 2619). An occasional paper prepared for the Government of India (1988: p.59, emphasis added) says, "It would thus be seen that the tenancy reform has not made much impact when judged by the incidence of informal or oral tenancies today. Such tenancies have come up even where leasing has been expressly prohibited. Thus providing some medium of security to this category of tenants is the most burning problem of agrarian reform today."
17. For evidence, see Sharma (1994) and a large number of studies cited there.
18. Detailed results of testing for agricultural practices on tenanted and owner operated lands have been presented in Sharma (1994) and Sharma and Dreze (1998). These tests were conducted to statistically test for differences in inputs as well as outputs (production) per unit area. The results of these tests support the hypothesis that there are no statistically significant differences in either use of inputs applied or in yields obtained.
19. Related the phenomenon of leasing-out of land by small landholders to big cultivators has been quite widely reported for quite some time now. This so-called reverse tenancy has been reviewed in detail in Sharma (1994). See Vyas (1970) for an early report and Murty (2004) for a recent paper discussing this problem at length based on all-India NSS data.
20. See Sharma (1994: 66-67) for further details.

13

Status and Relevance of Legalization of Land Leasing in Bihar

Dr. Pyare Lal

Background

Bihar was one of the first state to initiate agrarian reforms through the enactment of legislation to abolish the *Zamindari* system in the 1950s and probably thc least successful in implementing that and other agrarian reforms. In the wake of *zamindari* abolition and during the intervening years of land reform, large-scale evictions of occupancy tenants occurred, since the actual tenants could not prove their right of occupancy, they were evicted and the under-tenure holders such as the *bataidars* or sharecroppers, became tenants. The *batai* system existed in the past, but not to the extent that it does today. During the early twentieth century when the Survey Settlement operations took place, out of a total of 224.8 lakh acres, non-occupancy ryots and under-ryots accounted only for 6.6 lakh acres. Thus, the majority of the ryots enjoyed occupancy right. In Purnea, out of 19.8 lakh acres non-occupancy ryots and under-ryots accounted for only 1.43 lakh acres.

To form an idea of the growth of *bataidars*, examination of the data from the Survey Settlement of 1908 and from that of 1952-58 is revealing. In 1908, some 93,000 under-raiyati holdings were recorded; they owned 90,000 acres as homestead and *bari* lands and 35,000 acres of cultivated land. In the 1950s, the number of such under-raiyats jumped to 3.24 lakhs, holding only 2.64 lakh acres. But what is more interesting is that of these only one lakh had any land under cultivation, the rest having only homestead and/or *bari* land.

Thus, a majority of under-ryots had only homestead and *bari* land. These holdings were rent-free and they were tenants only in the sense that they have occupancy rights on these bits of lands averaging 0.04 acres. They were not tenants in the sense of being sharecroppers.

Since then, evictions have gone on at a very fast pace. In the 1950s, 30.50 per cent of operational holdings were recorded as partially or fully cultivated as undertenures. The 1970-71 agricultural census yielded a ridiculous figure of 0.40 per cent of operational holdings as the average for the whole state. The landlords had seen to it that undertenants are not recorded at all.

Thus, while only a fraction of *bataidars* are small and marginal farmers who take land on *batai* to survive as cultivators, the great majority of them are agricultural labourers. They stick to cultivation because of their peasant aspirations and the hope that they can get land through land reforms. The process of extensive land alienation from the occupancy tenants that began during the Great Depression, continued for all the following years: because of tenant-zamindar tension, and because of zamindars' precautionary moves against the consequences of the impending zamindari abolition, and later on ceiling legislation.

Present Tenurial Pattern

In Bihar the tenant landlord relationship is governed by Bihar Land Reforms Act, 1885 as amended in 1970. It provides that every tenant on ceiling surplus land may acquire the status of a raiyat (owner) on application within a specified period on payment of an amount. However, the landowner is entitled to resume for personal cultivation in an area not exceeding half of the total area held by the tenant.

Bihar Tenancy Act as amended in 1970 also provides that an under-raiyat in continuous possession of land for twelve years can acquire right of occupancy provided the landowners for whom the land leased-in own more than 5 acres (irrigated) or 10 acers (un-irrigated) of land.

Bataidars are liable to ejectment on expiry of the written lease or failure in payment of rent. However, there is no specific provision for either eviction or protection of an under-raiyat who holds an oral lease.

It is widely acknowledged that official figures of leased-in area are plagued by underreporting, and are gross underestimates of actual extent of tenancy. Nevertheless, the NSS figures provide us with at least a trend of change in the leasing-in of land in the state. Tenancy in Bihar is seen to fall quite sharply in the eighties, as compared to all-India trend. But quite interestingly, its incidence is seen to increase equally sharply during the nineties. In 1981-82, nearly 20 per cent of operational holdings were tenant holdings, and percentage of leased-in area in total operated area was 10.3 per cent (Table 13.1). In 1991-92, while the share of tenant holdings in total operated holdings fell to 5.6 per cent, the share of leased-in area fell to 3.9 per cent. Thereafter, there was a sharp increase during the nineties, with share of tenant holdings in total operational holdings rising to nearly 13 per cent, and area leased-in increasing to about 9 per cent of the operated area. This was in contrast to the all-India trend, which showed a secular decline of both the per cent of tenant holdings and per cent of leased-in area all through the eighties and the nineties.

Table 13.1: Per cent of Tenant Holdings and operated area leased-in Rural-Bihar

	% of Tenant Holdings			*% share of Leased-in Area*		
	1981-82	1991-92	2002-03	1981-82	1991-92	2002-03
Bihar	19.7	5.6	12.7	10.3	3.9	8.9
India	15.2	11	9.9	7.2	8.3	6.5

Source: NSS Report no. 492.

To know the present ground realities, about 200 households having possession of land either in capacity of owner, tenant or owner cum-tenant have been surveyed through schedule prepared for the purpose in four districts from different parts of the State, so that it can be generalized for the state as a whole. The survey seeks to understand the concealed tenancy in Bihar, the form of tenancy and the security of tenure particularly the tenants at will/sharecroppers. About 191 surveyed schedule were found to be filled properly and were analyzed further. Table 13.2 gives details of household covered for the field survey by their size of holdings. It shows that about 38.74 per cent of the household covered were marginal holders i.e. less than 2.5 acres followed by small holders i.e. between – 2.5 to 5 acres (22.5 acres), landless (21.47 per cent), Medium holders i.e. 5 to 10 Acres (12.57 per cent) and large holders having more than 10 acers of land (4.71 per cent). District-wise detail can be seen in Table 13.2. Similarly details of covered households by their Social Group is given in Table 13.3. About 47.64 per cent of such households were from OBC community while, 25.13 per cent were from others i.e. upper caste. Muslims community were classified separately and their coverage recorded 14.14 per cent while 13.09 per cent of such households were recorded as Scheduled Castes.

Table 13.2: District-wise coverage of Households by their Holding Size

(in %)

H.Hd/Distt.	*Landless*	*Marginal*	*Small*	*Medium*	*Large*	*N*
Nawada	24.49	42.86	16.32	12.25	4.08	49
Purnea	20.83	29.17	25.00	18.75	6.25	48
Samastipur	23.40	38.30	23.40	10.64	4.26	47
Bhojpur	17.02	44.68	25.53	8.51	4.26	47
All	**21.47**	**38.74**	**22.51**	**12.57**	**4.71**	**191**

Source: Primary Investigation.

Table 13.3: District-wise Coverage of Households by their Social Group

(in %)

Social Group/Distt.	*SC*	*OBC*	*Muslims*	*Others*	*N*
Nawada	20.41	44.90	18.37	16.32	49
Purnea	12.5	50.0	14.58	22.92	48
Samastipur	8.51	48.94	10.64	31.91	47
Bhojpur	10.64	46.81	12.76	29.79	47
All	**13.09**	**47.64**	**14.14**	**25.13**	**191**

Source: Primary Investigation.

When the households were mapped according to their landholding status, the biggest incidence of leasing-in was found to be among the marginal landholders (33.78%), while 24.39 per cent of landless too were found to be leasing-in land. Nearly 11.63 per cent of small land holders were leasing-in land. Thus, the incidence of leasing-in was found to fall uptil the semi-medium landholding category. As per the field survey 11.11 per cent in the large landholding group were also found to lease-in land, This was 8.33 per cent among the medium landholding category. (Table 13.4).

A somewhat larger figure for the leased-in area and proportion of households leasing-in land was reported by the A.N. Sinha Institute and ILO survey in the year 1982, and resurveyed of the same set of villages by IHD in 1999-2000 of Bihar. The survey also throws up interesting figures of social distribution of households leasing-in land. According to the findings of the survey, proportion of households leasing-in land considerably declined from about 36 per cent in 1982 to about 23 per cent in 1999-2000, but the proportion of the leased-in area to the total cultivated area has marginally increased from 24.5 per cent in 1982 to 25.5 per cent in 1999-2000. This had resulted in an increase in the average size of the leased-in land in general.

Table 13.4: Percentage of Household Leased-in and leased-in Area by their Holding Size

(in %)

Holding Size	*% of Leased in Household*	*% of Leased in Area to Total Cultivated Area*
Landless	24.39	100.00
Marginal	33.78	42.67
Small	11.63	5.26
Medium	8.33	4.23
Large	11.11	3.45
All	**22.51**	**33.57**

Source: Primary Investigation.

As per the survey conducted by A.N. Sinha Institute and ILO (1981) and IHD (1999-2000), the decade of nineties has seen the landless and the marginal farmers moving out of tenancy arrangements, but significantly enough big landed segments of the population moving into the land renting market. While in 1982, 34.53 per cent of landless were leasing-in land, in 1999-2000 only 22.63 per cent were leasing-in. Similarly, for marginal landholders (owning land of size less than 0-1.0 acres) while 53.5 per cent of these households were leasing-in land in 1982, in 1999-2000 only 36.09 were leasing-in. The share of leased-in area in total cultivated area of this segment also saw a fall from 62.65 per cent of area cultivated to 58.15 per cent in the same period. However, the present survey result shown in Table 13.4 indicates about the leased-in area out of the total cultivated area among different landholders groups. Apart from the landless groups who cultivate entire land in leased-in area of land, maximum percentage of leased-in land were found among marginal land holders i.e. 42.67 per cent of the total cultivated area by them, while small, medium and large landholders accounted for 5.26, 4.22 and 3.45 per cent respectively (Table 13.4). About 33.57 per cent out of the total cultivated area were found to be as leased-in area indicates further fall in the comparison of previous survey by A.N. Sinha Institute and IHD during 1982 and 1999-2000.

In fact, it is a general observation in the villages in Bihar that the households which take land on lease are those which are involved in cultivation and which are capable of making investments in cultivation to begin with and are able to pay rent even after low agricultural production.

Social Groups in Tenancy Arrangements

The importance of the leased-in land was found to be significantly high for the weaker sections of the population. Highest incidence of leasing-in was found to be among SC households (40%) followed by OBC farmer households (27.47%), Muslims Community (18.52%) and others included upper/middle caste farmers (6.25%).

Table 13.5: Percentage of Household Leased-in and Leased-in Area by their Social Group

(in %)

Social Group	*% of leased in Household*	*% of leased in Area to total cultivated area*
SC	40.00	75.90
OBC	27.47	36.42
Muslims	18.52	6.67
Others	6.25	3.45
All	**22.51**	**33.57**

Source: Primary Investigation.

As far as leased-in area is concerned, about 76 per cent of the total cultivated land possessed by SC household were the leased-in land whereas such percentage among OBC families was 36.42 and among Muslim Community about 6.67 per cent of the total cultivated land are found to be leased-in area. Only 3.45 per cent of the total cultivated area belonging to upper caste found to be leased-in area.

One of the dominant castes in the OBC group, the Kurmis witnessed an increase in the area leased-in despite percentage of household leasing-in land falling. This again could be a pointer to the emergence of capitalist production relations in agriculture wherein the better placed in the cultivating castes were producing for the market and seeking land in the tenancy arrangements. Similarly, for the Yadavs, though percentage of the leased-in land in the total cultivated area had fallen but still accounted for around 40 per cent of cultivated land.

Table 13.6 shows the distribution of household by the average yield in terms of Rupees per acre in their own and the leased-in land. It shows that productivity level was comparatively lower in the leased-in land rather than in their own cultivated land. However, it varies in different systems of tenure. Productivity was found to be comparatively higher in a few cases in terms of fixed money or produce than the share of produce.

Table 13.6: Per cent of Household

(in %)

Value of Yield (In Rs.) Per Year	*Own Land*	*Leased-in land*
Upto 15,000	58.78	67.44
15,000-20,000	17.57	18.60
20,000-25,000	14.19	11.63
25,000 & above	9.46	2.33
All	**100.00**	**43**

Source: Primary Investigation.

Overall the production level seemed to be lower in the leased-in land than in the own land cultivated.

Terms of Tenancy

NSS estimates show that while share-cropping still remains the predominant form of tenancy arrangement in Bihar, it has nevertheless been on decline over last two decades. In 1981-81, while 73.3 per cent of the leased-in area was on the terms of sharing of produce, it declined to 43.5 per cent in 1991-92, but rising thereafter to 67 per cent in 2002-03 (Table 13.7). There has been an increase in the area leased-in under fixed money and fixed produce tenurial arrangements. While percentage of area leased-in under fixed money contract increased from 6.5 per cent to 12 per cent over the period, the corresponding increase for fixed produce contract was from 3.6 per cent to 17.5 per cent.

Table 13.7: Per cent Distribution of Area Leased-in by Terms of Lease for Bihar and All-India

(in %)

		Terms of Lease Rural					
	Year	*Fixed Money*	*Fixed Produce*	*Share of Produce*	*From Relative: No terms*	*Others*	*All*
Bihar	2002-03	12	17.5	67	0.5	3	100
	1991-92	9.5	12.8	43.5	0.6	33.6	100
	1981-82	6.5	3.6	73.3	-	16.6	100
India	2002-03	29.5	20.3	40.3	4	5.9	100
	1991-92	19	14.5	34.4	7.4	24.7	100
	1981-82	10.9	6.3	41.9	-	40.9	100

Source: Table 3.15, NSS Report no. 492.

Increased incidence of sharecropping contract over the nineties together with increase in the incidence of tenancy is a pointer of the increased recourse to cultivation for subsistence by the population marginalized by the growth process during the nineties.

The result of present survey while corroborates the NSS estimates showing predominance of sharecropping as the predominant form of tenurial arrangement, but the figures for the fixed produce contract and cash/fixed money contract differ to quite an extent. Neverthess, fixed produce contract system is found to the second most prevalent system of contract, followed by fixed-money/cash rent system, which is the order of things emerging from the NSS estimates. Sharecropping emerged out to be the predominant form of tenurial arrangement between the landowners and the tenants, with 32.56 per cent of the households leasing-in getting it on a sharecropping basis. While 27.91 per cent of leasing-in was found to be on a fixed produce rental agreement and 25.58 per cent of the land leased-in was on the basis of fixed money.

Table 13.8: Terms of Land Leasing in Sample Districts

(in %)

District	*Terms of Leasing*					
	Share of input/output	*Share of output only*	*Fixed Money*	*Fixed output*	*Others including No term*	*N*
Nawada	7.69	53.86	7.69	15.38	15.38	13
Purnea	11.11	11.11	22.22	44.44	11.11	9
Samastipur	-	-	63.64	36.36	-	11
Bhojpur	10.00	60.00	10.00	20.00	-	10
All	**6.98**	**32.56**	**25.58**	**27.91**	**6.97**	**43**

Source: Primary Investigation.

Table 13.8 also shows that about 6.97 per cent each of the households leasing-in was on a share of input and output both and on some unidentified including no fixed term. District-wise details can also be seen in Table 13.8. Fixed money tenurial system was most popular in Samastipur district, while share of produce was popular in Bhojpur and Nawada districts.

Regarding duration of land leasing contracts, it considered data on percentage distribution of total leased-in land under different contracts periods—less than one year, one to less than 2 years, 2 years to less than 5years and 5 and more years. Such distribution for different categories of

Table 13.9: Duration of Leasing by Size of Operational Holdings

(in %)

Holding Size/ Duration	*Landless*	*Marginal*	*Small*	*Medium*	*Large*	*All*
Less than 1 year	(10.00)	(24.00)	(20.00)	(50.00)		(20.93)
1 to less than 2 year	(40.00)	(44.00)	(60.00)	-	-	(41.86)
2 to less than 5 years	(30.00)	(16.00)	(20.00)	-	-	(18.60)
5 & more years	(20.00)	(16.00)	-	(50.00)	(100.00)	(18.61)
N	10	25	2	5	1	43

Source: Primary Investigation.

operational holdings are given at Table 13.9. It shows that about 62 per cent land-lease contracts were for less than 2 years, while 18.60 per cent of the leased-in land families contracted for more than five years. Another important point to note is that there was tendency among the smaller tenants to get engaged in shorter duration lease contacts, while tenants from higher categories being engaged more in longer duration of lease contracts. Table 13.9 shows that while the percentage of lease-in land of tenants from small holdings under lease contract of less than two years has been 80 per cent, followed by marginal holdings (68%) and landless (50%). It seems that the land-lease contracts are dominated more by shorter duration contracts.

There is an important conclusion that the land and tenancy reforms measures adopted so far in the State are not satisfactory in establishing more equitable land distribution pattern and ensuring better access to tenanted lands by the lower categories of holdings. From the perspective of the peasants, the effect of liberalization of land-leasing might not be a happy one in all cases although supply of land for tenancy cultivation might improve.

We may now ponder about some possible consequences of liberalization of land-leasing:

1. It is possible that many households from higher categories who until now abstained from land-lease market due to legal prohibition of land leasing, but were eager to augment their scale of farming further, would like to engage in land-leasing-in a competitive situation created after withdrawal of restrictions of land-leasing.
2. Liberalization of land-leasing might result in emergence of a

new class of landlords particularly, absentee who harp on maximizing their earning through multiple channels including leasing-out of land.

3. The process of equalization of landholdings among the rural households through sale of land by big and purchase by small might shrink further with liberalization of land-leasing.

In the light of above observations following safeguard measures may be suggested to protect the interest of the group of landless marginal and small holdings in the market for land-lease as well as the landowners right:

- The landowner's right of resumption of leased-out land after the expiry of lease duration may be recognized to motivate more landowners to lease-out land.
- A mutual land-leased contracts between the landowner and the tenants with their terms and conditions should be made on non-judicial stamp paper as being practiced between house owner and the tenant. In the event of contracts violation by either party, the matter may be amicably settled locally with the help of *Panchayats* instead of pushing the matter to legal court.
- The contract document may be treated as valid document to obtain loans from financial institution.
- The duration of land-lease contracts needs to be appropriately specified. Longer-duration contracts would encourage the lessees to undertake investment on leased-in land which would raise agricultural productivity and production.

Conclusion

On the basis of the information gathered from different sources including field survey conducted in the selected districts of Bihar, it may be concluded that official figures of leased-in area are underreported and the gross underestimates of actual extent of tenancy. It is widely acknowledged that informal leasing is less productive and more exploitative.

Suggestions amd Recommendations

- Legalize land-leasing to promote both agricultural efficiency and equity. It will definitely increase occupational mobility of people.

- Legalisation of leasing. It will improve the availability of land in the lease market and also poor people's accessibility to land through leasing.
- Allow automatic resumption of land after the agreed lease period.
- Entitle and facilitate all tenants including sharecroppers to access short-term bank loan. It will discourage land being kept fallow and result in better utilization of land, labour and increase farm output.
- Remove the clause of adverse possession of land in the tenancy law.

14

Legalising Agricultural Tenancy: A Study in Odisha

Padmaja Mishra

Agriculture in India is in crisis today. Along with high rate of growth and the structural transformation that India is experiencing, agricultural sector has lagged behind. Its relative contribution to GDP has declined along with its own rate of growth and crop productivity is stagnant. The paradox is that it is still the biggest sector of employment and the major source of livelihood for millions. But, area under cultivation and input uses are also declining. To feed the growing population and to provide livelihood to millions, agriculture has to be productive and profitable. Widespread rural poverty and the recent crisis of high food inflation have led to the realization that any development effort will be nullified without due consideration for agricultural development. Therefore, increasing agricultural production and productivity is now an important objective before the policy-makers. Increasing public expenditure in agricultural infrastructure has been advocated and accepted in the 11th Plan by prescribing a 3-fold increase over the last plan. Increased public and private investment in agriculture is expected to revive the overall health of this sector with increased productivity.

Productivity in agriculture is mainly dependent on two sets of factors: technological and institutional. Traditionally, the older school of thought believes that agricultural productivity is purely a technological phenomenon and can be raised by the application of superior agricultural methods. Among the technological factors are the uses of agricultural inputs and methods such as improved seeds, fertilizers, improved ploughs, tractors, harvesters, irrigations etc. All these factors help to raise productivity, even if no land reforms are introduced. On the other hand, the institutional reforms, as advocated by the socialist school of thought, include the redistribution of landownership in favour of the cultivating classes so as to provide them a sense of participation in rural life, improving the size of farms, providing security of tenure, regulation of rent etc. Also these institutional factors, such as the existence of feudal relations, small size of farms, subdivision and fragmentation, insecurity of tenancy rights, high rents, etc. are the bottlenecks in raising production. However, as supported by practical experiences, both the schools of thought are converging and opinion has come to the idea that land reforms and technological changes are not mutually exclusive factors but are complementary in the process of agricultural development.

The present chapter argues that actually it is the agricultural institutions, particularly the agrarian relations, which determine all other factors necessary for productivity and profitability such as investments, technology adoption, input use, cropping pattern and marketing etc. Economic efficiency is a combination of technological and allocation efficiency. It is the institutions which combine the both.

Land reforms in India have been the major sources of institutional reforms. The purpose of land reform including tenancy reform is twofold. On one hand, it aims to make more rational use of the scarce land resources by affecting conditions of holdings, imposing ceilings and floors on holdings so that cultivation can be done in the most economical manner, i.e., without any waste of labour and capital. On the other hand, it is a means of redistributing agricultural land in the favour of less privileged classes, and of improving the terms and conditions on which the land is held for cultivation by the actual tillers, with a view to end exploitation. A very high correlation exists between poverty and access to land and land reforms have aimed at ensuring a fair and equitable access to land.

Land being the principal form of rural wealth, a representative of social status and political influence and a major source of household income, rural development revolve around land distribution pattern. Land ownership is highly valued for both economic and non-economic purposes such as social, cultural and political (Hill : 1986, Platteau : 1995). A clearly defined rights of access to land is at the centre of agricultural land use pattern and productivity. It is both a determinant of and determined by economic, social and political relationships in rural areas. Rights to access to land can be the most powerful resource to affect rural income and poverty. Combined with labour and other resources such as financial, physical and social, it can be the main factor behind sustainable livelihoods. The goals of economic efficiency and equity can be achieved with a proper land policy and can go a long way towards accelerated inclusive rural development. Agricultural land tenure is perhaps the most discussed aspect of agriculture and its reforms as it has a multidimensional importance.

The entire concept of tenurial reforms aims at the abolition of intermediaries and bringing the actual cultivator in direct contact with the state. The provisions of security of tenancy and rent regulation provide a congenial atmosphere in which the agriculturist feels sure of reaping the fruits of his labour. The scope of land reforms therefore entails abolition of intermediaries, tenancy reforms, i.e., regulation of rent, security of tenure for tenants and conferment of ownership on them. It also focuses on land ceiling and landholdings, agrarian reorganization including consolidation of holdings and preventing of subdivision and fragmentation and organization of cooperative farms.

Land tenure practices have varied over time and space. Impacts of different land tenure practices all-over the world have shown a strong bearing on land use pattern, cropping intensity, input use intensiveness and therefore, have a strong impact on productivity of the land. Different practices of tenure such as share tenancy or tenancy for fixed rent or fixed quantity have different implications on allocative efficiency and output. Various studies have indicated that sharecropping results in inefficient allocation of resources and reduced incentives for raising productivity of agricultural land (Hayami and Ostuka : 1993). Owner cultivated land and fixed rent lands are most productive as less restrictions are imposed by the landowners while sharecropping leads to loss of long-term investment incentives of the tillers (Ahmed *et al.*: 2002). But at the same time it is also observed that the above findings may vary with the size of the land. Small and marginal farmers have been found to have benefited by tenancy as they get land to cultivate and some support from the landowners which they could not have managed otherwise. At the same time, it also helps the landowners as they probably lack necessary manpower and time for self-cultivation.

The present chapter examines tenancy practices and its impacts on agriculture in the case of Orissa which is a pioneering state in India with regard to land reforms. The following part of the chapter is arranged in five sections. Section 2 introduces Orissa economy with special reference to its agricultural scenario. Section 3 discusses tenancy in Orissa and gives a theoretical discussion on the possible impacts of such practices. Section 4 is based on empirical observations gathered from a field survey undertaken in two different districts of Orissa and examines the theoretical arguments with the help of field level information with special reference to sharecropping and allocative efficiency. The final section presents policy implications which argue for careful legalization of tenancy for increasing agricultural productivity as derived from the empirical findings.

Orissa Economy and Agriculture

Orissa is an eastern coastal state of India having 4.9 per cent of the total area and 3.6 per cent of the total population of the country as per 2001 census with a density of population at 236 per sq.kms as against all-India average of 324. Its per capita NSDP is now 71 per cent of India's per capita NNP in 2008-09. 46 per cent of its total population live below poverty line while all-India average is only 27.5 per cent. 85 per cent of its population live in rural area as per 2001 census.

Orissa is an agriculturally predominant state with low productivity. The sectoral contributions to NSDP clearly reveal that the share of agriculture is declining (from a huge 56.58% to 23.45% in 2007-08). It is due to the poor performance of the agricultural sector and better performance of the other sectors which is indicated by the sectoral growth rates over the years along. The growth rate of agriculture has not picked up for a long period of time with a growth rate of –1.8 per cent annually at constant prices during the post-reform period. As per 2001 census, the proportion of workers to total population in Orissa is 38.9 per cent out of which the main workers constitute 67.1 and the marginal workers accounted for 32.9 per cent. But, the proportion of workers in rural Orissa is as high as 89 per cent where 29.7 per cent are cultivators, 35.1 per cent are agricultural labourers and a meager 4.8 per cent are others such as household industrial workers. It is noteworthy that the proportion of cultivators are declining over the decades whereas that of the agricultural labourers are increasing. This is a clear indication to the fact that landlessness is increasing day by day along with rural poverty. However, a total 65 per cent of the state's total workforce are engaged in this sector. The per capita availability of cultivated land was 0.39 ha. in 1950-51 and is 0.14 ha. in 2007-08. In view of increasing pressure on land, land use intensity in the state has been increasing and has reached 13.55 per cent in 2000-01 from 1.09 per cent in 1971-72 and land fragmentation has resulted in more number of marginal and small operational holdings. The share of marginal holdings are increasing both with regard to numbers and area covered. Net sown area is declining and it is now (2008-09) only 3.6 per cent of the total landmass. Share of net irrigated area to net sown area is only 31 per cent (44% for all-India) and fertilizer consumption is less than half and yield of paddy is less than 3/4th of the all-India average (15.29 qntl. per hect. as against 21.86 qntl/hect for all-India). Table 14.1 gives a bird's eye view of agriculture of the state.

Table 14.1: Agriculture in Orissa

Indicators	*1961*	*1971*	*1981*	*1991*	*2001*	*2008-09*
Share of agriculture in NSDP (%)	66.8	54.6	30		28.13	19.95
% of total workforce in agriculture	73.8	77.4	74.7	73.0	64.73	58.25
% of cultivators to main workers	56.8	49.2	46.9	44.3	44.2	N.A.
% of agricultural labourers to main workers	17.0	28.3	27.8	28.7	52.24	N.A.
Per capita availability of cultivated land	0.38	0.31	0.20	0.18	0.17	0.15

Source: Economic Survey, Government of Orissa and Statistical Abstracts of Orissa.

Tenancy in Orissa

During colonial administration, different parts of Orissa were under different provincial administration and after gradual expansion of the British rule, a number of land revenue systems emerged. While *Zamindari* system was prevalent in coastal area, *Raiyatwari* system existed in western parts. There was no law to protect the interests of the tenants before the Orissa States Order (1948) which conferred occupancy rights on tenants. Land legislations in independent India were motivated by 'land to the tiller' idea with an objective to end an exploitative and iniquitous system. Given the national mandate, individual states had their own land legislations as land is classified under the State list. The Government of Orissa has realized that access to and control over land resources hold the key to poverty alleviation and land reforms is a prerequisite for rural development. Therefore, the state has been a pioneering one in legislating land reforms. As a part of land reforms, land ceilings have been imposed. The average amount of land distributed per family in Orissa is 1.1 acres. A huge proportion of the agricultural holdings are extremely small and suffer from size disadvantages. Though agricultural policy of India has many a times emphasized the need for the development of lease market, tenancy legislations and the provisions at the state levels show that there is often complete ban or restrictive leasing laws. Concealed tenancy is a fact and nearly 22 per cent of cultivator households are tenants as supported by NSS survey and other micro surveys. A study in Bihar shows 34 per cent tenant households as against 4 per cent in NSS data. Available data on the number of tenants and acreage under tenancy is always underestimated because there is a fear amongst the landowners that the tenants might claim the legal occupancy right (as per adverse possession rule, after practical possession for 12 consecutive years) and they also want to escape the ceiling on land.

Orissa State Order, 1948 conferred occupancy rights on tenants. Orissa Land Reforms Act, 1960 and its amendments (1965, 1973, 1974), have banned leasing of land except by certain specific categories of *raiyats* such as widows, single women, minors, physically or mentally disabled etc. and has put a maximum limit of 1/4th of the gross produce on the rent. But, legislative ban of tenancy has only led to informal, shorter, seasonal and less secure tenancy than before. Moreover, by bearing explicitly on tenancy, the law has swept the problems of sharecropping under the carpet and no provision or efforts have been made to record concealed tenancy. Most of the tenurial agreements are verbal and most of the tenants don't even know about the Acts. Therefore, mostly the leasing practices have been informal and illegal. Usually, three forms of tenurial contracts are found in practice. One kind is that the landowners hire labourers to cultivate their land for a fixed wage. The second type is that of the fixed rate system where the landlords get a fixed amount as rent and the tenant gets the rest of the product. The third and the most prevalent type is the sharecropping pattern where the relative shares of the landlord and the tenants are determined as per a pre-fixed agreement.

As regards incidence of tenancy, state-wise tenancy statistics for rural India shows that Orissa belongs to the category of high tenancy states and has a very high percentage of leasing-out and leasing-in households compared to that of other states. It is estimated that nearly 20 per cent of the households participate in land lease market in rural Orissa, though with lots of inter- and intra-regional variations. Table 14.2 gives a picture of land-leasing in rural Orissa based on NSS survey reports.

As per the NSS reports, Orissa has high percentage of tenant households (nearly 25%) constituting 11.4 per cent of the total operated area and the average area leased is 0.449 hectare. As is evident from the

Table 14.2, the tenurial arrangements are clearly dominated by small and marginal farmers as well as small and marginal landowners.

Table 14.2: Distribution of Lessor and Lessee Households (HHs) and Leased-in and Leased-out Area by Size Class of Land Ownership in Rural Orissa

Size class of ownership (in Ha.)	*Total lessor HH (in %)*		*Total tenant HHs (in %)*		*Average leased-out Area (in ha.)*	*Average leased-in area (in ha)*
	1991-92	*2002-03*	*1991-92*	*2002-03*	*2002-03*	*2002-03*
Marginal (Below 1.01)	63.96	73.90	89.21	97.00	00.26	00.47
Small (1.01-2.00)	25.96	14.80	06.42	02.99	01.09	00.32
Medium (2.01-4.00)	07.01	08.70	03.56	00	01.77	0
Large (4.01-10.00)	02.38	01.90	00.81	00	03.34	0
Above 10.00	00.69	00.70	00	00	04.79	0
All sizes	100	100	100	100	0.689	0.449

Source: NSS reports 48th round and 59th round.

The major manifestation of tenancy in the state is sharecropping though fixed tenancy in cash and kind are common in western Orissa (Sarap : 1998). As per NSS 48th round statistics, leased-in area under share tenancy is increasing and constitutes more than 50 per cent of the total. The next prevalent form of tenancy contract are fixed rent and fixed produce type.

The vital question is: how such agrarian relations have affected our agriculture? How is the land used under various tenancy agreements? Indicators like land use intensity, cropping pattern and input use intensity can be studied to understand the productive use of the agricultural land. In a state like Orissa where more than 65 per cent of the households depend on agriculture for their livelihood and a fixed or diminishing supply of cultivable land is causing increase in the demand-supply gap associated with very skewed landownership, implications of tenancy is very crucial. Shortage of cultivable land, small and marginal size of the holdings and insecurity of tenancy have contributed to agricultural stagnation. Land use intensity is bound to increase to sustain a growing population. The Economic Survey of Orissa (2009-10) shows a decline in net area sown from 5845 thousand hectares in 2001-02 to 5604 thousand hectares in 2008-2009 and while the cropping intensity has grown from 135 per cent to 1625 during the same period. Orissa is one of the lagging states but with an increasing trend so far as input use intensity is concerned. Per hectare application of fertilizers and pesticides are one of the lowest but rapidly increasing. With this background, our intention is to study the implications of tenancy on production and productivity with the help of empirical findings.

The Empirical Study

An empirical study has been undertaken in two different districts of Orissa to cover different agro-climatic zones and socio-economic conditions. While Ganjam district has been chosen from the coastal plain land, the district of Bolangir has been taken from the central table land. In coastal Orissa there is more landlessness and more and more land are going to non-farm uses in comparison to other areas. As per yield rate of food-grains, the district of Ganjam occupies the highest position in Orissa, Bolangir occupies the 7th position. Two villages each from the two districts are taken. While the villages selected in Ganjam district are irrigated, the villages selected in Bolangir district are unirrigated and depend on tanks and tube-wells for water. These are Dura village from the Dura gram panchayat and B. Jaggannathpur from Bori gram panchayat in Ganjam and Khuntasamalei village and Daitarimunda village from Khuntasamalei gram panchayat in Bolangir. While the villages in Ganjam are well connected to urban municipal area, the villages in Bolangir are relatively more remote with poor connectivity to the nearest town Patnagarh.

Table 14.3 gives a summary picture of the two districts as per the information available in Economic Survey of Orissa (2009-10).

Table 14.3: A Comparative Picture of the Sample Districts

Sl. No.		*Ganjam*	*Bolangir*
1.	Per capita NDDP (in Rs. at 1999-2000 const. prices)	10994.00	11650.00
2.	Area under HYV Cultivation (in hect.) (2008-09)	275868	188328
3.	Yield rate (Qntl/hect.)	25.67	22.80
4.	% of irrigated area	42.11	27.02
5.	Consumption of fertilizer (N+P+K in thousand MT)	40136	19710
6.	Ratio of cultivators to agricultural labourers	1.52	1.82

Regarding the sample respondents, a total of 400 households have been selected randomly with 100 samples from each village. A questionnaire was administered to collect the necessary information with a focus on the agricultural activity under tenancy. A picture of the socio-economic profile of the sample is given in Table 14.4.

Table 14.4: Mean Socio-economic Profile of the Sample

Sl. No.	*Variables*	*Ganjam*	*Bolangir*
1.	Household income (Rs. per annum)	60356.00	55680.00
2.	Farm income	35,000.00	48,032.00
3.	Non-farm income	25,356.00	7, 648.00
4.	Yield level (HYV paddy) kg/hectre	1455.2	1025.4
5.	Operational holdings (hectares)	1.01	1.05
6.	Fertilisers used (kg per hect.)	36.5	25.1
7.	Educational qualification (average years of schooling)	07	05
8.	Earner-dependent ratio	4	3

Looking at the composition of the sample, it is found that the sample is dominated by cultivators who have some amount of own land as well as some leased-in land for cultivation. There are also some pure owner cultivators who cultivate their own land only and some pure tenant cultivators also who do not own any land of their own for cultivation. Therefore, the total sample has been divided into three categories: Owner cultivators (OC), partly owner and partly tenant cultivators (PT) and purely tenant cultivators (TC). Table 14.5 gives the distribution of the sample in both the districts according to their tenancy status and average size of holdings.

Table 14.5: Distribution of the Sample as per Tenancy Status and Size of Holdings

	Marginal	*Small*	*Medium*	*Large*	*Total*	*Average holding size (in acre)*
Ganjam					**200**	
OC	2	10	22	14	48	8.5
PT	22	44	27	02	95	6.5
TC	18	37	22	00	57	5.2
Bolangir					**200**	
OC	5	10	21	15	51	10.6
PT	15	47	22	–	85	6.8
TC	10	22	14	–	64	5.6

As is evident, in both the districts purely owner cultivator class is dominated by farmers having large size of holdings and purely tenant cultivators are obviously do not own any agricultural land and cultivate small size landholdings on lease or rent. The overall picture shows that the tenant cultivators mostly belong to small and medium farmer category. As the tenancy status and farm size seem to be correlated, farm size must be interacting with impacts of tenancy. This finding more or less supports the NSS findings. As per NSS statistics, nearly 84 per cent of the leased-out lands belong to marginal and small farmers and in Orissa it is still higher, i.e. 88.75 per cent while the landless and small and marginal farmers together account for 86.1 per cent of the tenants who lease-in. Again, it is the lower size class of holdings which dominate and the share of large holdings in leased-in land has been almost zero. It seems that lots of fragmentation and less number of large holdings have resulted in this kind of incidence of tenancy.

The question obviously arises why tenancy and leasing activity takes place at all? Why landowners do not self-cultivate and lease-out and why people lease-in? Is it by choice or by compulsion? Is the contract beneficial to both the parties and describe a positive sum game? Finally, is the arrangement beneficial for agricultural productivity?

In both the districts the reasons for leasing-in and out are found to be similar which are listed in Table 14.6.

Table 14.6: Reasons for Leasing-out and Leasing-in Order of Priority

Sl. No.	*Leasing-out*	*Leasing-in*
1.	Lack of manpower, labour and time	Insufficient land and availability of Labour, bullocks etc.
2.	Lack of interest for cultivation	Limited alternatives and non-farm options
3.	Uneconomic nature of holdings	Consolidate operational holdings

Source: Own sample survey.

The above picture clearly shows that tenancy is an agreement (informal) between labour scarce landlords and land scarce tenant farmers. The question is have such arrangements of trading of land between labour scarce owners/farmers and land scarce farmers been efficient and equitable? This is to be examined in the following discussion.

As mentioned earlier, tenancy can be of different forms as terms of tenancy differ. Though sharecropping is found to be prevalent form of practice in Ganjam, the sample in Bolangir shows a number of cases under fixed rent arrangement along with share tenancy. In the sample villages of Ganjam, we find only share tenancy and the product is shared on 50 : 50 basis if the owner makes investments. If they don't then 1/3rd of the produce is given in the form of rent. The dominant crops are paddy and black gram in this area. It is found that even if irrigation facilities are available, crop diversification is minimum as the tenants do not have the resources to adopt a diversified cropping pattern and take the advantages of irrigation. They don't want to move away from traditional crops as individual risk will be more. Moreover, it is the landowner who dictates what is to be produced.

In the sample villages taken from Bolangir district, the terms of tenancy are not confined to share tenancy as many cases of fixed rent tenancy have also been reported. It is invariably observed that for paddy cultivation which meets self-consumption needs, share tenancy prevails. But, in case of cash crops like cotton or sugarcane cultivation, fixed rent tenancy prevails. The sharing is on 50 : 50 basis and the rate for fixed rent is about Rs. 4000 per acre. The sharing ratio comes to 30:70 if the tenants undertake all the investments.

Therefore, it can be concluded that the terms of contract basically depend on the crop to be cultivated on which the risks and returns depend and the sharing ratio depends on the share of investments of both the parties. For low value paddy cultivation which is a subsistence crop and primarily meant for self-consumption, shared tenancy is noticed while for high value cash crops like cotton and sugarcane farming, fixed rent is preferred. It is also to be noted that paddy is labour-intensive and less risky in comparison to cotton which is capital-intensive and more risky. For such reasons probably, in agriculturally advanced states like Punjab, Haryana and Tamil Nadu fixed tenancy is more prominent than share tenancy.

Land-Leasing and Agricultural Productivity

Since there is rampant concealed land-leasing for agriculture in Orissa, we try to find out its effiiciency and equity outcomes. We focus on sharecropping since this is the most prevalent form of practice. There is a huge literature on the efficiency and equity implications of tenancy in general and sharecropping in particular. The opinion varies from classical disapproval and conditional approval of the neo-classicals to the Marxist disapproval on grounds of stagnation and exploitation.

Sharecropping is a form of land-leasing-in which the tenant and the landlord share the final output as compensation to labour supplied by the former and the land capital by the latter. It turns out as a method of sharing output as well as risk. Is it only sharing of risk or also provide incentives for both. Does it involve moral hazard? Two things are found to be important in the literature to answer such questions and to conclude on resource use or allocative efficiency. These two are: the contractual form and the contract length. In other words, the terms of contract and the degree of security of tenure are the two things those matter. The general arguments given in the literature are as follows.

Marshallians argue that land use will always be sub-optimal under share tenancy due to lack of incentives and inefficient production pattern. Because the sharecropper will always use the variable inputs up to the point where its marginal value product is equal to its marginal cost. Therefore, incentives for cultivation will depend on the sharing pattern. For a 50: 50 sharing arrangement, unless the return is double the cost of the factors of production. However, situation will be different depending on the sharing of the input cost as well. It is also argued that inefficiency can only prevail temporarily as competition among the tenants will make it unsustainable. Efficient contract between the landlords and the tenants can be negotiated to avoid inefficiencies (Cheung: 1969) conditional to the fact that land rights and lease markets are well defined. A counter-argument comes from another school of thought (Johnson: 1950) which says that insecurity of tenancy and threat of eviction may induce the sharecroppers to do their best and input use can be optimum. When the prospective tenants compete among each other for limited available cultivable land, the landlord would try to maximize his surplus and the tenants also try use their resource optimally. Therefore, due to competition in the market, an optimum is automatically arrived. As both the parties face an opportunity cost in cultivation of the land, that itself determines the incentive compatibility between both the agents and that is the reason for which share tenancy exists at all (Bardhan: 1983). Since landlords are risk neutral and the tenants are risk averse sharecropping is the second best as it trades off risk sharing and incentives.

There is a great deal of interlinkages between various submarkets in the underdeveloped rural areas. Both the landlords and the tenants face the markets for land, labour, credit etc. which are not independent of each other. Both the parties have to enter into various simultaneous informal agreements over labour, over credit, over risk hedging, over other inputs and so on. And, the scene is characterized by imperfections, asymmetry of information and moral hazard problems. As per the neo-classicals, sharecropping provides a method of interlinking of the submarkets to improve allocative efficiency in agriculture. When the landowner cannot monitor or supervise, provides incentives through consumption loan and other assistances to derive better work effort and commitment from the tenants (Braveman and Stiglitz: 1982). But, Marxists look at the whole situation as exploitative where the powerful landlords exploit the poor resource weak tenants. Where the weaker party gets the subsistence, the surplus ig whipped by the richer and powerful. Therefore, it violates the principle of efficiency and equity.

With these arguments in mind, we examine the productivity implications of land-leasing and tenancy in our empirical context.

Productivity Implications

In view of increased population and food insecurity, optimum use or most productive use of limited agricultural land is required. There are two ways to increase production or productivity of the land. Either expand the net area of cultivation or intensify cropping. Since independence, net sown area has already risen by more than 20 per cent and there is not much scope for further increase. Thus, raising cropping intensity is probably the only option at present for better land use.

Cropping pattern and cropping intensity are the two crucial indicators to understand the characteristics of a farm economy and also production and productivity. Cropping pattern should follow the land type and soil characteristics. In the irrigated villages of Ganjam district, paddy is cultivated during Kharif season and black gram and green gram during the Rabi season. Beyond this double cropping, no further crop diversification is seen except for very little vegetable cultivation. It is again observed that ownership and tenurial status of a household have not affected cropping pattern very much. Whatever little difference is found, we can indicate that diversification declines as one moves from ownership to tenancy status. As the sample villages are dominated by tenant households and marginal and small farmers with very few large farmers found as purely self-cultivators, risk aversion has been the main motive behind the cropping pattern. Land type and soil characteristics have not played any important role as farmers always prefer to do what others are doing to minimize individual risk of supervision, marketing etc. Data on cropping intensity as measured by the ratio of gross cropped area to net sown area is 1.57 on an average with slightly higher figure for the owner cultivators (1.61) and lower figure for pure tenants (1.30).

In the unirrigated villages of Bolangir, cropping pattern is different but not more diversified. Cash crop cultivaton like cotton and sugarcane are widely undertaken along with paddy. The large and medium farmers and owner cultivators have gone for cash crops. Some fixed rent tenants also do cash crops. Paddy cultivation is mostly observed under share-cropping arrangement. Cropping intensity is found to be lower in this area (1.25 average).

Though HYV cultivation using irrigation water and fertilizer is highly productive, if the inputs are not applied in time, in required quantity and in right combination at the appropriate stages of growth, yield reduces substantially. This is exactly what all the tenants have complained. Given their low risk absorption capacity, they cannot cope with production risk as well as price risk. Therefore, they don't go for much crop diversification. Cropping pattern is decided by the farming community in the village as a whole as no one wants to take independent risk for the input use or crop supervision or marketing. It is found that even if there is irrigation facility, tenants do not have the resources to adopt a diversified cropping pattern and to take advantage of irrigation. Land quality remains ignored which actually results in suboptimal use of the land as well as other resources.

Use of fertilser in improper ratio is a problem with Indian agriculture. This induces initial vegetative growth, susceptibility to pests and diseases, poor floral induction and delayed maturity and thereby reducing the yield. So also is the impact of indiscriminate use of pesticide. The field level information has indicated that the tenants are very much aware of these problems and have complained of the lack of timely availability of block level assistance on quality seeds, fertilizers and pesticides. At the same time, they feel the repeated pressure from the landlords to use fertilisers and pesticides for increased production on the one hand and their own resource constraint on the other hand.

The present study is also intended to look into any productivity difference between own cultivated and tenant cultivated lands. It has been much discussed in the literature that productivity is adversely affected under tenancy due to lack of incentives. Many empirical studies have been carried out to examine any productivity difference between owner operated farms and tenant operated farms. Rao (1971) finds that the former category produce more output per acre but the reverse is true if farm size is held constant. Similarly, Shaban (1987) by using broad based ICRISAT data finds that farmers are more productive in their own land in comparison to leased-in land and has noticed less use of family as well as bullock labour in leased-in land. Bhaumic (1991) also supports this as he finds more use of all sorts of inputs in owned plots. Hence, an incentive problem is indicated. But, on the other hand, Chattopadhyay and Sarkar (1997) observed no such significant difference except for fertiliser use and irrigabilty of the land. Therefore, the difference can also be attributed to land quality heterogenoty.

To examine the incentive implications, there are two ways to differentiate between own cultivation and tenant cultivation as implied by the categorization of the cultivators as given in Table 14.4. One way is to strictly differentiate between owner cultivators (OC) and tenant cultivators (TC) and the other way is to differentiate between owner cultivated land and tenant cultivated lands for the part tenants (PT) as maximum sample respondents fall into the part tenant category who also have some land of their own to be self-cultivated. Statistical *t*-tests have been undertaken to test the significance of average yield rate differences. Going by the two kind of differentiation, the following interesting observations are noted as follows:

(a) It is found that there is a difference between purely owner cultivators and purely tenant cultivators with regard to per hectare output though the difference is statistically insignificant (*t* value =1.87).

(b) A statistically significant difference (*t* value = 2.98) is observed between the productivity of own land and leased-in land in case of part tenants.

These findings lead us to conclude that leased-out lands have been less productive than the owner cultivated lands. The difference is less prominent in case of fixed rent contracts, mostly for cash crops as found in Bolangir district. Otherwise, the findings have been similar between the two sample districts. The dominant reasons for the productivity difference between owner cultivation and tenant cultivation as gathered from the respondents are listed as follows:

(i) The lands leased-out are inferior lands (unirrigated, unproductive, distant patchy lands and waste lands).

(ii) Tenants are resource constrained mainly due to lack of institutional finance.

(iii) They cannot take advantage of organized market (village *mandi*) and other government schemes for farmers due to lack of identity as farmers.

(iv) Input utilization and cropping intensity have been adversely affected because of lack of ownership and decision-making power of the tenants who are the real tillers.

(v) Tenants give more priority to resource use in their own pieces of land in comparison to the leased-in land due to the disincentives of benefit sharing and uncertainty of tenure.

It is invariably observed that the tenants give priority to their own land in terms of resource allocation be it family labour, bullock labour or other inputs such as seeds and fertilisers On the basis of the evidence gathered we have the following observations on tenants' incentive which impacts productivity under share cropping:

(1) In case of pure tenant farmers, incentive loss is minimized to some extent when they have no other options for better utilization of their resources and there is a constant fear of termination of lease.

(2) In case of part tenants there is a huge loss of incentive for leased-in lands and resource utilization priority goes to own land as leasing is temporary and product is shared.

Sustainability or Long-term Productivity Implications

Besides the above facts related to short-term productivity, one interesting finding has been made with regard to soil conservation and land developments which matters for sustainability and long-term productivity of the land. The owner cultivators are found using organic compost whereas chemical fertilizers are used in leased-in lands. Even small tenants prefer to sell their compost at a higher price and purchase fertilizers from block office at a lower price to be used in the leased-in lands. This is clearly observed in sampled cases from Ganjam district. This indicates that the tenants do undertake unsustainable farming practices which is harming the soil quality and reducing long-term productivity. There is a problem of trade-off between short-term incentives and long-term incentives also. They try to maximize short-term production to secure their tenancy but at the same time may not have the incentive to protect the value of the land and may not go for soil conservation because of insecurity of tenancy (contract length is usually one or two years) may hamper productivity in the long run. It can be concluded that care intensive activities like soil conservation and maintenance have been adversely affected due to lack of property rights of the actual tillers. In this context, expected life of the contract is important and therefore, security of tenancy is advocated. In this context, the importance of contract length has been much discussed in the literature. While Besley (1995) finds a positive correlation between land rights and investments and indicated better defined property rights and tenure stability avoids expropriation of long-term investments. Banerjee *et al.* (2002) find in context of West Bengal that increased tenure stability is associated with increased average land productivity. The landowners very often are found to have used the carrot and stick policy by taking advantage of concealed tenancy and make it more in secured by granting short-term leases. The tenants in such cases are bound to give satisfactory performance in fear of termination of tenancy. In such cases, there is less reason for tenancy being less productive. But, definitely becomes more feudal.

Equity Implications

As observed in case of our sample study, both landowners and the tenants have the freedom regarding the terms of contract and they can choose any tenant or any landowner. But, in practice, the tenants being resource poor have to depend on the landlord for many things, mostly financial loan. Therefore, interlinking of the factor markets leads to some form of coercion depending on the nature and degree of dependence—both economic and extra economic. Though the exploitative relation in feudal form of agriculture before independence is a past, it is still the prerogative of the landowners to decide on the terms of contract and the contract length. It is not coercion or exploitation by the landlords but the compulsion of the overall macro-economic realities and power imbalances implied by that. In sum, if share tenancy pays more than agricultural wages for a labourer then the poor farmers enter into leasing as per landowners' terms as they are not skilled enough to take other non-farm jobs.

Policy Conclusions

The prime pre-condition for inclusive growth, poverty reduction, food security and rural development in India is to revive agriculture. In this context, an appropriate institutional framework is a basic necessity to take advantage of any other pro-agriculture macro-economic policies undertaken by the government. There have been lots of policy dialogues around agricultural tenancy reforms. The existing reforms have been redundant and counter-productive. A proper designing for land tenancy is yet to be made which needs to combine both efficiency and equity concerns. Different stakeholders have conflicting views on legalization of tenancy which is often advocated for bringing a secured property right.

On efficiency grounds, it is often argued that the goal of increasing the size of holdings to allow scale economies and higher productivity can be accomplished though legalizing tenancy (Rao and Gulati: 1994, Vyas: 1994). Expansion of farm size is necessary to reduce production cost as small size farms with insecure tenancy have negative impact on production and productivity. It has been argued that increase of size holdings allowing scale economies and higher productivity can be accomplished through tenancy and may be more so by legalizing it. Removal of land ceilings have been advocated on such grounds of promoting viable size of the holdings, development of capitalist agriculture that leads to better allocation of resources and keeping parity of the agricultural sector with other sectors of the economy. Legalization of tenancy is also advocated on equity grounds as it can bring livelihood security and empowering land rights or user rights for the poor. Official agricultural policy documents also emphasise the need for the development of land lease markets. The tenancy laws aiming at ensuring security to tenants, fixing fair land rents, permitting owners to resume land for self-cultivation have been framed to make agriculture more attractive, non-exploitative and productive. However, due to many loopholes in the implementation, several deficiencies still remain because of which the goals have not been achieved. Rather, the uncertainties and lack of clarity have created a kind of fear and distrust among all the agencies which has adversely affected agriculture. For example, the landowners also have the fear of losing their land under the adverse possession rule and rotate the land among different tenants adding to concealed and insecure tenancy. Therefore, a well developed lease market is necessary.

In the backward rural economy of Orissa where wages are low, alternative employment opportunities are less, the small and marginal farmers can enlarge their holding size and reap the benefits of economies of scale if lease market is legalized. There is a relative consensus among economists that better defined property rights encourage investment decisions and a reliable legal system should positively impact land leasing efficiency. It is argued that security of tenure is necessary for increasing investments in the land as the tenants will have the incentives of receiving greater share of returns from their investments with reduced uncertainties. But, what about the resource constraints? Land rights, financial and other input provisions are other necessary complementary requirements. Improving access to other complementary resources can be provided by land titling. Therefore, legalization of tenancy is advocated for creating additional incentives for improving efficiency in production along with enabling better access to institutional credit and other service (Haque: 2001). Given a varied picture of agriculture across different regions of Orissa, region specific contract designs are needed given tenants' risk aversion, transaction costs and other exogeneous local conditions. Efficient land market helps a smooth and successful structural transformation of the economy with intersectoral balance. But, at the same time, there is a fear that in the absence of adequate non-farm employment opportunities, liberalization of lease market and tenancy may go against the poor and marginal farmers having low bargaining capacity with less capital. Increase in off-farm opportunities leading to increase in opportunity cost of labour will create incentives to rent out efficiently. However, there is a fear of reverse tenancy also where larger and richer farmers may lease-in land from small and marginal landowners to expand the size of operational holdings and maximize income. This may add to the number of agricultural labourers on the one hand and the number of large farmers on the other hand which will widen inequality. Orissa in its agricultural policy of 1996 has given agriculture the status of an industry and also created a space for contract farming and it has been observed that severe land alienations have taken place in tribal areas in the name of cash crop cultivation. Therefore, the efficiency and equity outcomes of a liberalized and legalized lease market will depend on the kind of options available to poorer tenants in the rural market. It is very often advocated that sufficient non-farm opportunities should be provided in rural areas so that income and employment will increase and will reduce overcrowding in agriculture. But, at the same time, agriculture should be sufficiently rewarding and attractive to keep it vibrantly alive with a healthy growth. It can only be achieved with dedicated and productive cultivators. Legalisation of tenancy can be one fundamental step towards this goal. To summarise, the specific recommendations are as follows:

(1) Tenancy should be legalized to provide security of livelihood to the tenant farmers and to increase land productivity also.

(2) Provision of user rights and land titling over agricultural land for the real tillers be considered on urgent basis so that they can be the real beneficiaries of all the pro-farmer government schemes.

(3) Specific care be taken to check reverse tenancy and dominance of money power so that poorer farmers will be preferred and protected in the lease market.

(4) Laws should be flexible enough to take account of the regional diversities within the state and therefore, contract designs need not be uniform over space and over crops.

(5) All kinds of inefficiencies and distorted incentives in the land policy and in revenue administration and implementation be removed.

(6) Lastly, sufficient non-farm opportunities be created so that rural life, both within and outside agriculture, will be rewarding.

References

Barlet, D. and M. Ali (2000): "Productivity Growth and Resource Degradation in Pakistan's Punjab: A decomposition analysis" Working paper 2480, World Bank.

Besley, T. (1995): "Property rights and investment incentives: Theory and Evidence from Ghana", *Journal of Political Economy*, Vol. 103(5), pp. 903-937.

Bhaumik, S.K. (1993): "Tenancy Relations and Agrarian Development: A Study of West Bengal", New Delhi, Sage Publications.

Chattopadhaya, M. (1979): "Relative Efficiency of Owner and Tenanat Cultivation: A Case Study", *EPW*, XIV (39), A93-96.

Cheung, S.N.S. (1969): "The Theory of Share Tenancy", University of Chicago Press, Chicago.

Economic Survey of Orissa, 2009-10, Government of Orissa.

Haque, T. (2001): "Impact of Tenancy reforms on productivity improvement and Socio-economic status of poor tenants", NCAP, New Delhi.

Marshall, A. (1961): "Principles of Economics", London, Macmillan.

Stiglitz, J.(1974): "Incentives and Risk sharing", *The Review of Economic Studies*, CXXVI, 219-225.

Rao, C.H. (1971): "Uncertainty, Entrepreneurship and sharecropping in India", *Journal of Political Economy*, 79(3), pp. 578-598.

Shaban R. Ali (1987): "Testing between competing models of sharecropping" *Journal of Political Economy*, 95(5), Chicago University Press.

15

Land Leasing by Women in Andhra Pradesh: Seeking Security and Reducing Risk

Karuna Vakati, Darryl Vhugen and *Tim Hanstad*

1. Introduction

Each year, often at the beginning of the growing seasons, landless rural women throughout Andhra Pradesh are entering into leases to cultivate agricultural land. The women are not alone: scores of agricultural labourers, current lessees, and landowners are leasing-in land each season. But these women are distinguished because they are among the poorest in their communities. With the support of community-based and non-governmental organizations (CBOs and NGOs), the women seek land to lease in order to provide a secure source of food for their families and to reduce their reliance on income from daily wage labour. Even though land leasing laws in Andhra Pradesh are highly restrictive, there is an informal landlease market. The present chapter analyses the mode of functioning of group leasing practices in Andhra Pradesh and assess their impact on the socio-economic status of poor women.

2. Study Description and Methodology

The study examines the actual leasing practices of poor women – cultivating leased land in groups or as households — in selected villages of five districts to help inform the following inquiries:

- Whether land leasing has a positive impact on the livelihoods of the rural poor;

- What conditions provide the best opportunities for the poor to experience positive impacts from tenancy;
- What factors are raising the costs of the poor's participation in the rural land lease market; and
- How can the state support the poor's participation in the land lease market and create opportunities for positive results from the participation.

Desk Research: The study's desk research involved analyzing the relevant tenancy, land reform, and land revenue laws in Andhra Pradesh and collecting prior state-level studies of rural tenancy.

Site Selection and Visits: The study identified local institutions (CBOs and NGOs) that are engaged in or otherwise facilitating agricultural land leasing projects for rural poor women in Andhra Pradesh, collected information about the activities of the institutions, and identified project sites for visits. The study team selected districts from several regions (one from Rayalaseema region, two coastal regions, and three districts in distinct Telangana areas). The team selected sites based on the known work of local institutions with women's groups on land activities and involvement with land leasing.

Between March 11-16, 2007, the team travelled to six districts and visited 13 villages. On May 18-19, 2007, the team visited an additional 6 villages, including a tribal area, in one district. Altogether, the team met with twenty-nine cultivation groups (289 members)[1] and 42 individuals engaged in leasing agricultural land in the visited villages and elsewhere. The team also met with directors of CBOs and NGOs with members engaged in land leasing, and with local land officials working with the IKP programme. At each location, the team reviewed any available documents related to the land leases and interviewed group members, individual lessees, NGO and CBO facilitators, government officials, and community members. The team interviewed individuals whom the team identified as lessees and where available, landowners who are engaged in leasing-out land. The team also interviewed several landowners who are not currently leasing-out land. In addition, the team visited some of the leased land and the land purchased by former lessees. Finally, the team interviewed officers and members of the Federation of Farmers Associations as well as representatives of the State Bank of Hyderabad and the State Bank of India to understand their perspectives on lending to tenants in the state.

3. Overview of Leasing Relationships Studied

District (region)	Leasing groups interviewed (group size)[2]	Leasing individuals/ households interviewed	Land owners interviewed	Extent of Leased land	Lease duration	Annual cash lease rates
Kurnool (Rayalaseema, Andhra)	2 (4, 10)	3	2 (not currently leasing out land)	0.60 - 20 acres	One season to 3 years	Rs. 2,000-4,000/acre
Nellore (coastal, Andhra)	2 (4, 10)	1 (with additional group of 11 individual lessees interviewed, one contemplating group lease)	1	1-7 acres	One season to 3 years	Rs. 3,000-5,500/acre
Medak (Telangana)	2 (24, 10, 30) 1 group representing 17 groups	1	2	4-15 acres	1-15 years	Rs. 0.00 (use rights) - 1,875/acre
Karimnagar (Telangana)	2 (26, 16)	0	1	3-5 acres	1-2 years	Rs. 1,500-6,000/year
Mahabubnagar (Telangana)	2 (50, 7)	1	0	3-7.5 acres	4-10 years	Rs. 200-2,000/acre
East Godavari Coastal Andhra	0	4	Multiple (all STs)	0.5 - 6 acres	1-3 years	Rs. 600-10,000/acre

4. Land

Availability: All but two of the villages visited or reported on to the team have land available for lease.[3] On average, residents interviewed estimate that 20-40 per cent of the agricultural land in most villages is leased or available for leasing. In a sample of eleven villages in Nellore District, each village had 800-1,000 acres of agricultural land, of which 200-300 acres is available for leasing or leased.[4] According to those interviewed, the amount of land available for lease fluctuates with the non-farm opportunities available to landowners, productivity of the land, availability of labour, and the landowners' need for cash. The data collected does not indicate any state-wide upward or downward trend in the availability of land for leasing.

Extent and Type: The lease relationships examined or reported on in the course of the study cover approximately 250 total acres of land. Of that total, approximately 90 acres is irrigated land and the balance is dry land.

Development: Most leased land is already developed and was being cultivated by landowners or other lessees prior to the subject lessees' lease. In several cases, lessees themselves cultivated the land as agricultural labourers prior to taking the land on lease. Except for the two cases noted below, none of the leases examined included any funds for (or expectation that) the lessees would engage in any significant improvement of the land, such as establishing irrigation.

- In R. Itikyal village, Medak District, the SHG took a 15-acre parcel of fallow land owned by the husband of an SHG member. The landowner agreed to allow the SHG to use his fallow land for 15 years in exchange for their agreement to clear the land, improve the soil, and cultivate the land.
- In Mahabubnager District, the NGO, Watershed Support Services and Activities Network ("WASSAN") assisted the cultivating SHG in Mukthipahad village with substantial improvements to the leased land, including piping in water, land clearing, and investment in upgrading the soil for organic farming. WASSAN and supporting NGOs note that the women are already obtaining returns on the development and improvements; they sell organic produce at a premium rate.

5. Lessees

Lessee Profiles: In its design and execution, the study focused on women SHG members, leasing as individual households and as cultivation groups. Accordingly, most lessees whom the team interviewed are female.[5] The majority of the female lessees are married; at least eleven are widowed, divorced, or abandoned. Most of the women are members of Scheduled Castes and other backward castes. Three individual lessees interviewed are Muslim (one in Kurnool District and two in Nellore District). In addition, in two villages, some of the lessees who are identified as members of Scheduled Castes also identify themselves as Christian. Some cultivating groups include members of Scheduled Tribes. The team also interviewed several male lessees in East Godavari District.

All women are members SHGs that have been in existence at least five years, and most SHGs have been in existence at least nine years. In all cases, the SHGs provide some measure of support for the leasing activity. The most common support derives from the SHG savings activity, which provides funds to engage in the leasing activity. In addition, the SHGs may serve as a forum for identifying and discussing land that might be available for lease. Household lessees may purchase seed and fertilizer in bulk through the SHG. Other SHGs undertake cultivation of the leased land as a group activity. In cases where the SHG is not also a cultivation group, the women form separate cultivation groups with members from other SHGs or lease and cultivate land as a household.

The majority of the interviewed women's household's are landless. The minority who own land have between one and two acres of dry land, often allocated by the government. The women's households all have sources of income or food grains other than as provided by the leased land. Most commonly, the women work as agricultural labourers, but in Mirzapur village, Medak District, the lessees and their husbands all work in a sugar cane factory. In Nellore District, some women leasing as a household are also professionals: one woman is a medical aide and another works for an NGO.

In the Scheduled Tribe areas of East Godavari District all of the lessees identified by the team were non-tribals. Perhaps due to the fact that non-tribal people may not legally lease-in land, few lessees were willing to speak to the team. Local revenue officials in Gangavaram village stated that the lessees they had identified were unwilling to be interviewed. However, a few lessees in Tantikonda and Rajavommangi villages attended large group meetings with the team and answered questions put to them. All of the non-tribal lessees who were willing to speak are landless. Many of the lessees have leased tribal land for many years. Indeed, in Tantikonda village, one lessee has leased the same land for more than twenty years on a series of one-year leases.[6]

***Decision to cultivate as a household or as a group*:** Many women made conscious decisions whether to lease land as an individual household or form (or join) a women's cultivation group. The women interviewed identify labour a consideration in determining whether to lease as a household or as a group. In general, those who have sufficient labour resources within their household are more likely to lease as a household; those with limited labour resources are more likely to seek group leasing opportunities.

As an example, in Nanore village, Kurnool District, a woman's family had previously leased three acres of land. They netted approximately Rs. 2,000 annually per acre. The woman joined a 10-member women's cultivation group because she believed they would be able to lease-in more land and obtain more production from each acre leased due to labour efficiencies within the group. The woman put in less time on the group leased land than on the land leased by the household. The group received Rs. 3,000 net per acre, and each member had access to fodder.[7]

Reflecting another perspective, in Kalva village, Kurnool District, a widow elected to lease land as an individual household in order to be able to receive the full benefits of her willingness to work hard. She believed that she would receive greater value from a household lease than if she shared the production with a group because as the head of household who had suffered significant adversity in the past, she was highly motivated to put in the time necessary on the land.

Another factor that influences the decision to lease as a household or collectively is the availability of funding. Some CBOs and NGOs have provided funds to groups specifically to engage in group leasing, and the funding is not available (in some cases) for household leasing. Women must elect group leasing-in order to qualify for the financial support.[8]

Other factors expressed by women that influence their decisions whether to lease-in land as a group or a household are: (1) the size of the parcels of land available for leasing-in the vicinity; (2) the landowner's willingness to lease-out land to a group; (3) the desires of the household (particularly the husband); and (4) the interest of other group members in group leasing.

A mix of these factors led a woman in Palakolan village (Kurnool District) to elect group leasing. The woman worked as an agricultural labourer on land planted with mango trees and saw lessees obtaining profits from the fruit. However, she did not believe her household could manage a lease of the entire orchard; the land would be too expensive and the labour requirements too intensive for a period of several months. The woman discussed the issue in her SHG and decided she would try an arrangement in which a group of three or four women lease the land for the mango season. She announced her desire to lease through the VO, which assisted her in finding women interested in joining her. She ultimately formed a cultivation group for the mango season. The same group (less one member) also leases agricultural land after the harvesting of the mango crop.[9]

Role of Local Institutions: In some cases, women began leasing land as a result of an NGO-initiated project. For example, in Medak District, the Deccan Development Society ("DDS") began a land leasing programme for landless Dalit women in the late 1980s as a complementary programme to the organization's food security programme. In other cases, the women themselves suggested land leasing as an activity to the CBOs and NGOs working in their communities. In general, in those cases where the leasing idea originates with the women members, the women appear to be more strategic in their decisions to engage in leasing and proactive in undertaking land-based activities. These women report taking into account the availability of certain types of land for leasing, labour availability within the household or potential cultivation group, and options for other income sources as factors in their decision-making.

Another potential source of financial support to female lessees is a new scheme offering joint liability agricultural production loans from commercial banks to groups of tenants organized as *Rythu Mithra Groups* (RMGs). In East Godavari district, RMG groups of 15 male tenant farmers receive commercial bank loans at 7 per cent interest facilitated by local Agricultural Department officials. None of those interviewed in any other districts appeared to be aware of this new credit programme available to tenants organized as RMGs and none had received loans from a commercial bank.

Women's Knowledge Relevant to Leasing: In general, the women interviewed exhibit adequate to substantial knowledge relevant to land leasing, including knowledge of the following:

- the availability of land for lease in their areas;
- the range of terms for leases or locally accepted standards;
- the productivity of various parcels of land and different crops;
- the costs and availability of necessary inputs; and
- the market prices for crops.

Several women take advantage of their knowledge and the flexibility of lease terms in making decisions about leasing. A woman in Kurnool District who wanted to determine if she could manage a lease initially opted for a small amount of land (60 cents). In Nanore village, a lessee opted for short-term lease after the landlord increased the lease rate; the lessee wanted to make sure the land would continue to be productive enough at the new rate. In Palakolan village, a four-member SHG has two leases to take advantage of the seasonal nature of various crops and spread risk: one lease is for mango trees; the second is for vegetable cultivation.

In short, in the majority of instances where lessees are supported by land-savvy CBOs and NGOs, the lessees interviewed articulated the knowledge necessary to make decisions about investing in land through leasing, and their responses suggest a logical, well considered approach to land matters. In several cases, in particular where the Kurnool District CBO, and NGOs Centre for Environmental Concerns ("CEC")[13] and WASSAN have assisted with capacity building and technical advice, the women are contemplating the future, planning how different land rights can assist them and their families, and developing steps to achieve their objectives.

***Desire for More Land to Lease/Engaging Labour*:** Some of the cultivation groups wish they had more land to lease-in. They believe they have sufficient members to cultivate additional land but they are constrained by: (1) lack of money to make the initial lease payment (and the inability to negotiate to pay some or all of the lease rate after the first crop); (2) lack of additional land available for lease in the area; or (3) the landowners' refusal to lease to the group. In contrast, those with household leases whom we asked about an optimum amount of land, such as a Muslim landowner (and former agricultural labourer, sharecropper, and lessee) in Kalva village (Kurnool District), generally are not desirous of leasing more land than they already have.

***Awareness of Legal Restrictions on Leasing*:** None of the women lessees interviewed is aware of any law regulating land leasing. However, several lessees, NGO representatives and Farmers Association representatives state that landowners' decisions about whether to lease, to whom to lease and for what term are based on the landowners' fear of the tenancy laws and the possibility that they might lose ownership rights to their land.

***Impact*:** All women interviewed report some measure of positive impact from their leasing arrangements in one or more of the following areas:

- **Increased and Diversified Sources of Food and Nutrition:** Many women state that leasing-in land increased their food security.[10] In Nellore District, women note that they are more interested in ensuring their families are eating better as a result of the leased land than they are in obtaining additional income. One woman who plants her leased land with a diversity of vegetables for home consumption states that her family is healthier as a direct result of the use of leased land. In Thakapalen and Manganmuru villages (Nellore District), and in Mirzapur village (Medak District), the cultivation groups sharecrop, and their focus is on ensuring sufficient production for consumption. Many of those sharecropping assign access to food grains priority over income.
- **Availability of Fodder:** Leasing provides lessees with fodder for livestock. The access to fodder is of particular value in summer, when fodder is difficult to obtain.
- **Increased Social Status in the Community:** All lessees report they experienced a positive change in their status within their communities when they transitioned from agricultural labourers to lessees. Even those women in Manganmuru village, Nellore District, who sharecrop land of landowners for whom they are obliged to provide day agricultural labour state that they believe their positions are slightly better than if they were simply agricultural labourers. Some women report having an easier time obtaining loans because of their status as lessees.

  Not surprisingly, the positive change in status is most significant for women who have transitioned from lessee to owner. As one Scheduled Caste woman in Palakolan village, Kurnool District, reported, prior to her evolution from labourer to sharecropper to cash lessee to landowner, she could not have sat in a chair during our interview. The widowed, divorced, and abandoned women who lease-in land in Karimnager District specifically mention the importance of their social status. They would like to purchase land in order to provide for their children but also to ensure a place for themselves in the community. They note that their status had increased as lessees, and they seek to solidify their status further as landowners.

  The converse is also true. The women in Galipalli village (Karimnager District) have not experienced a rise in their social status within the village. They believe their status is unchanged because they have not been able to convince farmers to lease land to them. Instead, the responses of the women suggest that as the result of their involvement with the NGO, Andhra Pradesh Mahila Samata Society ("APMS"), and APMS' reputation, the women may have been marginalized within the community. Women lessees in the DDS project (Medak District) who are witnessing rising lease rates express their dissatisfaction with their anticipated transition back to a total reliance on wage labour and the loss of their status as lessees of land.

- **Increased Control:** Women frequently report expressly or impliedly that leasing land gives them more control over their lives. In Kurnool District, women from a small cultivation group state that leasing land frees them from total reliance on wage labour for income. As lessees, they can work together with their families and schedule the work on the land in accordance with family needs. If a member of a group or family is ill and unable to work for a period, or if someone needs to travel to the city, the group or family can adjust so no production is lost.
- **Increased Sense of Security:** Leasing-in land provides some women with a feeling of greater economic security than they experienced when they were totally dependent on agricultural wage labour. The women lessees in Mahabubnager District seek landownership as a means to increase the security for themselves and their families.
- **Increased Cash Income:** In many cases, the lessees report earning a modest cash income from their lease. Annual net income per person ranges from Rs. 375-20,000. Far less frequently, the leases result in net losses.
- **Significant Step Toward Land Ownership:** At a majority of sites, women understand that land leasing is on a path to land ownership and is an important intermediate step. In Palakolan village, the women interviewed sought land to purchase after two or three years of leasing because they recognized that: (1) leasing was ultimately less secure than landownership because the landowner could decide not to renew the lease; (2) leasing provided the experience with farm operations and confidence to undertake landownership; and (3) landownership gave them the best control over land use, production, and profits.

6. Landowner/Lessor Profiles

In the non-tribal areas, two of the landowners are male – one from general caste, and one from backward caste. Two women landowners are Scheduled Caste members and two are Muslim. The two Muslim landowners interviewed are not currently leasing-out land. In the tribal areas, all of the landowners are members of Scheduled Tribes, as required by law.

The lessees, village residents, landowners, and CBO and NGO representatives interviewed report that the majority of the landowners who lease-out land own between one and twenty acres of agricultural land in addition to the leased-out land. The landowners select land to be leased-out based on a variety of factors, including absentee status (one farmer leases-out the land owned by his absentee brother), location (one farmer selected the land most distant from his house to lease-out), lessees' knowledge of land (many lessees worked as agricultural labourers on the leased land), and quality of soil/extent of development (one farmer granted an SHG use rights to fallow land that he did not want to invest in developing himself).

In three villages, the landowners who lease land to the SHG have household members who belong to the SHG or are themselves SHG members. In two of the three leases the landowners received a cash payment from the SHG. In one case, the landowner granted the SHG use rights to fallow land in exchange for development of the land by the SHG.

Reasons for Leasing-Out Land: In general, landowners (or those speaking on behalf of landowners) favour leases because the landowner can retain the asset while also receiving a lump sum payment of cash. Leasing allows them to respond to local conditions, such as a shrinking labour market or a drought, while maintaining status as a landowner.

More specific reasons for leasing-out land that landowners, lessees, NGOs, CBOs, and village residents identify are:

- Lack of labour in area or difficulty in managing labour[11];
- Inconvenient to manage land because distant from owner's residence or business;
- Owns another, more profitable business;
- Desire to invest time in non-agricultural enterprises;
- Needs cash for daughter's wedding and dowry, to support children, pay-off debts, or for undisclosed reasons;[12]

- Children are not interested in farming;
- Owner is physically unable to farm some or all of the land;[13]
- Land is fallow and must be cleared and developed before it will be cultivable (and lessees will develop land);
- Lessees will improve soil with manure and their labour and they will return the land in better shape; and
- No water on the land.

Preference for Sharecropping: In Nellore District, lessees in three villages report that landowners only agreed to lease-out land for a per cent of the production. According to the lessees and NGO representatives interviewed in those villages, fixed-rate leases are not available.[14] However, the SHG representatives from 11 villages in Nellore (all of whom are lessees) report that while landowners tend to favour sharecropping arrangements in that district, each of them has a fixed-rate cash lease for horticulture.

Share Leases Conditioned on Labour: In two villages near Kavali (Nellore District), lessees report that landowners will lease a portion of their land for sharecropping only if the lessees agree to work as agricultural labourers on other parts of the land. Lessees in one of the villages state that their children are also required to work in the landowner's house as a condition of the lease.

Landowner Awareness of Legal Restrictions on Tenancy: None of the landowners interviewed has a clear understanding of the law applicable to leasing. Without mentioning any specific legal provision or constraint, two landowners state that they would consider extending lease durations only to lessees whom they trusted. In several instances, lessees report that landowners' fear of losing their rights to the land affected the willingness of landowners in their areas to: (a) enter into any leases at all; (b) lease to SHGs; (c) enter into leases in excess of one year; or (d) execute written leases. Landowning officers and members of the Federation of Farmers Association confirmed that the law results in landowners to leasing less land, on shorter leases, and to a smaller pool or potential lessees. However, no one could identify the exact nature of the legal restriction that they believe impacts the landowner's leasing practices. Rather, the landowners generally understand that the legislation favours lessees and could lead to a loss of their land rights in favour of lessees.

Impact of Personal Knowledge of Relationship with Lessee: The landowner's knowledge of the lessee plays a role in lease terms. In

R. Itikyal village (Medak District) a landowner who regularly leases five acres of land for one-year periods states that he would be willing to consider a longer duration lease if he knows and trusts the lessee. In Anemadugu village (Nellore District), a landowner agreed to lower the lease rate by Rs. 500 for the second year in part because she knows the lessee is taking care of her land and improving the soil. Similarly, in the tribal village of Tantikonda, landowners stated that they preferred to lease only to those they new well. The team learned of several leases in the tribal areas in which landowner had leased-out to the same lessee for as much as twenty years.

Familiarity also can operate to the disadvantage of a lessee, however. In Kachireddypalli village (Karminager District), the lessees report that none of the farmers will lease land to them for one or more of the following reasons: (1) the farmers know they are affiliated with the NGO, APMS; (2) they are women, and (3) they require written leases in which the farmer must identify all landholdings. This situation appears to be the result of particular circumstances with the village and NGO. Elsewhere, landowners are far less concerned about the SHG status of lessees: in Kalva village in Kurnool District, lessees report that farmers will lease to the person offering the highest rate for the land, without regard to relationship or status. Similarly, the 11 women engaged in household leasing-in 11 different villages in Nellore District confirm that landowners will select the highest bidder, even if the individual is unknown or a non-resident.

7. CBOs and NGOs

All of the lessees interviewed are members of SHGs, some of which are part of organized and functioning community based organizations (CBOs) that have established tiers: the SHG, the village organization (federation), and the Mahilya Samitrya ("MS"). Lessees at some sites are also aided by local NGOs, which often work in conjunction with the CBO. Depending on the site, the CBOs and NGOs provide one or more of the following functions for SHG members:

- A funding source for leasing activities (through loans and grants);
- Capacity building;
- Community sensitization and education on issues affecting women farmers;

- Financial advice;
- Bank linkages;
- Technical advice and extension services; and
- Peer group support.

Community Based Organizations: The Orvakal CBO operating in Kurnool District is in its 11 year, and its members have been engaged in leasing land since 1998. The CBO has grant funds from the Turner Foundation earmarked by the MS for loans for land leasing and purchase. The idea for the land programme originated with SHG members who were working as agricultural labourers and witnessed "outsiders" leasing and purchasing land and reaping profits and benefits that were beyond the reach of agricultural labourers. The CBO's land programme initially supported land leasing, expanding to providing support for land purchase. Since 1998, 110 women in 30 villages have purchased approximately 200 acres of land.[20]

Over the years, the CBO has developed a comprehensive structure for guiding and supporting its members' land activities. The VO reviews all applications for land projects. Members must submit micro plans with any application, and a five-person land committee at the MS level reviews all applications, the micro plans, and checks the land title. The committee focuses its attention on whether it believes the land can be productive enough to warrant the member's investment of time and financial resources. The MS requires members seeking loans for land leasing and land purchase to show some evidence of agricultural expertise, and often suggests that those seeking land purchase begin with land leasing to gain experience and confidence.

Lessees can decide to cultivate the land as a household or as a group, and the CBO is often instrumental in helping women to form cultivation groups. Leases must be reduced to writing and the MS office maintains copies of the originals. The CBO loans funds for leasing at 12 per cent annual interest and for 15-month durations. The SHG retains three per cent for its revolving fund, the VO keeps three per cent, and six per cent goes to the MS. Lessees tend not to repay early to save interest; they elect to keep the cash for consumption purposes.

In other areas, CBOs are less proactive in relation to land activities. In Nellore District, a ten-member SHG (*Priyanka*) in Thakapalen village was formed 11 years ago. The SHG members have sharecropped approximately 10-11 acres of land collectively since its inception. Each household negotiates its agreement separately, but they collectively farm the land. Each agreement lasts only one year, and each year the SHG sharecrops different land. The terms of the leases, the productivity of the land, and the profitability of the lease for group members have not altered in the last 11 years. The SHG has been involved with IKP the last five years but, to date, the SHG has not explored land purchase possibilities.

Non-governmental Organizations: Several NGOs in Andhra Pradesh support land leasing activities of SHG group members, with assistance ranging from technical advice and capacity building to financial support. The following is an alphabetical listing of the NGOs visited in the course of the fieldwork:

- **Andhra Pradesh Mahila Samata (Society)**[16] **("APMS")** works in nine districts of Andhra Pradesh and has its roots in women's empowerment and promotion of social justice issues. APMS received funding through UNDP and the Sustainable Dryland Agricultural Project for Women (*Samatha Dharini*) during the 2000-2003 time period. The project expanded the reach of APMSS to include agriculture projects in addition to its social justice initiatives. In Medak District, the project covered seven mandals and 125 villages and provided funding for development on land owned or leased by sangha women. The project provided the women farmers with technical training and farm implements, and assisted groups in establishing community grain storage facilities and seed banks.
  Examples of APMS' involvement includes work with a 24-member SHG formed 13 years ago in R. Itikyal village. With support of APMS, the group is developing and cultivating 15 acres of land in exchange for use rights; after 15 years they will return the land to the landowner. In Manthur village, 38 SHG members farm ancestral land held by 15 of the SHG members.[17] In Kachireddypalli village, Karimnager District, an SHG is cultivating SHG members' land. The NGO reports that its members have been denied opportunities for cash leases from non-member farmers because (1) the project has taken much of the available agricultural land in some areas; (2) farmers discriminate against the women because of their affiliation with APMS and their sex; and (3) farmers do not want to sign the lease form required by APMS, which requires survey numbers for the land and a disclosure of all land owned by the farmer.[18]
- **The Centre for Environmental Concerns ("CEC")** is a Hyderabad-based organization that was founded in 1984 by a

group dedicated to sustainable development and environmental awareness as a means of improving rural livelihoods. The organization works in semi-arid topical regions where the natural resources on which the rural poor depend are highly degraded. CEC works with SHGs, providing capacity building, technical advice on matters ranging from farming technologies to drought mitigation, and advocacy on economic policies, gender, health, and poverty.

CEC has been successful in developing leasing relationships as a means for SHG members to access land. In Mirzapur village landowners approach members of the SHG inquiring about the group's interest in leasing because the SHG has cash available. The SHG has a considerable amount of bargaining power because they have the cash reserve and a good reputation. The combination pays off in savings on lease rates: the group negotiated a reduction of the rate sought by the landowner from Rs. 10,000 to Rs. 7,500 for six acres. Landowners seek out the group with offers of leases and land purchase.[19]

- **The Deccan Development Society ("DDS")** is a Medak-district based NGO that works with approximately 5,000 sangham women in 75 villages surrounding DDS' headquarters in Zaheerabad. The organization's programmes include creating seed banks, providing for food security through development of degraded agricultural land, protecting indigenous crops, and establishing a market for agricultural produce. DDS began its land leasing programme in 1988, and at its height in the early-mid-1990s, 70 SHGs (500-600 members total) were leasing land. DDS gave loans to the groups for the lease payments. Lease payments are made in a lump sum at the beginning of the term, and written leases are required. About ten years ago, 30-35 members became landowners through the SC Corporation land purchase programme.

  The numbers of DDS leasing groups have been declining since the late 1990s, with many members returning to agricultural labour. Currently 17 SHGs (about 200 members) are leasing land, and the number is still declining. DDS states the reason is the high lease rates demanded by landowners. Leases that were originally Rs. 250-400/acre in the late 1980s are now Rs. 2,000-2,500/acre. Lease rates for well sited, irrigated land in the district range from Rs. 6,000 to 7,000/acre/year.

Most of the land leasing members interviewed cannot visualize themselves as landowners, and DDS has no apparent plan to provide the skills and financial support for such a transition. The group interviewed (representatives of 16 different leasing groups) report having problems with group decision making. The leadership of groups is not elected; the same women have held positions for years because "no one else wants the responsibility." Many report concern with the impact on their lives if rising land rates force them back to reliance solely on agricultural labour.

- **The Watershed Support Services and Activities Network ("WASSAN")** was formed in 1995 as an informal network of NGOs engaged with the economic improvement of the poor and marginalized sections of the state living in drought-prone areas. WASSAN provides capacity building and technical assistance to farmers, sharing expertise in areas of preventing land degradation, growing organic produce, and developing sustainable agricultural practices. The NGO works closely with the mandal mahila samakhas in building capacity of women farmers.

  In Mukthipahad village, Mahabubnager District, WASSAN selected seven women who are widowed, divorced, or abandoned to begin savings and livestock programmes, ultimately graduating to land leasing, nursery development, and related activities. Using APDAI and CBO funds, WASSAN provides loans to the women and pays them for their labour on their leased land. Repayment of the loans does not begin for two years, and the repayment goes to the VO's revolving fund. The consultants providing technical assistance believe that the leases will be self-sustaining after about three to five years. In Chaudapalli village WASSAN worked with the VO to organize a group of owners of 7.5 acres of assigned land to lease the land to the SHG (which includes the landowners) for a nominal sum. The SHG will clear the land and grow fodder.

Success factors for Institutional Support: The CBOs and NGOs that have successfully supported land leasing efforts to the significant advantage of the SHG members share some of the following qualities in their organizational or programme structure:

- The idea for the land activity originates with the individual members or SHG;

- Land projects are one of several options for activities and are consciously and intentionally selected by the individual or group (who have some amount of agricultural experience), usually after some successful and sustained experience with activities such as savings programmes and livestock projects. Within the CBO, other SHG members who are less interested or skilled in land cultivation have opportunities to pursue other economically viable activities such as a small shop or tailoring business;
- The NGO or CBO has access to funds (in the form of grants or loans) sufficient to help provide members with all or part of the cost of a lease and necessary inputs;
- The NGO or CBO has a respected presence in the communities in which the SHG members reside and work;
- The NGO or CBO has some knowledge of land issues, understands the role land leasing and land purchase can play in diversifying livelihoods of the rural poor, provides leadership to its members on land issues, provides (or provides linkages to) technical knowledge and training relevant to land selection, cultivation, and economics of leasing, and prioritizes capacity building for SHG members.

The institutions are less successful in supporting land leasing activities where:

- The CBO or NGO has not educated itself on land matters, including land values, land availability, and the interests and needs of landowners and lessees in a particular community;
- The CBO or NGO leadership does not tend to think strategically about land activities (e.g., how they can be funded, how they can be used to diversify livelihood options for their members, how they can be combined and stepped, etc.);
- The CBO or NGO provides limited (or no) financial advice and education to its members and has not succeeded in building the capacity of its members in that area;
- The CBO or NGO does not spend time in its members' communities to build relationships with the elders, political leaders, elites, and civil society members, to investigate the economic climate and income opportunities, and to understand the needs and interests of all groups within the community; and
- The idea for the land activity originates from an external source

and is imposed on SHG members as the only activity (or the only economically viable activity) offered.

Lease Rates: The median lease rates for the small study sample are Rs. 1,000 per acre per year for dryland, Rs. 5,555 for irrigated land, and Rs. 4,762 for orchard land. Annual lease rates paid by lessees interviewed range from Rs. 200 per acre (for dry fallow land owned by SHG members in Mahabubnager District) to Rs. 10,000 per acre for irrigated land in Mahbubnagar District and in East Godavari District.

Within villages, rates tend to be relatively consistent within categories of land, varying by not more than Rs. 1,000 per acre. Factors that influence lease rates within a category of land (e.g., dry land, double crop land, mature orchards, roadside land) include the quality of the soil, maturity of fruit trees (in the case of orchards), the location of the land, and the land's recent productivity.

Change in Rate: The majority of landowners appear to raise the lease rate at the conclusion of a lease period. In areas such as Anemadugu village (Nellore District) and Kalva village (Kurnool District), residents report that lease rates are increasing at an average rate of approximately Rs. 500 per acre per year. In some areas of Nellore District, women report that lease rates are increasing by Rs. 1,000 per acre per year. Increases are not, however, guaranteed. As noted above, in Anemadugu village a woman reduced the lease rate on her land by Rs. 500 per acre because the rains had been poor and the landowner recognized that the lessee was improving the soil.

Sharecropping: Interviewees report that sharecropping is the dominant form of tenancy in the areas of Nellore District that we visited and is commonplace for certain crops such as paddy and black gram in parts of Kurnool, Medak, East Godavari and Karimnager districts. In some cases, women prefer leasing for a per cent of the production (i.e., sharecropping) because: (1) they do not have to take a loan or use savings to make the lease payment upfront; and (2) they will likely receive some output and thus at least have something to eat. A woman landowner in Kalva village intentionally began as a sharecropper before committing to a fixed-rate lease. From the perspective of these women, sharecropping reduces some of the risks inherent in leasing land. In Kachireddypalli village, APMS representatives state that the NGO favours sharecropping arrangements because women had high default rates on loans taken for fixed-rate leases.

In Nellore District, however, some of the sharecropping arrangements described are less than favourable for the lessees. The lessees receive access to a small amount of land, provide inputs, and receive only one-quarter of the production after paying the landowner's share and a share for plowing. Moreover, in many cases the leases are conditioned on the women's willingness to perform agricultural labour on the remainder of landowners'/lessors' land, often at below market rates.

***Lease Payment Schedule*:** In the majority of the fixed-rate leases encountered, lessees make advance payments to hold a lease or pay the entire lease amount upfront. In Kurnool District, landowners of orchards hold a public auction once a year for lease rights to the fruit. After a lessee's bid is accepted, the lessee will make an advance payment, called a *samskara*, to hold the lease rate. If the lessee is an SHG member, the member will then determine how to pay the lease price and make payment within one month. In some cases of multiple year leases, lessees pay one year's rent in advance and pay for the second year after they harvest the crop.

***Negotiation Process*:** The lessees uniformly conduct negotiations with landowners themselves. Where a small cultivation group seeks the lease, all members generally attend the negotiation. In the case of large cultivation groups, a subset of members attends the negotiation. Only one case involved a third party. In Mahabubnager District, the lessees refused to pay the landowner's requested rate. The landowner met with the village elders and VO to ask for assistance in convincing the lessees to accept his rate, but the VO and elders did not support his position. The landowner was ultimately unsuccessful in his attempt for a higher rate.

In most cases the negotiations over lease rates requires only one meeting. In three cases, lessees report making an offer and then simply waiting some time for the landowner to accept the offer.

***Breach*:** One cultivation group reports breaching two lease agreements. The 16-member cultivating SHG in Galipalli village, Karimnager District, entered a one-year lease of five acres of dry land for Rs. 5,500, which was payable after they harvested the crop. The crop was poor, and the group paid the landowner Rs. 2,000. On another occasion, the same cultivation group entered a two-year lease for four acres for a total of Rs. 8,000, payable in installments. The group cultivated the land for one year, was dissatisfied with the production, paid the rate for one year, and broke the lease. The NGO, APMS, states that neither landowner objected to the group's decision to break either lease.

9. Lease Documents

Formal and Informal Leases: Of the 44 lessees and lessee groups questioned about the form of their lease[20], 24 have (or had, in the case of a prior lease) written lease agreements memorializing the terms and conditions of one or more lease. The team did not view any written lease for a sharecropping arrangement. Most written leases are (or are described as) one page documents. None of the leases viewed was longer than two pages. Approximately half the written leases are prepared on stamp paper.[21] Written, plain paper leases were especially prevalent in East Godavari District. In Chaudapalli village, Mahabubnager District, the parties to a four-year lease elected to use white paper to save money. In Kachireddypalli village, Karimnager District, the parties who had written leases for two lease periods elected to continue the lease on an informal basis because they trust each other and thus believe no writing is necessary.

Lessees who wish to have a formal lease appear to be responsible for initiating the requirement of a formal lease and for providing the form. Most lessees who use form of leases obtain the form from the CBO or NGO. In Nanore village, the village accountant has land lease forms.

Contents and Form of Lease: Most formal written leases include the name of the landowner, the name of the lessee, a description of the land, the lease rate, and the duration of the lease. Where written forms are used with a group lease, the form identifies the SHG or cultivation group as the lessee. Individual members may or may not be identified on the form. In some cases, both parties sign the lease, but in most cases, only the landowner signs.

Breach/Dispute Resolution: None of the lease agreements reviewed on site or described to the team contains dispute resolution clauses or penalties for breach of the agreement by either party.

Maps and Survey Numbers: None of the leases we reviewed required maps to be attached. One NGO, APMS, requires leases to have survey numbers and requires landowners to identify all landholdings on the lease form.[22]

Registration and ROR: The team found no examples of a written lease being registered, and no one reported advising the panchayat secretary or mandal revenue officer of the need to enter a lease arrangement in the record of rights and make passbook notations.

10. Dispute Resolution

No one who is or has been party to a lease reports having any disputes regarding a lease, lease duration, lease rate payment, or any other aspect of the leasing relationship. All lessees and landowners questioned about hypothetical disputes state that they would take any problems to the village elders. No one identifies the panchayat secretary, MRO, RDO, or Collector as persons to approach with a leasing dispute. No one mentions the civil court system or private lawyers. The PD in Nellore District comments that informal leasing arrangements are confirmed in front of witnesses and will be enforced by the village elders and supported by the community.[23]

In addition to reliance on the village elders and customary ways of witnessing agreements and seeking enforcement, some disputes are addressed through NGO and CBO processes. In Chaudapalli, Mahabubnager District, representatives of WASSAN and the VO worked with landowners, who are also members of the project cultivation group, to reach agreement on who owned what land within a 7.5 acre parcel – a matter that the parties had contested for years. In Mukthipahad village, Mahabubnager District, the VO was instrumental in mediating an impasse in lease rate negotiations.

11. Economics of Leasing

Whether leasing is economically feasible depends on the cost of capital, the lease rate and terms, input costs, the market price for cash crops, the value of production consumed by the family and their livestock, and the availability of food and fodder for consumption through other sources.[24] As this study was limited in time and scope and its objectives were broadly focused, the team did not attempt to collect all of the information necessary to assess and evaluate the economics of leasing by SHG members. However, some general information is noted.

In most of the cases examined, women are engaged in leasing dry land and growing low risk crops, such as pulses and ground nut, which require low investment. In those cases, the output is also low. In the majority of cases, lessees use the production for consumption. In a few cases, most notably the women working with WASSAN in Mahabubnager District, the lessees are making substantial investments in the land, including making it suitable for organic farming, with the expectation that in the future they will reap significant profits from their investment.

In the majority of cases, financing for cash lease payments comes from some combination of the SHG (at 12% interest), a bank and/or private moneylenders (at 36% interest). Usually, the lessee needs financing in part to satisfy the general requirement that the entire lease amount be paid at the beginning of the lease term. (Less often the landowner permits the lessee to make annual lease payments at the beginning of each year of a multi-year lease.)

Most, if not all, SHGs seem to be unaware of the availability of low-cost cropping loans from the Rythu Mithra Group (RMG) programme, an initiative of the Reserve Bank of India through the Andhra Pradesh Department of Agriculture. The team interacted with a functioning RMG in Peddapalam Village, East Godavari District. In this programme, owner or tenant farmers can form joint liability groups of 15 and access commercial bank loans at 7 per cent interest. RMG formation and loan acquisition is facilitated by local Agricultural Department officials. The banks will finance even oral leases so long as the existence of the lease arrangement is confirmed by the local officials.[25]

The team interviewed representatives of the State Bank of India and the State Bank of Hyderabad whose duties include the bank's participation in joint-liability lending programme to groups of tenants organized as RMGs. The bankers were very enthusiastic about the programme and wish to see it expanded. They report that the repayment history has been very good to date with results exceeding even the 98 per cent repayment rate achieved by SHGs.

Over the past year, statewide lending by commercial banks to groups of tenants organized as RMGs under this new programme amounted to about Rs. 100 crores. However, the bankers see a potential state-wide market for lending to tenants of Rs. 1,000 crores to Rs. 4,000 crores (currently, the total volume of agricultural lending in Andhra Pradesh by commercial banks is Rs. 20,000 crore). *Importantly, the bankers interviewed noted that the biggest constraint for increasing lending to tenants is the lack of capacity in organizing the tenants into RMGs and certifying them.* The bankers noted that this was a logical activity for IKP as the programme staff were well situated for organizing the tenants into groups, especially women tenants.[26]

The parcel size leased per household or SHG member encountered does not vary widely and is relatively small. Median acreage per household is 0.75 acres for dry land, 0.60 acres for wet land and 2.75 acres for orchard land.

Lease rates range from Rs. 200 to Rs. 10,000 per acre for dry land, from Rs. 1,000 to Rs. 10,000[27] for wet land and from Rs. 1,000 to Rs. 10,625 per acre for orchard land. The median lease rates for the small study sample are Rs. 1,000, Rs. 5,555 and Rs. 4,762, respectively. The rates depend on a variety of factors, including the length of the lease, prevailing rate in the area, whether the land was wet or dry land, the landowner's need for cash and the lessee's ability to engage in meaningful negotiations. In some areas lessees perceive that investors from the coastal areas are coming in and bidding up the lease and purchase rates, either for speculative or commercial purposes, thus making it less economically feasible for the local people to lease or purchase land.

Not all lessees prefer cash leases. Some prefer less risky sharecropping arrangements.[28] In some areas, landowners are unwilling to enter into cash leases leaving sharecropping as the only alternative.

We lack sufficient information to assess the profitability of sharecropping leases. In Mangamaru village (Nellore) the women generally believe that access to sharecropping gives them a slightly better standard of living than reliance solely on wage labour. Such improvement is derived primarily from the share of food grains to which the lessees are entitled which enhance food security for the family. In addition, the lessees could also count on receiving the straw from the crop for their livestock.

Most lessees continued to work as agricultural labourers when not occupied farming their leased land. Daily wage rates for female agricultural labourers were generally Rs. 20-40[29], except for the AP Rural Employment Guarantee Scheme projects, for which a daily rate of Rs. 80 was in effect[30] and prevailing rates as high as Rs. 120 in areas of East Godavari District experiencing a labour shortage.

In some cases, lessees report that by leasing-in land, they were not able to work as many days as wage labourers. However, in the majority of cases, group and household lessees report working on their leased-in land on a schedule that allows them to continue to work as much daily wage labour as available. In some cases, lessees select certain land to lease-in and choose certain crops in order to cultivate the leased-in land during periods in which no agricultural labour is available. In other cases, household or group members who could not for some reason go for wage labour work on the leased-in land. In other words, in many cases, leasing-in land adds to the lessees' total number of days of labour rather than substituting for day labour days.[31]

No one reports that the wage labour wages foregone was greater than the net income from the leasing activities. In most instances the data gathered is insufficient to calculate the amount of foregone income. In Nanore village in Kurnool District, SHG group lessees estimate that as a group they had given up Rs. 3,600 in annual wage income as a result of farming leased-in land, but their net annual income from the leasing activity was Rs. 16,400.

Overall, the findings suggest that leasing can be a means to improved standards of living for the lessees and their families. In addition, in some cases land leasing can be a step on the path to landownership. The head of the Orvakal CBO reports that after a few seasons of leasing, most of the lessees in the CBO are ready to take the next step to purchasing land. Indeed, interviews revealed some marked success stories, where women had parlayed lease profits into buying land. Leasing appears to be giving them the necessary confidence to consider and evaluate whether they can undertake ownership or accumulate the capital necessary to purchase land.

12. Impact of Law on Leasing

As stated above, none of the lessees, other than those in the Scheduled Tribe areas, report any knowledge of the laws regulating leasing. Local practice governs their decisions to lease, the duration of the lease, lease rate, and understanding of the rights and responsibilities of the parties to the lease. None appear aware of any minimum requirements (or other restrictions) on lease durations.

Similarly, none of the institutions assisting the lessees demonstrate a clear and comprehensive knowledge of the laws governing leasing, although some requirements (such as APMS' requirement that landowners' identify all landholdings and WASSAN's leases with smallholder landowners in the Telangana area) suggest that some knowledge of some legal provisions may have informed the institutions' practices.

In addition, even certain government institutions appear willing to ignore legal restrictions on leasing. For example, the programme of joint liability lending to groups of tenants organized as RMGs requires the identification or even active facilitation of technically illegal leases by officials of the Agricultural Department. Moreover, other local officials are fully aware that Scheduled Tribe landowners are leasing their land to non-tribals in clear violation of the law and even provided written details of scores of such leases to the study team.

Lessees and NGO and CBO representatives attribute some

knowledge of the law to landowners. Several individuals state that landowners make certain decisions (such as refusing to lease to groups, leasing only for short time periods, and refusing written leases) because they fear they will lose their land rights to lessees through some application of an unidentified law.

While most of the small sample of landowners interviewed know that laws regulating leasing exist, none have a clear understanding of the law. This appears to be the case even for individual landowning members of the Federation of Farmers' Associations. Our interviews suggest that the landowners have only a vague understanding of the nature of the tenancy law governing new leases, and they have misconceptions about the rights granted to tenants (often apparently confusing them with ownership-like rights granted in Andhra Pradesh and other states at the time the laws were enacted).

For example, according to APMS, landowners in Medak and Karimnager Districts state they are unwilling to lease land to SHGs or individually to members of the SHG because they believe the group will somehow obtain ownership rights to the land. The landowners state that individual lessees (who are not involved with the NGO) do not carry a similar threat to landownership rights.

Despite landowners's lack of familiarity with the details of the tenancy legislation, they are generally aware that the laws favour lessees and that their own land rights are at some risk when they lease their land. Thus, the laws do have an impact on their decisions about leasing land, including whether to lease, to whom, and for how long. The most common impact of the law (or the landowners' perception of the law) is the short lease durations. Landowners limit the duration of leases at least in part in response to a perceived risk under the law.[32] All of the 11 lessees interviewed as a group in Nellore District (who had leases in 11 different villages) have one-year leases; all state a desire for longer (3-4 year) lease periods, but that landowners are unwilling to lease for longer periods. In Medak District, the sarpanch reports that landowners in the area lease-out land to a group of lessees, but none lease for more than a year because lessees might gain some rights to remain on the land. As a result, approximately 20-30 lessees move from parcel to parcel each year.[33]

Officers of the Federation of Farmers' Associations stated definitely that tenancy laws dissuade many landowners from leasing-out land, cause them to give shorter leases when they do lease out, and make them more careful about to whom they lease. Moreover, they observed that the requirement that lessees be identified in "Column 16" of village revenue records (*pahani* in Telangana areas and *adangal* in Andhra areas) is a barrier to leasing. It is their view that landowners will be more comfortable leasing-out land if the village revenue records completed by local revenue officials do not have a column for reporting non-owning occupiers or lessees.[34] In fact, they stated that if landowners had no fears concerning the risks of losing land rights to lessees, much more land would be given on lease.

13. Conclusions

Even in this small study of land leasing activities, the possibilities for significant economic and social change through leasing stand in high relief: with the help of strong CBOs and NGOs, women in rural Andhra Pradesh are using land leasing to escape some of the constraints imposed by caste, gender, and economic circumstance. The women are entering into agreements that are governed by contract, not status. The women SHG members and landowners are freely negotiating the leases, and the agreed terms and conditions of their leases are, at least in part, economically driven.

In most cases, the leasing relationships created are mutually beneficial, although in most cases landowners have more bargaining power than lessees. Some relationships, particularly sharecropping arrangements, continue to smack of feudalistic oppression. All are outside the formal law.

The informality of the leases creates risks, distorts contracting behaviour, and ultimately dilutes the benefits of leasing. The restrictions in the formal law drive the arrangements underground, where no protections against exploitive relationships exist and the market cannot correct inefficiencies. The informal, local systems of contract enforcement and dispute resolution are powerless against the invocation of the jurisdiction of civil court and formal law.

In addition, the extent to which land leasing must proceed outside the existing legal framework limits its effectiveness as a means for the rural poor to access land. Only women fortunate enough to live in areas served by CBOs and NGOs that have: (1) a commitment to improving rural livelihoods through land-based programmes; and (2) independent sources of funds (so that they need not rely on state funding for their land-based programmes) are likely to benefit. The vast majority of rural women in Andhra Pradesh lack access to financial resources, committed organizations with land expertise, and sustained opportunities for capacity building and training. Leasing-in land is typically not, therefore, a viable option.

The state has an opportunity to make that option viable. If the state is able to create an appropriate policy and legislative framework that legitimatizes leasing by SHG members at market rates and terms, and if the state can extend the existing IKP land activities to include support for land leasing activities taken up by SHG members, the achievements of women witnessed in Kurnool District and elsewhere in the course of this study can become the reality statewide. Moreover, improving credit access for women tenant farmers by organizing them into RMGs or SHGs can make this livelihood option available to many more women.

Notes

1. This number includes the 16 DDS cultivation groups in Medak District represented by one or more member.
2. This column identifies groups that cultivate the leased land together.
3. In Manthur village, Medak village, the residents reported no land available for leasing because a development project took a substantial portion of farmland. In Nellore District, residents of Gowavaram village state that no land is leased or available for lease in the village.
4. We obtained these estimates from a group of women meeting in Nellore, each of whom was from one village in the district.
5. In the course of the fieldwork, the team heard about other lessees operating in the area who are not affiliated with an SHG. We did not inquire about the profiles of these lessees, although description suggest they are household leases. We did not hear of any male cultivation groups, and we did not hear of any women leasing-individually, as a household, or as a women's cultivation group, who were not affiliated with a CBO.
6. Generally, the leases of tribal land arise from a visit to the area by a non-tribal person who makes contact with tribal landowners to determine who has land to lease. This contact person then brings non-tribals to the area who negotiate leases directly with the tribal owners. The lease rates in tribal areas appeared generally to be much lower than in non-tribal areas.
7. We did not determine whether the woman's belief that she was ahead with the group lease because she worked less and the group got more production per acre (and fodder) pencils out. We would need to know the inputs into the household lease, her share of the inputs into the group lease, her share of the production from the group lease, and whether the household unit was identical in the household lease and the group lease.
8. In some cases this financial support also creates the lease market for the

groups: landowners make land available for leasing because they know that the groups have the financial support of the CBO or NGO.

9. A woman in Nellore District tells a similar story. She was working as a labourer in a mango orchard, "doing all the work" while the absentee lessee of the orchard appeared to do little. She saw the lessee with money for presents and some lifestyle comforts and approached the VO about the criteria for qualifying as a lessee.
10. This impact (and the women's recognition of the impact) is due in part to the focus of NGOs such as CEC, APMSS, and DDS on food security.
11. In some coastal areas reported on, landowners lease-out a small portion of their land in order to obtain labour. Lessees sharecrop the small leased plots and provide the landowner with labour on the landowners' land.
12. In some cases that resemble use mortgages, the landowner leases out land for a lump sum until the landowner can repay the lessees the lease amount. The landowner retains an amount calculated as interest as the lease rate.
13. This reason was especially common in the tribal areas.
14. One factor influencing the number of sharecropping agreements examined is the preference of one NGO, APMS, for sharecropping in some areas. One NGO representative told the team that they had difficulty obtaining loan repayments from the lessees so preferred them to engage in sharecropping.
15. The MS does not have statistics regarding how many women have leased land.
16. In its written materials, the NGO identifies itself as Andhra Pradesh Mahila Samata Society, although it is commonly referred to as Andhra Pradesh Mahila Samata.
17. The NGO refers to this arrangement as a lease, but the landowners receive no monetary consideration for their ownership; all SHG members work on the land and take equitable shares of any production.
18. We did not have an opportunity to discuss the issue with any of the landowners. However, an in-depth discussion with members of the NGO suggests that the NGO has a reputation for successfully advocating women's rights and has had difficulty combining that role with the role of building relationships with community members (such as landowners) that will assist their members economically.
19. The group is currently considering a five-acre parcel that the landowner would like to lease for Rs. 50,000. The group has offered Rs. 20,000. The group has also been approached by a landowner wishing to sell his ten-acre parcel of wet land at the rate of one lakh/acre; the group is considering the purchase.
20. The 44 includes 11 individuals questioned in a group in Nellore District, all of whom are household lessees of land from different villages. Of the 11, only one lessee had a written lease, which she had requested because she leased mango orchards and the rate is high.
21. The Kurnool District CBO requires stamp paper.

22. Officials from APMS, which requires this information before making grants or loans for the lease, were unable to state why that information is required other than it is "security" for the lessee. The Telangana Tenancy Act limits leasing to small holders and it may be that someone advised the NGO that leasing landowners must disclose their landholdings in order to determine if the lease is legal.
23. He also states that lessees are able to get loans from banks for inputs without written lease agreements. This is also true of commercial bank crop loans made through the RMG programme.
24. In addition to cash income, leasing generally provided the lessee households with food for consumption purposes. Any comprehensive analysis of the overall economics of these leases must consider the value of such consumption. The data gathered is insufficient for such an analysis.
25. The banks have a schedule of cropping economics that they use for these loans. Lease payment expense is not included. Forms used for RMG formation and loan applications accompany this report as Appendix-D.
26. Thus far, it appears that RMG groups consist largely or perhaps entirely of men. The programme is available to female farmers, including members of Shags.
27. The highest reported rate for wet land is Rs. 10,000. One such report came from Kurnool District, but is not used for the calculation because the rate was reported by a resident of the area, who while knowledgeable, is not a party to an actual lease. The other such lease was found in East Godavari District. In that transaction, the Scheduled Caste Corporation leased 38 acres of temple Endowment Land and then sub-leased the land to an SHG group consisting of 64 members. That lease is included in the database.
28. In addition, in some villages (such as Mirzapur village in Medak District) use mortgages are common.
29. Men tended to be paid 50-100 per cent more than the women.
30. The team did not receive evidence that the EGS wage rate had actually influenced the rates for non-EGS work, although there are opinions that it will do so as the volume of EGS work increases. To date, the number of days of EGS work available in the areas visited is limited (30 days or less per worker per year, and in one area less than ten days per worker per year).
31. The cultivation group in Manthur village takes the opposite approach in cultivating the ancestral land of the members. The 38-member SHG farms a total of fifteen acres, which is comprised of parcels owned by 15 members of the SHG. All members of the group work on the land, using the production solely for consumption. The group prioritizes (or at least schedules) work on this land over agricultural wage labour. They decide how many days each member must put in on the land and what tasks they should perform. If the member is not required to perform work on the land, the member is free to go for wage labour.

32. While information on landowners is limited, there are some indications that they are also limiting the duration of leases in order to control the lease rate to take advantage of increasing land values.
33. In Nellore District, many of the 11 women interviewed with one-year household leases are leasing-in the land for consecutive one-year periods.
34. In practice, it is rare that leases are reported to the local authorities.

Part III

LAND RIGHTS FOR WOMEN AND TRIBAL

16

Empowering Landless and Voiceless Women in Rural India

Hina Haque

Introduction

Women in India, especially those living in rural areas, are considered to be both resource poor and voiceless. This is especially so because the majority of them do not possess either any landed property or education and skill for gainful employment which can result in their economic independence and upward mobility. Unfortunately, there is no adequate realisation by India's planners and policy-makers that economic deprivation of women has a backwash effect and slows the process of overall economic growth. There are studies to indicate that more equitable distribution of resources and opportunities between men and women leads directly to higher economic growth and productivity (The World Bank: 2001). Since agricultural land is the most important form of property and productive asset in rural India, the present chapter intends to examine the issue of land rights for rural women at length. The main objectives of the chapter are to (i) highlight the importance of women's land rights in rural India, (ii) identify the constraints to women's land rights, and (iii) suggest policy options and strategies for improving women's access to land rights for their accelerated economic and social development.

Socio-Economic Status of Rural Women

India has one of the highest concentration of poor people in the world. Also, the incidence of rural poverty is relatively more among landless households. Since rural women have limited access to land right, the incidence of poverty and that of malnutrition is also high among women.

According to India's Eleventh Five Year Plan, (Govt. of India: 2007), the incidence of income poverty among women tended to be higher in rural areas. The percentage of female persons living in poor households in rural India increased from 28 per cent in 1993-94 to 29 per cent in 2004-05. In the age group of 15 to 49 years when women are in their most productive years, 73 per cent of SC women, 79 per cent of ST women, 61 per cent of OBC women and 61 per cent of Muslim women were illiterate. Also about 42 per cent of SC women, 46 per cent of ST women and 30 per cent of other women were malnourished. Nearly 56 per cent of women in the country suffered from anemia. In states like Bihar, Jharkhand, Andhra Pradesh, Assam, West Bengal and Tripura this was as high as 62 per cent to 70 per cent (Govt. of India: 2007).

It is indeed a sorry state of affairs that nearly 86 per cent of women agricultural labourers in the country and 74 per cent of women farmers were either illiterate or had education below the primary level, the average education of women agricultural labourers being less than one year (Govt. of India: 2006). Consequently, the wage rate of women labourers also was relatively lower. The women-men wage ratio in agriculture was 0.69 for casual workers and 0.79 for regular workers. In non-agriculture, it was 0.65 and 0.57, respectively. According to current daily status, about 8.1 per cent women workers in rural areas were unemployed and 17.0 per cent underemployed (Govt. of India: 2010).

Importance of Land Rights for Women

Several research scholars (Roy *et al.*: 1990) have pointed out that landlessness is at the root of some of the world's most serious and persistent problems with consequences frequently extending to severe exploitation and deprivation of minimal political rights and basic human needs. In most developing countries, titles to land are normally vested in men, even though women are the primary cultivators in many places. Since the great majority of the world's poor people live in agrarian settings, this is a fundamental source of vulnerability for poor women (The World Bank: 2001). Marginalized women constitute the majority of those living in poverty. Unless, land is allocated to poor women who have no professional skills to acquire a decent purchasing power for food, their freedom from poverty and hunger will remain a distant dream (CWLR: 2005). Gender equality in land rights is both a livelihood objective and a powerful means of eradicating poverty. The amount of land women own has a significant and positive impact on food expenditure as well as on children's educational attainment (Katz and Chamorro: 2002). Women's ability to have independent access to and exercise control over assets is a critical determinant of their welfare and income earning capacity (Fate, Champs and Quinsunbing: 1999). Besides, when women directly own land and control income, they spend mostly on basic family needs, whereas the men spend a significant percentage on personal items. Women and children also face much less risk of poverty if women have some direct access to land. Even a small plot of land can reduce the risk of poverty by acting as a bargaining point for attracting further resources from the state and from within the household (Agarwal: 1994).

In India, the majority of the rural population is dependent on agriculture for livelihood and therefore, access to land has an important bearing on any individual's economic and social status. Apart from being one of the most durable assets that can be used as collateral security, land holds a broader social value by instilling a sense of identity and belonging within the village setting and Providing Political Power (R. Mearns: 1999). Roughly about 35 per cent rural households in India are *de facto* female headed, due to reasons ranging from widowhood, marital breakdown or male migration (Govt. of India: 1998). As more men shift to urban or rural non-farm livelihoods, more and more households depend on women for managing farms and bearing the burden of family assistance. According to the latest NSS data (NSSO: 2006), women farmers constitute nearly 41.8 per cent of the total farmers. It would be seen from Table 16.1 that

Table 16.1: Number and Percentage of Farmers Among Agricultural Workers

	1993-94	*1999-2000*	*2004-05*
A. Number of Farmers (Million)			
Male	88.0	85.3	96.8
Female	55.2	51.9	69.4
Total	143.2	137.3	166.2
B. Percentage to Total Farmers			
Male	61.5	62.1	58.2
Female	38.5	37.8	41.8
Total	100.0	100.0	100.0
C. Percentage of Rural Agricultural Workers in each Sex Category			
Male	74.1	71.3	66.5
Female	86.1	85.2	83.2

Source: NSSO, 2006, quoted from Srivastava and Srivastava, EPW, July 10, 2010, vol. XLV, No. 28, pp. 51-52.

per cent share of women farmers in total number of farmers marginally increased from 38.5 in 1993-94 to 41.8 in 2004-05. The absolute number of women farmers increased from 55.2 million in 1993-94 to 69.4 million in 2004-05.

Table 16.1 further shows that about 83 per cent women workers in rural India were engaged in agriculture as either cultivators or labourers, while only 67 per cent of men workers were engaged in agriculture. Of the total agricultural workers, about 35 per cent were casual labour and 65 per cent were self-employed. However, in the absence of land right, women farmers fail to cultivate land efficiently because they lack access to institutional credit facilities for want of collateral. Besides, they lack access to other institutional support services such as extension, marketing etc.

Status of Women's Land Rights

According to Agricultural Census, 2005-06 (Govt. of India), of the total individual operational holdings in India, about 12.4 per cent were operated by women, covering 9.9 per cent of the total land. In the lowest size class (below 0.5 hectare), the percentage of individual landholdings operated by women was relatively higher at 13.8, covering 12.5 per cent of the area operated by them (Table 16.2). It would be seen from Table 16.2 that there is a kind of inverse relationship between size class of holding and women's access to operational land. Although this does not clearly say whether the land operated by women also owned by them, it more or less indicates the same. Table 16.3 further shows that in states like Andhra Pradesh, Goa, Karnataka, Kerala and Tamil Nadu, the percentage share of women in total number and area of operational holdings was comparatively higher than other states. In some Eastern states, namely Assam, West Bengal and Orissa and also in Punjab in the north, the percentage share of women in the total number of operational holdings as well as area was relatively very low. In this respect, some region specific studies throw some more light.

A recent study of 4110 households in Karnataka by Swaminathan *et al.* (IIM-B, 2011) shows that in rural Karnataka, 71 per cent of all agricultural land was owned individually by men, while only 14 per cent was owned by women individually, 2 per cent was owned by principal couple and 8 per cent was under various forms of joint ownership. Only 9 per cent of women owned agricultural land as against 39 per cent of men. Women constituted about 20 per cent of all agricultural landowners

Table 16.2: Per cent Distribution of Total Number and Area of Individual Operational Holdings in India by Sex Category

Size of holdings	Gender	*1995-96*		*2000-01*		*2005-06*	
		Number	*Area*	*Number*	*Area*	*Number*	*Area*
Below 0.5	Male	87.87	89.24	13.41	12.01	86.19	87.54
	Female	12.13	10.76	86.59	87.99	13.81	12.46
0.5-1.0	Male	90.25	90.23	11.40	11.48	87.53	87.62
	Female	9.75	9.77	88.60	88.52	12.47	12.38
1.0-2.0	Male	90.75	90.78	10.92	10.85	88.37	88.54
	Female	9.26	9.22	89.08	89.14	11.63	11.46
2.0-3.0	Male	91.98	92.06	9.55	9.48	89.71	89.86
	Female	8.02	7.94	90.45	90.52	10.29	10.14
3.0-4.0	Male	92.92	92.95	8.21	8.20	91.03	91.06
	Female	7.06	7.05	91.79	91.80	8.97	8.94
4.0-5.0	Male	93.62	93.61	7.53	7.52	91.75	91.78
	Female	6.38	6.39	92.47	92.48	8.25	8.22
5.0-7.5	Male	93.94	93.96	7.00	6.99	92.32	92.36
	Female	6.06	6.04	93.00	93.02	7.68	7.64
7.5-10.0	Male	94.47	94.41	6.33	6.39	92.93	92.94
	Female	5.63	5.59	93.67	93.61	7.07	7.06
10.0-20.0	Male	94.93	94.95	5.82	5.77	93.48	93.57
	Female	5.07	5.05	94.18	94.23	6.52	6.43
20 & above	Male	95.81	94.65	5.07	5.27	93.99	93.86
	Female	4.19	5.35	94.93	94.73	6.01	6.14
All Sizes	Male	89.85	92.32	11.63	9.07	87.63	90.12
	Female	10.15	7.68	88.37	90.93	12.37	9.88

Source: Government of India, Ministry of Agriculture, Agricultural Census Division, Website.

Table 16.3: Per cent distribution of total number and area of individual operational holdings in various states by sex category

State	Gender	1995-96		2000-01		2005-06	
		Number	*Area*	*Number*	*Area*	*Number*	*Area*
Andhra Pradesh	Male	81.81	85.85	79.63	83.37	78.01	81.90
	Female	18.19	14.15	20.37	16.63	21.99	18.10
Assam	Male	100.00	100.00	98.04	98.21	98.20	97.67
	Female	0.00	0.00	1.96	1.79	1.80	2.33
Bihar	Male	92.39	93.57	NA	NA	88.00	88.42
	Female	7.61	6.43	NA	NA	12.00	11.58
Goa	Male	77.43	83.88	77.63	84.56	78.53	85.45
	Female	22.57	16.12	22.37	15.44	21.47	14.55
Gujarat	Male	91.19	92.57	90.57	91.80	89.94	90.78
	Female	8.81	7.43	9.43	8.20	10.06	9.22
Haryana	Male	93.91	95.47	92.58	93.32	90.86	92.16
	Female	6.09	4.53	7.42	6.68	9.14	7.84
Himachal Pradesh	Male	93.95	95.75	93.11	95.17	93.21	95.37
	Female	6.05	4.25	6.89	4.83	6.79	4.63
Jammu & Kashmir	Male	95.81	97.02	92.18	93.95	92.32	94.02
	Female	4.19	2.98	7.82	6.05	7.68	5.98
Karnataka	Male	85.57	88.27	83.91	86.76	82.63	85.43
	Female	14.43	11.73	16.09	13.24	17.37	14.57
Kerala	Male	76.58	82.03	79.13	83.83	80.59	84.76
	Female	23.42	17.97	20.87	16.17	19.41	15.24
Madhya Pradesh	Male	93.21	95.17	93.45	94.91	92.83	94.44
	Female	6.79	4.83	6.55	5.09	7.17	5.56
Maharashtra	Male	87.02	88.93	84.47	86.44	84.41	86.32
	Female	12.98	11.07	15.53	13.56	15.59	13.68

		1995-96		2000-01		2005-06	
State	*Gender*	*Number*	*Area*	*Number*	*Area*	*Number*	*Area*
Orissa	Male	98.67	98.77	96.99	97.26	97.26	97.50
	Female	1.33	1.23	3.01	2.74	2.74	2.50
Punjab	Male	99.34	99.49	99.22	99.38	99.29	99.47
	Female	0.66	0.51	0.78	0.62	0.71	0.53
Rajasthan	Male	96.40	96.22	95.88	95.65	93.96	94.28
	Female	3.60	3.78	4.12	4.35	6.04	5.72
Tamil Nadu	Male	83.51	85.79	81.86	84.87	81.28	84.04
	Female	16.49	14.21	18.14	15.13	18.72	15.96
Uttar Pradesh	Male	93.57	94.24	92.40	93.49	91.74	92.93
	Female	6.43	5.76	7.60	6.51	8.26	7.07
West Bengal	Male	96.87	97.52	96.67	97.88	96.60	97.94
	Female	3.13	2.48	3.33	2.12	3.40	2.06
Chhattisgarh	Male	NA	NA	89.54	92.03	88.73	91.33
	Female	NA	NA	10.46	7.97	11.27	8.67
Uttaranchal	Male	NA	NA	92.19	92.06	91.09	91.61
	Female	NA	NA	7.81	7.94	8.91	8.39
All India	**Male**	**89.85**	**92.32**	**11.63**	**9.07**	**87.63**	**90.12**
	Female	**10.15**	**7.68**	**88.37**	**90.93**	**12.37**	**9.88**

Source: Government of India, Ministry of Agriculture, Agricultural Census Division—Website.

in rural Karnataka. Also only 17 per cent of adult women had ownership of home against 47 per cent of adult men having ownership of home in rural areas. The situation may not be better in other states.

Gupta's survey of 870 households in Midnapore and 24 Parganas districts of West Bengal (Gupta: 2002) revealed that 34 women had inherited property as daughters, 07 women as widows; one as a daughter in law and there were seven cases of joint ownerships of various combinations. Of the 34 cases of inheritance by daughters, 19 cases were in situations where there were no sons to inherit. In two cases, only house was inherited. A study of Gender, property rights and responsibility for farming in Kerala (Kodoth: 2004), revealed that dowry is replacing inheritance rights as a mode of transfer of property to or on account of women. Both among Nairs and matrilineal Muslims, there has been a gradual shift towards dowry. Even though gender inequality in the ownership and control of land was found to be worse in North India than in South India, women had unequal access to landrights in almost all parts of the country. As Chen (1998: 2000) observed, only about 18 per cent women in North India inherited some land. On average, 87 per cent of the women did not receive any land as daughters. Among widows, about half of them with claims did not inherit any land.

Constraints to Women's Land Rights

In India, about 15 million rural households are reported to be landless. But the position of women is more vulnerable. Even if they belong to land-owning families, there is hardly any land in their names. As a result, they lack independent status and basic economic and social security. In fact, there are several legal, religious and socio-cultural barriers to women's land rights in India. Therefore, it would be worthwhile to identify and examine some of the major constraints to women's land rights.

Legal Constraints: The religious and customary laws which govern property rights relating to inheritance, marriage and divorce in India are highly gender biased against women.

Hindu Personal Law: Till recently, under the Hindu Succession Act, 1956 which governs the right of inheritance in the absence of a 'will' for the Hindus who account for nearly 82 per cent of India's population, only male family members became co-parcenars and received an undivided share of the joint family property at birth. Female family members were excluded from sharing the joint family property (Debekar: 1999), as quoted by Tim *et al.*: 2004). It was only in relation to the self-acquired property that the Hindu Succession Act provided for equal shares to male and female children as well as the widow of the deceased, or if the deceased was male, his mother (Gopal: 1999). However, the Hindu Succession Act has been amended in 2005 which gives equal rights to men and women in matters of inheritance of both self-acquired property and joint family property. Nevertheless, the injustice caused to women in this matter over the past several decades cannot be washed away, as the law will not have any retrospective effect. Also, there is lack of adequate awareness and sensitivity on the part of both women and men in Indian society at large to enable effective implementation of the new law.

Muslim Personal Law

Muslims constitute about twelve per cent of total population in India. The inheritance of property by Muslims is dictated by Muslim Personal Law (Shariat) Application Act of 1937. According to this Act, widows and daughters who are in the same relation as a male family member to the deceased, receive half the share of property received by the male family member. Also, a Muslim man is not permitted to bequeath more than one-third of his property by will which provides a safeguard to his spouse and female children against complete disinheritance. However, the Muslim Personal Act does not apply to agricultural land, the devolution of which is decided by state law or if the state law is silent, then custom (Aggarwal: 1995). In the southern states, the provisions of Shariat Act has been extended to agricultural land; while in other regions, the devolution of agricultural land through inheritance is governed by customs and tenurial laws.

Tribal Land Law

Land relations in tribal areas are generally governed by customary land laws that are often uncoded. In many tribal communities, lands were customarily considered as the Clans' common property and transferred to only clan male or female members as custodian in most cases. According to Santhal law, a daughter has no inheritance rights in land as the goal for every woman is supposed to get married and move to her husband's house. However, when a Santhal girl gets married and her father dies leaving no widow or sons, she either shares his land with her sisters or if there are no sisters, she inherits it entirely (Archer: 1984, Rao: 2008). Besides, in some areas, **'taben jom'** is a small piece of land kept for the married daughter by her parental households which could support her in any unwanted situations. But the women's children cannot inherit this land, as they represent a different lineage. Also '**taben jom**' is considered to be more in the nature of a gift than a right. In addition, under '**ghar jamai**' form of marriage wherein the son-in-law gives up his claims on the property of his own father and lives in the wife's parental house, the girl inherits the land on which they live. This right is recognised by Santhal Parganas Tenancy Act (Section 20-C).

In a quite similar form of marriage, called '**ghardi jamai**' where in the girl's parents have no sons and their daughter and son-in-law look after them until death, the couple gets the house, all the lands and half the cattle and goats for a temporary period. In the absence of a male adult child, it is the daughter who becomes the inheritor but as she cannot perform certain tasks, her husband becomes a representative. Moreover, after marriage, a Santhal woman has almost total shared control on the use of agricultural land and total control over the use of homestead plots (Rao: p. 215). But this continues, as long as the marriage stays.

The status of a tribal woman in relation to land right in north-eastern region is quite different. In Khasi and Jaintia hills, the position of women is comparatively better than their counterparts in other regions. It is the youngest daughter and not the sons who inherit the ancestral property. If there is no daughter of the deceased mother, the property goes to the mother's sister, the mother's sisters' daughter and so on. In Garo Hills, the leading mother has the right to nominate any of her daughters as the custodian of land, and not necessarily the youngest one as is the case of Khasi hills. Conversely, the Mizo customary law allows only the youngest son to acquire the father's entire property and the other brothers and sisters do not acquire such rights, although the newly codified Mizo customary law in 1963 provided that the daughters/wife of a deceased male would get some share in the immovable properties of the husband/father. In Nagaland also, women are not entitled to inherit ancestral property and can only have a share in movable property (Das: 1990, Karna: 1990). The **Tiwa** in Assam used to be a matrilineal group, but are now in the process of change to patriliny (Krishna: 2005).

Land Reforms Policy

During the past few decades, a number of redistributive land reform measures were undertaken, namely (i) imposition of ceilings on land holdings and redistribution of ceiling surplus land, (ii) allotment of government land to the poor, (iii) allotment of homestead plot to the landless families, and (iv) conferring ownership right to erstwhile tenants, in both agriculture and homesteads. Up to 2005, nearly 7.34 million acres of land were declared ceiling surplus and 5.40 million acres of such land were distributed among 5.75 million beneficiaries. Similarly, about 14.7 million acres of government wasteland were distributed in the country up to 2003-04. In addition, about 12.4 million tenants were conferred ownership right over about 15.0 million acres of land. In some states, homestead plots were given to landless families for constructing house. Unfortunately, in all these land reform schemes, women did not find place, as they were thought to automatically benefit as being members of beneficiary households.

During the past three decades India's policy-makers recognized the importance of land rights for women. It was the Sixth Five Year Plan (1980) that for the first time, stated: "Government shall endeavour to provide joint titles to husband and wife, initially in cases of transfer of agricultural land and house sites. Perspective Plan for Women (1988-2000) stated that women's undiluted access to land would bestow on her necessary economic independence and power and would improve her social position in the family. (Govt. of India: 1988). The Eighth Five Year Plan (1990-95) underscored the need to increase women's control over economic resources for improving agricultural production. It recommended that for married women, joint titles would be desirable for productive assets, houses and house sites. In 1992, the Government of India organized a Chief Ministers' Conference in new Delhi on Land Reform; one of the recommendations was that joint title should be given to both husband and wife in all land redistribution measures. It pointed out that due to both gender biased land reform legislations and redistribution of land, women have traditionally been losers. This is partly reflection on the lack of gender sensitivity on the part of central as well as state governments, despite the fact that policy announcement for gender-equal land reform has been made from time to time.

Following the recommendation of the Chief Ministers' Conference held in New Delhi in 1992 that 'joint title' should be given to both husband and wife in all land distribution measures, states like West Bengal and Assam passed executive orders to give joint title to both husband and wife. But in practice, nothing substantial happened. Even in a state like West Bengal, which claims to have achieved significant amount of success in land reforms, studies by Gupta (1997: 2002) show that 98 per cent of the land distributed in Midnapore district went to men. Only in a few cases, some single women received land. None of the married women received joint titles. In Bankura district, 800 *pattas* were distributed after the said circular, but in no case the name of the spouse was included. A recent statewide survey (SIPRD: 2003), further reveals that less than 10 per cent of the *pattas* were in the joint names of the spouses, ranging from 3 per cent in Darjeeling and 24 Parganas to 19 per cent in Bankura. Single *pattas* in women's name constituted hardly 5 per cent of the total *pattadars*. Also under 'operation barga' in the late 1970s and 1980s, men were the beneficiaries as land was recorded in the name of male members of the tenant household. There was not a single woman registered as a sharecropper. In three rounds of field research in West Bengal, Jennifer Brown and Sujata Das Chowdhary (RDI: Nov., 2002), observed that there was no evidence that the joint rights policy has been implemented in West Bengal.

Kerala is another state which is reported to have implemented land reform relatively more effectively. But the Land Reforms Act, 1969, did not address the gender question in land reform. As Sardamoni (1983) points out, in the early 1980s, women were the major casualty in land reforms. More than one-fourth of those who lost land were widows. Moreover, under ceiling laws in almost all states, adult sons get special consideration. In Haryana, Punjab, Uttar Pradesh and Delhi, the household can hold additional land on account of each adult son, but not adult daughters. In other states like Bihar, Gujarat, Madhya Pradesh, Maharashtra, Andhra Pradesh, Himachal Pradesh and Tamil Nadu, each adult son counts as a separate unit to hold specified extent of the land, but not the daughters, excepting in Kerala where both adult sons and adult daughters were entitled to a separate specified unit. Also the definition of tenants under tenancy laws of various states are vague and ambiguous which curtails women's leasing right. Tenancy follows generally male lineage.

In October, 2000, the Revenue Secretaries' Conference held in New Delhi emphasized that states may identify the causes of the failure of the revenue machinery to allot ceiling surplus land, *bhoodan* land and government wasteland in the name of women beneficiaries and take corrective action in a time bound manner. At least 40 per cent land should be exclusively assigned to women and in the rest of the cases, the allotment may be jointly in the name of husband and wife. The Agricultural Policy-2000 mentioned that efforts will be made to grant property right in land to women wherever possible and self-help groups of women may be encouraged to take up activities like the regeneration of wastelands.

Women will be given preference in the allotment of ceiling surplus lands (Govt. of India: 2003). India's Ninth Five Year Plan (1997-2002) made a policy statement that the government would endeavour to give joint titles to both husband and wife, when distributing agricultural land and house sites. The Eleventh Five Year Plan (Govt. of India: 2007) did not mention about women's' accessing ceiling surplus land, but pointed out that all landless families with no homestead land as well as those without regularized homestead should be allotted 10-15 cents of land each, priority being given to female headed households. Also all regularized homestead should be in the name of the wife. But in the ground, not much has been done to empower rural women through redistributive land reform measures, thus far. On the whole, women did not gain much access to land by way of Government's land reform measures, as state governments seemed to be insensitive to the issue in most cases.

Gender Bias in the Implementation of Land Resettlement Schemes

The gender bias in land allotment has been observed even in resettlement schemes in all major project areas, such as Sardar Sarovar Project, covering Gujarat, Madhya Pradesh and Maharashtra, Tehri Project in Uttarakhand, the Upper Krishna Project in Orissa. In each of these projects, the resettlement packages for landed families were male biased (Aggarwal: 2004).

It hardly requires to be emphasized that redistributive land reform, if effectively and properly implemented in a gender unbiased manner, could empower the rural women through land rights and enhance their social, economic and political status. Unfortunately, redistributive land reform which was once considered to be a key factor in the socio-economic transformation of rural India, has been relegated to the background of development effort to-day.

Women's Access to Land Market

Women's access to land right remains restricted, firstly because not much land transactions take place at affordable prices and secondly due to the fact that dependent women have hardly any money to purchase land in their own name. During the thirty year period between 1950 and 1980, only about 4.1 per cent of owned agricultural land had been sold (Aggarwal: 2003). However, the practice of dowry feeds the land market in several places. Brown and Das Chowdhary (RDI: 2002) observed that about 50 per cent of land transactions in West Bengal are due to dowry or wedding costs, followed by illness (13.4%), other distress related reasons (13.0%), migration due to employment, change of place etc. (10%) and other factors respectively. Even though dowry is illegal throughout India, the law considers the recipient of the dowry as having received the dowry on behalf of the bride and he or she is legally required to transfer, it to the bride (Brown and Chowdhary, RDI: 2002). However, the RDI's field research (2002), in West Bengal indicates that some women held ownership right to purchased land. Some women held independent rights to a portion of their households' purchased land and some held joint rights to purchased land although women rarely purchased land on their own because of culturally constrained gender role and lack of financial resources. Some other women owned land because they supplied the funds to purchase the land through their labour, by bringing money to the marriage or by selling the jewelry. Besides, the concept of co-ownership of marital property is a means of assets within marriage. The husband and wife own an undivided one half interest in the property acquired during the marriage unless the property is spouses' separate property obtained prior to the marriage or received by one spouse as a result of gift or inheritance during the marriage (RDI: 2004).

Socio-cultural Barriers to Women's Land Rights

It is one thing to have a law that entitles women to own land and quite another to effectively allow them to own and control land. The main socio-cultural constraints to women's land rights include (i) a widespread notion that daughters are transitory members for whom shares in parental property cannot be allowed without destroying brother-sister ties, (ii) post-marital residence location, (iii) the myths that women should not plough. Marriage in the form of Karewa, as practised in Punjab and Haryana lead to the forfeiture of the women's estate (P. Chaudhary: 1993). Moreover, women's subordination to men in agriculture should be seen in the background of gendered division of labour and the invention of myths to create ritual taboos against women ploughing (M. Kishwar: 1987). It is indeed strange that when cultivation of land by hired labour is permitted by law, women not ploughing stand against their having land ownership right, as there is a saying that land ownership belongs to one who ploughs the land. In fact, the denial of property rights to women is a significant instrument of patriarchy, as historically developed, cultural and ideological force giving gender relations their specific character in each situation (N. Banerjee:

1996). The patriarchal ideology dominates India's rural society and therefore, necessary interventions for enabling women to own and control land must aim at changing the traditional social, religious, political and legal institution's that reflect and reaffirm such patriarchal ideology. The attitude of the Indian people towards gender unequal land rights has not changed over time. Even though the Hindu Succession Act, 1956 was amended in 2005 to give equal right to men and women in matter of devolution of all kinds of landed property, a recent study (Roy: 2006) in Bihar reveals that only 10 per cent of the people cherished the positive aspect of the legislation. Nearly 43.5 per cent respondents reacted saying that a heavy amount was spent on dowry and therefore, there was no question of sharing of father's property by the daughters. About 14.5 per cent said that this would destroy the family ties. Five per cent respondents mentioned that it was against Indian culture and tradition. Others responded in a mixed manner, saying that property right would help women against a rainy day or against the system of dowry and also some saying that the amended Act would never be implemented. It also needs to be ascertained whether rural women in general and in backward regions, in particular are aware of the new law. While amendments in the land laws would be necessary, the social, cultural and religious norms which dictate women's land rights should be changed through education and awareness building.

It would be especially necessary to sensitize the local revenue and land reform officials who enforce the land laws. It is well known that revenue officials in India, (especially the village *patwaris*) have a strong, bias against women as women's shares are not generally recorded.

Conclusions

It becomes evident from the foregoing discussion that women in India have very limited access to land rights. In some cases, even if they legally own small pieces of land, their effective control is with male members of the family. This has implications not only for agricultural efficiency and equity, but also for food and livelihood security of a vast majority of rural women. Over the years, various national (Ninth Five Year Plan: 1997-2002) and international concerns (WCAARD: 1979) and ICAARD (2006) have hardly resulted in improving the situation. While appropriate legal changes, especially in the case of Muslim Personal Law for its extension to agricultural land would be necessary, it would be all the more important to create social awareness and mobilize public opinion in favour of women's effective land rights. The civil society organizations, women's self-help groups, committed bureaucrats, university and school teachers and panchayat members have an important role to play in this respect. Also, there is need for a clear road map and action plan for achieving gender equality in land rights and all other spheres of life. The possible areas of intervention would include the following:

(i) There should be a national policy that every landless family should be provided with at least ten to fifteen cents of land in women's name.
(ii) There should be a nationwide campaign and social mobilization for effective implementation of Hindu Succession Act, 1956, as amended in 2005.
(iii) The application of Muslim Personal (Shariat) Act, should be extended to agricultural land, as agricultural land continues to be the most important form of productive asset in rural India.
(iv) Rural women should be assisted through subsidized credit to purchase and manage land and other resources either individually or collectively. Group leasing and cultivation of land has already shown positive results in some parts of the country (DDS Experience in Zahirabad, A.P., APMSS in A.P. and Kudumbashree in Kerala).
(v) All the state governments should initiate a scheme of land purchase and allotment of land to landless poor either jointly in the name of men and women or individually in the name of women.
(vi) All rural families who do not own any homestead should be provided a homestead plot in the name of women.
(vii) Create local level pressure to ensure recording the women's share in land through both inheritance and land redistribution programme of the government;
(viii) Provide legal aid to women in all villages.
(ix) Create Women's Club and Resources Centre in every village panchayat. The fact that 30 per cent of total seats are now reserved in all panchayat bodies for women, the women's empowerment committee should be set up at the village level to address the issue of land rights for women, along with other issues of general welfare.
(x) Government should ensure that women cultivators get adequate facilities of credit, extension, technical inputs and marketing.

References

Aggarwal, Bina (2003): "Gender and Land Rights Revisited", *Journal of Agrarian Change*, Vol. 3 Nos. 1 and 2, pp. 184-224.

Aggarwal, Bina (1998): "Disinherited Peasants, Disadvantaged Workers", *EPW*, March 28, 1998, A-5.

Aggarwal, Bina (1994): "Gender and Legal Rights in Agricultural Land in India", *EPW*, Vol. XXX, No. 12.

Archer, W.G. (1984): "*Tribal Law and Justice: A Report on the Santal*", Concept Publishing Co. New Delhi.

Biswas, Devika (2006): 'Working Towards Women's Land Rights through Land Records and Livelihood Programmes', Paper Presented at the Seminar organized by BIES, Patna and RDI U.S.A., April 29, 2006.

Banerjee, N. (1996): 'The Structural Adjustment Programme and Women's Economic Empowerment', in N. Rao, L. Kurup and R. Sudershan (ed), *Sites of Change*, FES and UNDP, New Delhi.

Brown, Jennifer and Sujata Das Chaudhary, Rural Development Institute, U.S.A., November, 2002 (mimeo).

Chaudhary, P. (1993): "Conjugality, Law and State inheritance Right as a Pivot of control in Northern India", *National Law School Journal*, Special Issue.

Chen, M.A. ed. (1998): "Widow's in India: Social neglect and Public Action", Sage, New Delhi.

Chen, M.A. (2000): *Perpetual Mourning: Widowhood in Rural India*, Oxford University Press, New Delhi.

Consult for Women and Land Rights, Indian Case Study Working Group for ICAARD, New Delhi Dec., 15-16, 2005.

Das, J.N. (1990): *Land Relations in North Eastern Region* – Paper presented in a Seminar held at NIRD North-Eastern Regional Centre, Guwahati, March, 22-24.

Dibekar, G.M. (1999): *Law of Hindu Co-parcenary*, 2nd Edn., Hindu-Law House, Pune.

Fatchamps, M. and A.R. Quisum (1995): "Human Capital, Productivity and Labour Allocation in Rural Pakistan", *Journal of Human Resources*, 34(2), pp. 369-406.

Food and Agricultural Organisation, (2006): International Conference on Agrarian Reform and Rural Developments; Draft Declaration, Porto Alegre, Brazil.

Food and Agriculture Organisation (1979): World Conference on Agrarian Reform and Rural Development, Rome, Italy (Report).

Gopal, Gita (1993): "Gender and Economic Inequality in India: The Legal connection", *Boston College Third World Law Journal*, Vol. 13.63, as quoted in Tim *et al.*, Land and Livelihoods, FAO, May, 2004.

Govt. of India, (2010): *Employment-Unemployment Situation in India*, 2007-08, NSSO, Ministry of Statistics and Programme Implementation, New Delhi.

Govt. of India, (2007): *Eleventh Five Year Plan*, Planning Commission, New Delhi.

Govt. of India, (2006): *Employment-Unemployment Situation in India*, 61st Round, 2004-05, NSSO, Ministry of Statistics and Programme Implementation, New Delhi.

Govt. of India (2003): Ministry of Agriculture, *Agricultural Census, 2000-01*, All India, New Delhi.

Govt. of India (2003): *India's Five Year Plans*, Academic Foundation, New Delhi.

Govt. of India (1998): Ministry of Rural Areas and Employment, Report of the Committee for Gender Equality in Land Devolution in Tenurial Laws, New Delhi, p. 2.

Govt. of India (1998): *Sixth Five Year Plan*, Planning Commission, New Delhi.

Govt. of India (1998): *National Perspective Plan for Women*, Department of Women and Child Development, New Delhi.

Govt. of India, (1997): *Ninth Five Year Plan*, Planning Commission, New Delhi.

Govt. of India (1992): *Eighth Five Year Plan*, Planning Commission, New Delhi.

Gupta, Jayoti (1997): 'Voices Brake the Silence' in N. Rao and L. Kurup (eds.) *A Just Right: Women's Ownership and Natural Resources and Livelihood Security*, RES, New Delhi, pp. 135-69.

Gupta, Jayoti (2000): "Women Second in the Land Agenda", *Economic and Political Weekly*, May 4, 2002.

Hanstad, Tim, Robin Nielsen and Jennifer Brown, RDI, *U.S.A.—Report on Land and Livelihoods*, submitted to FAO, May, 2004.

Karna M.N. (1990): *The Agrarian Scene in North-East*—Paper presented in a seminar held at NIRD, North-Eastern Regional Centre, Guwahati, March 22-24.

Katz, E. and J.S. Chamorro (2002): 'Gender, Land Rights and the Household Economy in Rural Nicaragua and Honduras', Paper presented at the Regional Workshop on Land Issues in Latin America and the Caribbean, May 19-22, Pachna, Mexico as quoted in *Land Policies for Growth and Poverty Reduction*, the World Bank, 2003.

Kishwar, M. (1987): "Toiling without Rights: The Women of Singhbhom", *EPW*, Bombay, January 31.

Krishna, Sumi (2005): "Gendered Price of Rice in North Eastern India", *EPW*, June 18.

Rao, Nitya (2008): "Good Women do not Inherit Land" Social Science Press and Orient Black Swan, New Delhi.

Roy, Anubha (2006): "Recent Amendment in Hindu Law of Inheritance", Paper Presented at the Seminar, organized by BIES, Patna and RDI, USA on April, 29 at Patna.

Mearns, R. (1999): "Access to Land in Rural India", *Policy Research Working Paper 2123*, World Bank, Washington, DC.

Sardamani, K. (1983): "Changing Land Relations and Women: A Case study of Palghat District, Kerala" in V. Majumdar (ed) *Women and Rural Transformation*, Concept, New Delhi p. 173.

Srivastava, Nisha and Ravi Srivastava (2010): "Women, Work and Employment outcomes in Rural India", *Economic and Political Weekly*, July 10, vol. XLV. No. 28.

State Institute of Panchayats and Rural Development. (2003): "Beneficiaries of Land Reforms: The West Bengal Scenario", Kalyani, Spandon, Calcutta.

Swaminathan, Hema, Suchitra J.Y. and Rahul Lahoti (2011): "Karnataka Household Asset Survey-Measuring the Gender Asset Group", IIM-B, Monograph.

Roy, L. Prosterman, Mary N. Temple and Timothy M. Hanstad (1990): "Agrarian Reform and Grassroots Development: Ten Case Studies", Lynne Rienner Publishers, USA.

The World Bank (2001): *World Development Report, 2000-01: Attacking Poverty*, pp. 117-122, Washington DC, U.S.A.

17

Tribal Customary and Formal Law Interface in North-Eastern India: Implications for Land Relations

Walter Fernandes

The North-East is known in the rest of India mainly for its conflicts. One cannot deny that this home of many ethnic groups and tribes has for five decades witnessed armed conflicts that are integral to its people's search for a new identity amid the economic and cultural crises they face. One of the reasons is the interface between modernisation and their traditions with no preparation. It has had an adverse impact on many tribes. An area in which this interface has made a major impact is the legal framework governing land relations. Most tribes of the region run their civil affairs according to their community based customary law but the individual based land law of the country is superimposed on them. Because of the disruption it has caused in their lives one of the demands of many tribes is recognition of their customary law. It has been recognised through constitutional amendments in Nagaland and Mizoram. Community ownership also gives women some control over land, but this chapter will limit itself to the land laws and will only allude to its gender and class implications.

The community ownership tradition has not remained unchanged but has been modified over time. Even when their customary law is recognised, the elite among them tend to interpret it in their own favour. For example, the Sixth Schedule was meant to be a protective measure but it has not always gone in favour of the community, especially women. For example, the Garo of Meghalaya who come under the Sixth Schedule have experienced a changeover to commercial crops and individual ownership that goes against women and leads to class formation. Signs of its beginnings are visible among the Dimasa of Assam, the Aka of

Arunachal Pradesh and others. This chapter will discuss some of these issues, especially the interface of their customary law with the formal law and its implications for land relations

Land Laws and Tribal Communities

Though the complex phenomenon of the conflicts has often been oversimplified as secessionist or terrorist, in reality an identity search is central to it. Their land and customary laws are closely linked to their identity which is in fact built around them. Today land relations are being modified by immigration, encroachment and the changes that the modern legal system introduces in their tradition. They do not begin the conflicts but exacerbate those existing already. The conflict was initially against "outsiders" who, the local people felt, were controlling their economy, alienating their livelihood and were attacking their culture. Because of this combination of causes, the conflicts combined economic demands with sub-nationalism and cultural resurgence (Datta 1990: 36-37) including the customary law.

In other words, commercialisation of their resources is only one of the causes of the conflict against "outsiders". Central to it is the homogenising tendency of the Indian State. In attempting to turn itself into a nation, the Indian State rarely respects the cultural and ethnic identity of different groups. It often fails to recognise that the tribals have a culture and a religion of their own (Singh 1990: 234). The tribals react to the effort to homogenise cultures and monopolise their livelihood. Thus, most struggles in the north-east are in reaction to the tendency of the dominant "one State one nation" thinking of the Indian State and "to take the degree of Aryanisation as the measure of Indianisation" (Datta 1990: 41).

An example is the meeting of A.Z. Phizo, the Naga Nationalist leader with Mahatma Gandhi in 1946. There are reasons to believe that the Mahatma was sympathetic to the idea of the type of autonomy Phizo demanded. It was based on the specificity of their culture, customary law and livelihood resources that the Nagas wanted to safeguard. However, most leaders of Independent India failed to understand the specificity of the North-east that is different from the rest of India (Sanyu 1996: 116-118). Most of them seem to view the region as the extreme end of India whose territory has to be secured. As such their national belonging is suspected and the region is considered a security risk. Very little thinking goes into the possibility of considering their difference from the rest of India an asset or as a link with South-east Asia. Most leaders seem to judge the nation from the homogenising rather than the pluralistic perspective. Diversity suffers because of it (Mishra: 2000). In reaction to it, many communities proclaim their "sovereignty" which others call anti-national.

That is one of the reasons why in most struggles a major demand is recognition of their identity linked to their land and customary laws. Their laws have changed in many ways in response to formal education and the new economic, religious and political relations. In some cases individual land ownership has come to be accepted as the norm. However, most tribes continue to treat their customary laws and the community ethos as intrinsic to their identity. That is also the reason why the negotiations leading to the formation of Nagaland and Mizoram States had recognition of their customary laws as one of the conditions. It was granted to these two States through Articles 371A and G respectively of the Constitution.

Implications for Land Laws

The tendency to view the nation from the point of view of the Centre rather than the periphery also has implications for the land laws and land relations in the region. Most tribes are community based but the formal land laws are individual based, and are founded on the principle of the State's eminent domain. In this view land is only a commodity for cultivation and construction (NCHSE: 1986) while to the tribals it is an ecosystem with the local community at its centre. For centuries their communities have treated the resources as renewable and have built a culture and economy of their sustainable use (Iyer 1996: 375-377).

The 19th century land laws made to suit the colonial need of exploiting the resources to the benefit of the British Industrial Revolution ignored the latter view. The colonial regime needed to change the Indian economy and turn the colony into a supplier of capital and raw material and a captive market for its products. It required monopoly over land for coal mines, coffee and tea plantations, roads, railways and other schemes. New land laws were enacted to facilitate the process of land transfer to the profit of British plantation and mine owners. The effort to turn the livelihood of the local communities into a commodity began with the *Permanent Settlement Act, 1793*. It continued in the *Assam Land Rules, 1838,* the Calcutta law of 1824 and others meant to make land available for purposes like salt pans and culminated in the *Land Acquisition Act, 1894* that remains in force today (Upadhyay and Raman 1998).

The colonial regime based these laws on the principle of eminent domain. In Australia it is called *terra nullius* (nobody's land). White colonisation of native land there and in the Americas is based on the principle that land without an individual title belongs to none as such anyone can occupy it. In 1992, the Australian judiciary declared it unconstitutional (Brennan: 1995: 4-5) but Indian land laws continue to be based on its American version of eminent domain. Its first facet is that land without an individual *patta* is State property. The second is that the State alone has the right to decide a public purpose and deprive even individual owners of their assets. This State power is overriding (Ramanathan: 1999: 19-20).

The Law and the North-Eastern Communities

That is the background of the interface of the customary and formal laws that has modified many tribal laws but its extent depends on the nature of their contact with the world outside their own, the application of the Sixth Schedule and of their customary law. The Constitution was amended in 1963 and 1987 to recognise the law in Nagaland and Mizoram under Articles 371A and 371G respectively. The Sixth Schedule that recognises community ownership of land and forests is applicable to parts of Tripura, two districts of Assam and to the whole of Meghalaya (Fernandes, Pereira and Khatso: 2005: 22-23). In the remaining States, the tribes live according to their community based customary law but the formal law recognises only individual ownership. It has changed their customary law to various degrees.

The Aka of West Kameng district in Arunachal Pradesh are close to their tradition and govern themselves according to their customary law. The Sixth Schedule does not apply to them but it has not had much impact on them because their contact with the outside world is recent (Sinha: 1962). This *jhum* practising tribe lacks the very concept of landownership and only has the tradition of community control and of usufruct right over the CPRs. In the *jhum* season every family cultivates as much land as it needs. After it the land reverts to the community. However, slow change is visible among them. Today some claim individual ownership but others with a salaried job call themselves landless since they have lost their right over the CPRs by not practising *jhum* any more (Fernandes and Bharali: 2002: 7-8).

The Dimasa of North Cachar Hills in Assam have been exposed to the dominant cultures since their Hinduisation by the Bengali administrators who accompanied the British colonial rulers but have retained their internal autonomy, continue to be governed by their CPR based customary law, come under the Sixth Schedule and have a district autonomous council. Many of their clans have dual descent with property inheritance through the male line and the clan and family name coming through the female line (Bordoloi: 1984). However, their elite are moving towards individual *pattas.* One of their leaders owns over 200 acres (Barbora: 2002: 1287). On the other side, a voluntary agency has introduced oranges and other commercial crops in some villages without changing the ownership pattern. Some families that have accepted individual ownership have conferred inheritance rights also on women. Thus the trend towards individual ownership goes hand in hand with change of their land use without changing the landownership pattern drastically (Fernandes and Pereira: 2005: 37).

The matrilineal but patriarchal Garo of Meghalaya are governed by their customary law and the Sixth Schedule. Many of them in the East Garo Hills have begun to show signs of class formation and of strengthening patriarchy. An important reason of this change is the introduction of rubber plantations. It got them to interact with the administration that gives loans and subsidies only to individual owners and "heads of families" that most financial institutions interpret as male. The families we studied in West Garo Hills continue their CPR culture and have not planted commercial crops. Thus, they have not combined the modem with the traditional that the Dimasa families can become if more accept recent changes.

On the other side, the *Adibasi* of Jharkhand origin whom the British brought to Assam in the 19th and 20th centuries as indentured labour to work in its tea gardens are not included in the Schedule though they speak of a customary law that has very little value today. So they represent the type of modernisation that results in impoverishment and exploitation (Fernandes, Barbora and Bharali: 2003: 33-34). Landlessness is the highest among them and literacy very low (Toppo: 1999: 133-134). Though they have lived in Assam since the 1850s, they continue to be considered non-indigenous. As late as the May 2004 general elections a candidate appealed to the voters to recognise himself as indigenous and reject his opponent from the plantation labourer community as non-indigenous (*The Telegraph*: May 5, 2004).

The Boro, a plains tribe not under the Sixth Schedule, have won a Boro Territorial Council (BTC) after a struggle but till recently their community based customary law was not recognised (Roy: 1995: 16-

17). As a result they have almost fully internalised the ideology of individual ownership in the sense that those living on the CPRs call themselves encroachers while others like the Aka and Dimasa consider themselves CPR dependants. The BTC has recognised their right partially but has not granted them the Sixth Schedule status. Also the Rongmei, a Scheduled Tribe of Bishnupur district in Manipur are not governed by the Sixth Schedule. Several have lost their land to ethnic conflicts and some to the Loktak project but have not even been compensated since much of what they sustained themselves on was CPRs that the law does not recognise as their livelihood. These two tribes represent the interaction of their CPR based customary law with the individual based administration that can deprive them of their livelihood with no right to any alternative (Fernandes and Bharali: 2002: 16-17).

The Angami, a major tribe living mainly in the Kohima district of Nagaland numbered 97,433 in 1991. Terrace cultivation based agriculture was their main economic activity till recent years many took up salaried jobs. They were in the forefront of resistance to the British regime and after 1947 they led the Naga Nationalist Movement and played a major role in the ethnic movement under the leadership of Z.A. Phizo. In the process they underwent rapid social change and gained access to modem education and political systems. They have the Village Development Board that also has women among its members. Thus, tradition and modernity live side by side (D'Souza, Kekrieseno and Nokhwenu: 2002: 26-27).

Land Laws and Ethnic Conflicts

We have mentioned these tribes as examples of various degrees of interaction of their communities and tradition with the formal law. Its implications have come out clearly in most of our studies. We give in the Table 17.1 an example from two of them. One can see from it that landlessness is all but non-existent among the Aka whose interaction with the external world is recent. The "landless" are persons with a salaried job and have stopped *jhum* cultivation and have thus given up their usufruct right over the CPRs. The slow change is visible in the two persons who claim to own one or two hectares each of individual land. They belong to the elite that have appropriated to itself irrigated land near the river.

The extent of landlessness among the *Adivasi* shows their exploited state. Alienated from their land in Jharkhand and later from their customary

Table 17.1: Land Ownership Pattern among Some Tribes in the North-east (Acres)

Tribe	*Landless*	*Jhum*	*Below l1*	*1.01-2.5*	*2.51-5.0*	*5.01-10.0*	*10+*	*Total*
Adibasi	128	0	23	12	6	0	0	169
Aka	3	38	0	0	2	0	0	43
Angami	13	1	64	21	18	24	29	170
Boro	32	0	3	14	6	3	0	58
Chiru	0	0	8	13	6	0	3	30
Chotei	1	0	8	9	6	0	3	27
Dimasa	1	79	0	21	4	1	0	106
Garo	84	0	57	41	28	8	5	223
Rongmei	4	0	54	31	19	1	1	110
Total	266	118	217	162	95	37	41	936
%	28.42	12.61	23.18	17.31	10.15	03.95	04.38	100.00

Source: Fernandes and Barbora: 2002a: 116; Fernandes and Bharali: 2002: 23.

law and community, they have slowly lost their tribal identity itself and have internalised a non-tribal psyche of subordination. Such internalisation has been intensified by the regimented work structure of the tea gardens and by the denial of a Scheduled Tribe status to them. Among them we chose 50 per cent of the respondents from the *bastis* outside the gardens. They are not regular workers in the garden and are expected to be cultivators but their landownership remains low.

Of importance is the extent of landlessness among the Garo who were till recently a CPR dependent matrilineal tribe. When rubber plantation was introduced among them, the Rubber Board and the financial bodies insisted on individual *pattas* as preconditions for loans and subsidies. *Patta* ownership was usually understood as male. It has started a move away from the community to the individual male owner. Today individual ownership is a fact and male ownership has become a demand of middle class men. This move has been in existence for many decades on a low scale (Kar: 1982: 254). Rubber plantation has intensified the demand and its catalytic role was crucial to it (Fernandes, Pereira and Khatso: 2005: 10-11). Thus, even where the Sixth Schedule recognises community ownership, the administrative system goes by individual values and sets in motion processes that can weaken the tribe.

That the Dimasa too are close to their tradition but not to the same degree as the Aka is seen in the nature of landownership among them. Three-fourths of them continue to depend on *jhum* land which is their livelihood. Some of them have started moving towards individual ownership because of their exposure to the commercial forces and the administration. They want individual *pattas*. If done without any preparation it can take them towards the extent of landlessness seen among the Garo (Fernandes and Pereira: 2005: 93-95).

Landlessness is very low among the Rongmei but 50 per cent of them own less than one acre each and nearly 30 per cent own less than 2.5 acres (one hectare) considered the minimum a family requires for its sustenance. They lost much of their land in their conflict with the Kuki. The law could not protect their land. Some others lost their land to the Loktak Project but were not counted among the displaced or compensated for it. The reason for both is that the formal law did not recognise community ownership according to which they lived. Most of them have now occupied the CPRs of some neighbouring villages and have declared them their own.

There is very little landlessness among the Angami whose customary law has been recognised. It bans land transfer to persons not belonging to their tribe. They also have a tradition of combining individual with community ownership on one side and terrace cultivation with *jhum* on the other. That tradition combined with the political processes of the nationalist struggle and access to education to make it easier for them to deal with the changes than what the others have done. However, some tension is felt today in the form of land alienation within the tribe and the consequent class formation. One of the forms it has taken is hidden absentee landlordism (Fernandes and Barbora: 2002a: 48-49).

Land Relations and Conflicts

We have discussed till now the impact of imposition of the formal land laws on tribes that have lived by their tradition. The formal law was imposed on them with no preparation. Where there was some preparation, they were able to deal with the changes more than those who came face to face with them with no preparation. The Garo belong to the latter category while the Angami and Dimasa have had some preparation in different forms. Such an intervention has disrupted their societies and their reaction was initially against outsiders like the immigrants. Because of it the region has witnessed killings of Biharis in Karbi Anglong, anti-Bengali conflicts in Tripura and others elsewhere (Bhaumik: 2003: 84-85).

This interface and land alienation that results from it has also resulted in shortages. They have led to conflicts between various communities of the region and most of them are around land. On one side, the local people view the immigrants as a threat to themselves. On the other, the shortages that result from this encounter force them to compete among themselves for the scarce resources. One of its results is hardened identities and exclusive claims over the resources of an area. Each community rewrites its history to declare itself the original inhabitants of a given region, as such the sole claimants to its resources. Be it the Naga-Kuki conflict in Manipur (Fernandes and Bharali: 2002: 52-55), the Bodo-Santhal (Roy 1995) and Dimasa-Hmar tension in Assam (*The Telegraph:* 23rd April 2003) or the Tripura tribal demand for a homeland (Bhaumik: 2003) all have their origin in competition for land. These conflicts between communities of the region arise because in the context of shortages caused by encroachment and the failure to create productive jobs every group views the limited land and jobs as its exclusive right. Given their symbiotic relationship with land, the community views the resultant conflicts as defence of its culture, identity and livelihood. Thus, the changes begin with outside intervention. Once they result in shortages they both intensify traditional rivalries and create new ones within the region (Baruah: 1999: 29-32).

A crucial issue in them is commercialisation of land and forests and the resultant transition of most ethnic groups from subsistence to a commercial economy and of their land from livelihood to a commodity. The administration supports the transition to a commercial commodity by insisting on individual *pattas* for subsidies and loans. The reaction of the local communities is ambiguous. Many of them want to retain control over land but at the same time move away from it as a livelihood. Most Aka and Dimasa families sustain themselves on land but it is not the case with the Rongmei, *Adibasi* and to some extent Boro who have lost much of their land to ethnic conflicts. Some of the Rongmei sell their land in order to earn money to bribe officials and get a job in the administration or for their children's education. They have lost hope in land as sustenance and are ready to part with it for an alternative but are not to abandon their cultural attachment to it as an essential ingredient of their identity. This effort to keep control over it while moving beyond it is another cause of conflicts.

Displacement and Land Laws

One can thus see that a crucial factor in the transition to an individual *patta* that is alien to the tribal CPR culture is encouragement that the individual-oriented administration gives to processes leading to it. Besides, the State alienates land for development projects and while so doing, it recognises only individual ownership. It has become a crucial issue in the North-east in the context of the thinking that the region should be turned into the powerhouse of India. 48 possible major dams are being planned in the region and a list of more than 100 others is being (Menon *et al.*: 2003). The former Prime Minister Mr. Atal Behari Vajpayee gave recognition to this process by launching on 24th May 2003, the 50,000 MW hydro-electrical initiative for the Northeast (*The Telegraph*: 25th May, 2003).

These initiatives have to be studied in the context of past experience because most dams being planned are in the tribal areas where the CPRs are the norm. Many of them are in Arunachal Pradesh where the tribes live according to their community ownership ethos but the individual based law does not recognise their land as their livelihood (Menon: 2003). Many other dams too are in the tribal areas. For example, according to official sources, the proposed Pagladia dam in the Nalbari district of Assam will displace 18,347 persons while field data show that it will affect 1,05,000 persons, 80 per cent of them Bodo and other tribals who live on their CPRs that the law does not recognise as their livelihood (Bharali: 2004).

That has been the case also in the past. We have stated above that the Rongmei in Manipur lost their land to the Loktak Project without even being included among the displaced and being compensated. Impoverishment has been its consequence. By the late 1960s, the indigenous tribes of Tripura had lost more than 60 per cent of their land to Bengali Hindu immigrants from the erstwhile East Pakistan. It resulted in conflicts between them and the settlers. That is when the Gurnti or Dumbur dam was announced in the 1970s. They protested against it but were forced out of their land. It submerged 46.34 sq. km, most of it level land that is only 28 per cent of the State's total. By official count it displaced 2,558 families that had *pattas.* Another 5,500 to 6,500 CPR dependent families were not even counted. Their only alternative is *jhum* cultivation in its catchment area or on other common lands. It causes environmental degradation and they come to be considered enemies of nature (Bhaumik: 2003: 84-85). In other words, where the Schedule is not recognised, people can be displaced without being counted among the displaced.

We have noticed it in the ongoing study on development-induced displacement in Assam 1947-2000 (Fernandes and Bharali forthcoming). All the projects together have used around 15 lakh acres of land, two-thirds of it CPRs. Around 20 lakh persons have been deprived of their livelihood. However, the official records account for a little over 4,00,000 acres of land and 3,00,000 persons displaced or deprived by it. The rest of the land is not accounted for and the people whose livelihood it is have not been counted because the law does not recognise their community ownership. So in the eyes of the State, the communities that have inhabited it for centuries before the colonial law was enacted are encroachers on it.

The situation becomes easier if the law recognises community rights and the people are aware of it. That, for example, is the case at Domiasiat in the Khasi Hills of Meghalaya. The State has been trying to acquire their land for uranium mining but the local community has been opposing what it considers loss of its livelihood. There has been a stalemate on this issue for seven years with the State speaking of the need to get hold of that land and the people refusing to part with it by asserting their right to security of life and livelihood. The State is unable to deprive them of their land since the area comes under the Sixth Schedule.

Search for Solutions

We have seen in this paper that conflicts are intrinsic to the contradiction between the tribal customary laws and the formal law because of their contradictory value systems. Another crucial issue in the process their livelihood alienation is the transition of many communities of the region to a commercial economy. Commercialisation of land and forests is thus closely linked to the transition of land from livelihood to commodity. One cannot begin by stating that a commercial approach is bad in itself. One only states that such a transition without any preparation or imposition of another system on them for the benefit of another class does them harm. The shortages it causes result in ethnic conflicts. Linked to it is the changeover from an egalitarian to a class society. That weakens the community based tribal culture with no alternative to take its place. These are among the processes leading to ethnic conflicts most of which are around competition for land.

That has to be situated in the context of the proposed dams and other projects in the North-east that has a backlog of at least 3 million persons which is about 25 per cent of the active workforce. The fear is that it will deprive a large number of persons of their livelihood with no alternative to take its place. The experience in the rest of India shows that it results in the people's impoverishment and eventual marginalisation. That is where alternatives have to be found to the system that impoverishes them in the name of national development. Change of the law is only one step in it but it is essential. One has to begin by recognising their customary law as basic to their identity formation. The law has then to be updated instead of being replaced by another system that can weaken them as a community. Recognition of the CPRs as their livelihood is basic to it. Other changes can follow from it.

Conclusion

We have studied in this paper the role that the formal individual based law plays in causing shortages, impoverishing the communities and thus causing ethnic conflicts in the North-east. The law and land alienation are not their only causes but are crucial to them. The main issue is the process that makes land encroachment possible or legitimises land alienation. That becomes an attack on the livelihood of the communities of the region in their predominantly agrarian economy with *jhum* as its main form in the hills. Conflicts arise because different groups vie for limited land. Given their symbiotic relationship with it, they come to view conflicts as defence of their culture, identity and livelihood. That shows the need to search for alternatives that combine their community ethos with commercial needs.

References

Barbora, Sanjay. (2002): "Ethnic Politics and Land Use: Genesis of Conflicts in India's North-East", *Economic and Political Weekly,* 37 (n. 13, March 30), pp. 1285-1292.

Baruah, Sanjib. (1999): *India Against Itself: Assam and the Politics of Nationality,* New Delhi: Oxford University Press.

Bharali, Gita. (2004): "Dams a Threat to Livelihood", *The Telegraph,* November 29.

Bhaumick, Subir. (2003): "Tripura's Gumti Dam Must Go", *The Ecologist Asia* 11 (n. 1, Jan.-Mar), pp. 84-89.

Bordoloi, B.N. (1984): *The Dimasa Kacharis of Assam.* Guwahati: Tribal Research Institute.

Brennan, Frank. (1995): "Parliamentary Responses to the Mabo Decision", in M.A. Stephenson (ed). *Mabo: The Native Title Legislation: A Legislative Response to the High Court Decision.* St Lucia: Queensland University Press, pp. 1-25.

Datta, Brijendranath. (1990): "Ethnicity, Nationalism and Sub-Nationalism, with Special Reference to North-East India". In D. Pakem (ed), *Nationality, Ethnicity and Cultural Identity in North-East India.* New Delhi, Omsons Publications, pp. 36-44.

D'Souza, Alphonsus, Christina Kekrieseno and Veronica Nokhwenu. (2002): *Status of Women in North-East India: A Comparative Study of Three Ethnic Groups in North-East India, Aka, Angami and Dimasa.* Jakhama: St Joseph's College (mimeo).

Fernandes, Walter and Sanjay Barbora. 2002a. *Modernisation and Women's Status in North-Eastern India: A Comparative Study of Six Tribes.* Guwahati: North-Eastern Social Research Centre.

Fernandes, Walter and Sanjay Barbora. 2002b. *The Socio-Economic Situation of Nagaon District: A Study of Its Economy, Demography and Immigration.* Guwahati: North-Eastern Social Research Centre (mimeo).

Fernandes, Walter and Gita Bharali. (2002): *The Socio-Economic Situation of Some Tribes of Bishnupur and Palizi.* Guwahati: North-Eastern Social Research Centre (mimeo).

Fernandes, Walter, Sanjay Barbora and Gita Bharali. (2003): *Primary Education of Plantation Labourers in Assam.* Guwahati: North-Eastern Social Research Centre (mimeo).

Fernandes, Walter and Melville Pereira. (2005): *Land Relations and Ethnic Conflicts: The Case of North-east India.* Guwahati: North-Eastern Social Research Centre.

Fernandes, Walter, Melville Pereira and Vizalenu Khatso. (2005): *Customary Laws in North-East India: Impact on Women.* Guwahati, North-Eastern Social Research Centre (mimeo).

Fernandes, Walter and Gita Bharali. Forthcoming. *Development-Induced Displacement in Assam 1947-2000.* Guwahati: North-Eastern Social Research Centre (mimeo).

Iyer, K. Gopal. (1996): "Land Alienation, Deforestation and Predicament of Tribals: A Comparative Study of Bihar and Tripura", in K. Gopal Iyer (ed). *Sustainable Development: Ecological and Socio-Cultural Dimensions.* New Delhi: Vikas Publishing House, pp. 393-403.

Kar P.C. (1982): *The Garos in Transition.* New Delhi: Cosmo Publication.

Menon, Manju. (2003): *Large Dams for Hydropower in North-east India: A Dossier* Pune: Kalpavriksh.

Menon, Manju, Neeraaj Vagholikar, Kanchi Kohli and Ashish Fernandes. "Large Dams in the North-east: A Bright Future?" *The Ecologist Asia* 11 (n. 1, Jan-March), pp. 3-8.

Misra, Udayon. (2000): *The Periphery Strikes Back.* Shimla: Indian Institute of Advanced Studies.

NCHSE. (1986): *Rehabilitation of Displaced Persons Due to Construction of Major Dams: Volume I,* New Delhi: National Centre for Human Settlements and Environment.

Roy, Ajay. (1995): *The Boro Imbroglio.* Guwahati: Spectrum Publishers.

Ramanathan, Usha. (1999): "Public Purpose: Points for Discussion", in Walter Fernandes (ed). *The Land Acquisition (Amendment) Bill, 1998: For Liberalisation or for the Poor?* New Delhi: Indian Social Institute, pp. 19-24.

Sanyu, Visier. (1996): *A History of Nagas and Nagaland: Dynamics of Tradition Village Formation.* New Delhi: Commonwealth Publishers.

Singh, A.K. Ranjit. (1990): "Emergent Ethnic Processes in Manipur: A Reappraisal." In D. Pakem (ed). *Nationality, Ethnicity and Cultural Identity in North-East India.* New Delhi, Omsons Publications, pp. 233-251.

Sinha, Raghuvir. (1962): *The Akas.* Itanatar, North-Eastern Frontier Agency.

Toppo, Hippolytus. (1999): "Violation of Human Rights in the Tea Plantations of Assam and West Bengal", in Thomas Pullopillil (ed). *Identity of Adivasis of Assam.* Delhi: Indian Publishers Distributors, pp. 129-156.

Upadhyay, Sanjay and Bhavani Raman. (1998): *Land Acquisition and Public Purpose.* New Delhi, The Other Media.

Part IV

LAND ACQUISITION AND INCLUSIVE DEVELOPMENT

18

Land Acquisition and Peasant Resistance Critique of Policy Interventions

K.B. Saxena

The government undertakes a large number of development activities which require land. This land is acquired from landowners under various land acquisition laws and the cause is justified 'Public Purpose'. The most prominent and widely used of these laws is the Land Acquisition Act, 1894 enacted during the colonial period. The power of the government under the Land Acquisition Act is absolute and the landowner has no choice to opt out or refuse to part with land. This is the reason why people who are dependent on land for their subsistence are so agitated about the prospects of its acquisition. The changes made in the Act in 1984 extended its application to the land required for private companies as well which enlarged the scope of acquisition. The shift to neo-liberal economy and its integration with global economic order has widened the scale and accelerated the pace of acquisition of land. The enactment of SEZ Act, 2005 has introduced a new dimension to the demand for acquisition of land in terms of area, location, size and geographical spread. This pressure on land and the level of displacement it has unleashed has given rise to widespread resistance from the affected peasants. The land Acquisition Act, 1894 being the chief instrument of acquisition has come in for a very sharp attack on grounds of its inconsistency with democratic governance and principles of equity and social justice. There has been a long pending demand for drastic changes in it and even of its abrogation altogether. This demand was ignored by the government until recently. But the ongoing opposition to acquisition of land in several places in the country compelled the government to take a call. The Land Acquisition (Amendment Bill), 2007) (MORD: 2007) a (hereinafter referred to as the Bill, 2007) was a response to this demand.

With a view to appreciating what the changes proposed in the Bill, 2007 imply, it is necessary to have an idea of what the principal (existing) Act contains and why its provisions cause so much hostility. The conceptual framework of laws on acquisition of land is heavily loaded in favour of the state which has been vested with absolute power to forcibly acquire private land while conceding the affected persons virtually no right or choice to prevent or stop it. The instrument of land acquisition presents a chaotic picture of numerous statutes, mostly catering to the needs of specific sectors (Atomic Energy, Railways, Highways, Coal etc.) with no commonality forged among them regarding procedure of acquisition and compensation entitlements of the persons whose lands are acquired though, in the late 1980s, the Central Government did advise the concerned agencies that the package of entitlements enshrined in the L A Act, 1894 should be followed. The peasantry, lacking in legal literacy, cannot comprehend the rationale of using different laws with differing acquisition procedures and norms of compensation. Multiplicity of laws confuses and bewilders them making difficult the exercise of limited rights available to protect their interests. They also fail to obtain relevant information clarification from land acquiring agencies on their quarries and apprehensive. As a result, they are unable to make use of specific legal provisions contained therein to raise their objectives and articulate their interests in the limited space that may be available to them for this purpose. State agencies are least interested in making the acquisition process people friendly with information, facilitation and conscious efforts at eliciting their participation. The acquisition process is entirely bureaucracy-centric and focuses on adherence to the letter of the law unconcerned with its human dimensions. This aggravates the traumatic impact of acquisition on the beleaguered farmers and other affected poor dependent upon land.

The diversity of legal architecture is only one (though not the most significant) among the affected people's concerns. The other features of law on Acquisition hurt their interests more pronouncedly. The important among these concerns are discussed as follows: [for the purpose of this paper, only the L A Act, 1894 as amended in 1984 is being referred to here as major part of acquisition is carried out under it]. These concerns have been clubbed under the following heads:

(a) Conceptual framework
(b) Definitional limitations
(c) Processual matters

(d) Substantive issues
(e) Implementational practices

Conceptual Frame

Title of the Act: The Act as the title suggests restricts itself to acquisition of land and its processes. Its primary focus is on regulating state action to achieve this objective. What happens to the persons from whom land is acquired or those who were dependent on the land for livelihood and other life supporting benefits is not its concern. The scope of the law is exhausted once the land is taken possession of after acquisition process has been completed. The limited concern shown for the land losers in the Act is with regard to payment of compensation. The severest indictment against the law is that it is overwhelmingly focused on the interests of the state and ignores the consequences of this action on the displaced persons. It has therefore been advocated that changes may be made in the law to incorporate state's obligation towards persons affected by this acquisition. So far, whatever the government has done by way of attending to the problems of displaced persons is largely in terms of relocation usually called resettlement. This process has been accomplished by executive action though some state governments have enacted laws for rehabilitation of persons displaced as a result of specific projects (usually eternally funded) which have laid down broad norms for this purpose. But the larger acquisition of land is carried out for projects of the Central Government which has no such law and until recently no uniform policy either. In the absence of any statutory obligation with regard to their rehabilitation, the displaced persons find it difficult to get even the limited entitlements promised to them by the Central or state governments under the project specific package enforced. The incorporation of an unequivocal obligation of the state to rehabilitate and resettle persons displaced as a result of acquisition of land and the norms of their entitlement within the ambit of the acquisition law has therefore been stressed to strike a balance between the power of the State and the interests of the citizens.

Eminent Domain: The Land Acquisition Laws (the colonial version and the later amendments carried out since independence) are embedded in the doctrine of 'Eminent Domain'. This doctrine rests on the premise that the 'right of the state or the sovereign to its or his own property is absolute while that of the subject or citizen to this property is only paramount. The citizen holds his property subject always to the right of the state to take it for public purpose'. (CPCNR: 2007) The state is not even required to prove that acquisition is for public welfare. A mere declaration that acquisition of land is for public purpose is sufficient to meet the legal requirement. The doctrine was a part of the common law in England which was applied to the colonies and later to India coming under colonial occupation irrespective of the prevailing indigenous practices regarding use of and control over land here. Though the LA Act, 1894 has made acquisition of land conditional to the serving a 'public purpose', the 1984 amendment has extended the ambit of 'public purpose' to cover acquisition of land for private companies also. Besides, the law, as interpreted by the apex court, does not cast an obligation on the state to establish the public good involved in the specific proposal of land acquisition. In fact, it has even held that acquisition of land for private enterprises for earning profit can also serve 'public good' if the state has so declared. (Upadhyay *et al.*: 1988) Thus, the 'absolute' nature of state power has emerged from the application of the law against which affected citizen has no relief. Since right to property is not a fundamental right, but only a constitutional right, its abridgment by state through a procedure established by law cannot be challenged on grounds of infringement of the basic structure of the constitution. Though acquisition of land deprives people of their established modes of livelihood, the right to life under Art.14 has not been creatively interpreted by courts to limit the power of the state in this regard or even to cast an obligation on it to provide a viable alternative source of income and a dignified standard of living to the displaced persons. The absolute and arbitrary nature of state power based on the doctrine of 'Eminent Domain' is sharply criticized as being anachronistic in a democratic polity besides being inhumane and needs to be drastically curtailed in a manner that makes acquisition of land conditional to the consent of the persons affected by it.

Definitional Limitations

'Public Purpose': The need for a public purpose is cited as the Justification for acquisition of privately owned land. But the 'public purpose' has not been defined. It has only been explained by listing a few such purposes as illustrative examples of 'public purpose'. The judicial review of acquisition laws has also not required that it need be defined or elaborated or even its use justified in specific cases. State's declaration that the acquisition in a particular case is for a 'Public Purpose' is considered sufficient for meeting the requirement of law. The flexibility in the application of 'Public Purpose' given to the state has rendered its use arbitrary and the affected peasantry powerless to stem its onslaught. There is a strong demand that 'public purpose' should be clearly and precisely defined and it should be made justiciable.

'Land': 'Land' has not been expressly defined. The existing definition of Land in the Act has only elaborated the expression through an 'inclusive' explanation. This explanation does not make any reference to the ownership, occupation or use. But the practice of land acquisition has expressly linked it to the private ownership with a valid documentary proof of title. This conceptual understanding of land tends to exclude the category of land from its ambit which is not privately owned but extensively used by the local community in a defined territory. The sanctity of such use is founded in customary practices and traditions. The current interpretation of the expression 'land' therefore does not take into account different categories of land, i.e., common land, forest land, government/ public land and other common property resources like water bodies, forest produce etc. even though these resources provide a wide range of vital life supporting benefits. These common property resources, therefore, get excluded from any claims by the user community and the individuals to be compensated for their loss either financially or by way of replacement in the state provided habitat resettlement plan. Since these resources are designated as 'government property' in land records, the government determines the terms of their transfer to the project agency. The compensation, if any, for loss of such CPRs, therefore, accrues to the government. Besides, there are also plots in these lands which are under individual occupation but have no valid authorization from the competent authority though some of the occupants may possess certain documents to support their claims. Where the occupants are *dalits* or *adivasis* or other rural landless poor, eligible for receiving allotment of land under the land reforms policies of state governments, such occupancies are entitled to regularization. But in the absence of formal validation of such claims by the government, these occupants are considered 'encroachers' and, therefore, not entitled to any compensation. A comprehensive definition of land to include all such lands which people use is, therefore, sought to accommodate all such affected interests.

'Person Interested': This expression has also not been defined but refers to a person who claims an interest in compensation for acquisition of his/her land. But the practice of acquisition has applied this term to only those whose rights are enshrined in the record of rights of the concerned village. Usually, this gets confined to owners/title holders of land as the land records, by and large, do not reflect the existence of tenancies and other interests, privileges, usages, concessions, occupancies and benefits arising out of land. Similarly, it excludes from consideration agricultural labourers, artisans, and fishermen etc. who use private land of others for cultivation under different arrangements, homestead facilities, and other benefits either on customary basis or by way of oral contracts. Such interests rarely get reflected in record of rights or backed by legal sanction or documentary proof to support them. Similar is the case with occupants of plots on government/public/forest land referred to above. The definition, therefore, needs to be widened to include these persons.

'Person Affected': This expression has not been incorporated in the law. As a result, in official practice, 'person interested' as per land records alone is taken as 'Person affected' for the purpose of compensation payment and rehabilitation, if any. This makes a very large section of displaced persons totally invisible in law and official action even though they suffer most from the acquisition of land. This section includes all those who are deprived of their sources of livelihood and access to life supporting facilities. The only consideration they receive from the government is a small residential accommodation in the resettlement colony where they are physically shifted from their existing house. The ambit of the term is, in fact, even wider than the dimensions of displaced persons conveyed by the extended definition above. The acquisition of land affects persons in terms of loss of land, assets, livelihood and quality of life much beyond the area acquired. The pressures generated by the project and the influx of persons from outside the project area makes ever increasing demands on private and public land in the vicinity and meager life supporting resources and facilities used by the local community. This leads to loss of ownership of assets and access to common property resources and tends to de-stabilize livelihood patterns and social networks in the area and restricts opportunities for peaceful ownership and access to land and dignified survival. This phenomenon is usually described as 'secondary displacement' which the government policy and practices on rehabilitation and resettlement do not take into account. This inadequacy needs to make up by inclusion of an appropriate provision.

'Project Authority': The law has recognized only two authorities for the purpose of acquisition of land. One is the 'appropriate government', which takes the decision to acquire land. The other is the 'collector' who carries out the acquisition proceedings on behalf of the appropriate government and also executes the task of relocating the displaced persons as per the policy of the concerned government. But the agency requiring the land for a project or activity may not be the 'appropriate government' itself or even an agency controlled by the government. In such cases, the land is transferred after acquisition to the concerned agency for utilization. This agency is nowhere in the picture as far as the law is concerned and therefore bears no responsibility for action taken under it. The project agency deposits the cost of acquisition with the 'appropriate government' and provides assistance, if any, sought by the latter in accomplishing the task. The 'Collector', however, relies on the agency to carry out most of the tasks which fall in his domain, particularly those relating to the relocation of the persons to be displaced as a result of land acquisition. This is because the project agency is endowed with better resources financial and manpower. The omission, therefore, enables project authority to escape responsibility for its action/inaction and get the cover of protection from the 'appropriate government' to the detriment of displaced persons. Where the project agency itself has been delegated the power of acquisition of land by the 'appropriate government' under the LA Act directly or is so empowered under a separate law specific to its sectoral area, [e.g. Coal Areas (Regulation and Development) Act] this contingency may not exist. A suitable provision needs to be inserted to delineate the responsibility of the 'Project Authority'.

'Entitled to Act': The expression in the Act refers to persons who have been authorized to represent the 'interested persons' suffering from disability before public authorities for various purposes. The definition restricts this privilege to unmarried women less than the full age, minors, lunatics and idiots. However, certain restrictions have been placed and conditions laid down on the competence of such persons to act in this regard in the matter of receiving compensation amount. This provision does not include persons claiming to represent those who are excluded from the ambit of compensation payment due to omission of their status and entitlements in land records or communities which have no legal identity or resources less persons lacking in knowledge, financial capacity and competence though otherwise free from disability and mature in age. NGOs, social activists and organizations who represent the interests of displaced persons or sections thereof are also deprived of this entitlement in the absence of clear and express authorization in the law. The orientation of officials involved in the acquisition of land being usually conservative, the operation of process of acquisition turns out to be restrictive and exclusionary and, therefore, does not bridge the gap. It is necessary therefore to provide a role for civil society organizations in the process of acquisition.

'Market Value': The amount of compensation for the loss of land included in the award made under Act is to be assessed on the basis of the 'market value' of such land. The market value has not been defined in the Act. Only matters to be considered and excluded for awarding compensation have been outlined. This leaves the government with freedom to devise its own method of determination of 'market value'.

The practice of government in this regard is based on the average of registered transactions (sale/purchase) of a similar quality of land in the vicinity of the land to be acquired over a period of 3-5 years. The 'market value' so determined is unrealistic and suffers from many distortions on account of several factors – undervaluation to evade registration fee, restrictions on transfer of land, unequal status of buyers and sellers etc. As a result the compensation awarded to the land loser on this basis is inadequate fails to obtain for the awardee a similar quality of alternative land in the area. This turns the erstwhile landowners into landless labourers. There has been a strong demand for 'Market Value' to be substituted by 'Replacement Value' to remove this distortion.

Processual Matters

Initial Screening: The Act does not provide for a mandatory initial screening of the project proposal for land acquisition in terms of certain crucial parameters from the viewpoint of the affected persons before the process of acquisition is initiated. For instance, it is not examined whether there are non-displacing or least displacing alternatives for making land available which would reduce the hardship of the affected persons and at the same time make land available for the project. The rationale for quantum of land sought as also choice of proposed location are also not scrutinized. The status of land records of the area is not checked with reference to the changes that have taken place in the ownership, use and character of land on the ground. This creates a huge disconnect between persons interested on the basis of land records and those who are so otherwise and, therefore, deprives people with valid rights, though not reflected in land records, of the meager benefit of compensation and rehabilitation, if any. Land transactions induced by vested interests intent on grabbing land of poor people cheaply for accessing future benefits are not freezed before the process of acquisition starts. This has serious implications for identification of the 'Persons interested' and their entitlements. Also, it is not ensured whether the environmental and social impact assessment has been duly carried out and taken into consideration while preparing the project, choosing its site and working out the matrix of cost-benefit analysis. The extent of displacement likely to be caused and the financial and other measures required for resettlement and rehabilitation are also not meticulously gone into to see if information and cost reflected in the project is adequate and realistic. Usually, the scrutiny into these dimensions, if any, starts only after the preliminary notification about the intention to acquire land for the project is issued. The result is that virtually no meaningful scrutiny covering various dimensions of the proposal takes place because the time available is too short and no expertise is available with the 'collector' to undertake it. In such circumstances, the project proposal gets endorsed. This leads to serious problems during the process of implantation. The absence of rigorous scrutiny jeopardizes the interest of affected persons and intensifies their opposition to the project. The acquisition also gets delayed because serious problems not anticipated earlier come to the surface from time to time which needs to be resolved. In some cases, irreparable damage is caused to the affected persons and the ecology of the area on account of insufficient scrutiny since judicial review rarely goes into these aspects. A provision for expert scrutiny prior to the initiation of acquisition process has, therefore, been sought by the critics.

Issue of Notification: The issue of preliminary notification conveying the need for specified land and the interest of the 'appropriate government' to acquire it for 'public purpose' to the persons likely to be affected is envisaged under the Land Acquisition Law/laws. But the nature of information and the mode of communication to the affected persons specified therein do not ensure that the information reaches them and is sufficient to enable a decision to be made. The provision regarding publication of the notification in two newspapers fails to take note of the large scale illiteracy of the affected peasantry and lack of circulation of newspapers in rural areas beyond the district headquarters or major towns. The requirement that the notification be displayed in prominent places does not taken to account that there may not necessarily be such prominent places in the concerned areas which all people frequently visit and the concerned persons may not have the capacity to comprehend the information conveyed to enable them to form a view on this basis. These provisions also seem to ignore the official behaviour and practices known for their disinclination to share information, lack of transparency in dealings, denial of access to official documents and records through myriad ways, delays in publication of the Gazette and unhelpful attitude of the functionaries to respond to the quarries and apprehensions of the local people. The provisions have also become outdated since they do not provide a role to the local panchayats for dissemination of this information and collective deliberation by it in the Gramsabha on its implications. This deficiency in the ambit of information supplied and its communication to ensure its reach to the target group results in valuable opportunity being lost to raise concerns about the proposed acquisition by people likely to be affected. There has been concerted demand to meet this deficiency.

Reference: The current provision for seeking 'reference' to the competent court by persons who are aggrieved by the award is restricted to those persons who have not accepted the award of compensation through a written application to the Collector. This entitlement is also extended to those who are forced to accept the award under pressure of officials but record their protest. The latter situation conveys that they are aggrieved by it. The land acquisition authorities have interpreted this provision to imply that if persons interested have agreed in writing to the terms of the award or taken its payment, such an award cannot be sent to the reference court even if they make a request in writing to the Collector. This is unfair to the aggrieved persons because most such persons particularly the poorer among them do not know what constitutes the terms of award. They accept the 'award' due to pressure exerted by the land acquisition authority. They have little resistance capacity. Besides, agreeing to the terms of award or what should be included in the award before the Collector does not take away the right of the person to go for reference against the awarded amount and the basis on which it has been assessed etc. A clarificatory provision in law would protect the interests of such persons.

Redetermination of Compensation: In the cases, which are sent for reference, the court order where higher compensation has been fixed is rarely known to persons other than the party/parties who sought this reference. Even NGOs/social activists engaged in assisting project affected persons do not get to know this information. The decisions in such cases come long after the process of acquisition has been completed and the affected persons have been relocated. There is no institutional arrangement to disseminate such information to others who may benefit from this information. This prevents other persons similarly placed from approaching the Collector for seeking redetermination of compensation on this basis. There is no effort by the land acquisition authority to convey this information to enable such persons to file a claim within the time limit prescribed in the law. The deserving persons thus fail to take advantage of the legal provision and lose enhanced compensation by default. Lack of knowledge and resources to approach the concerned authority for this purpose is a common disability of poor people which the law does not take into account. The duty should be cast on the 'Collector' in the law to disseminate this information to the concerned persons in the quickest possible time to enable them to file a petition for seeking reconsideration of compensation within the time limit prescribed for this purpose.

Taking Possession of Land: There is no requirement in the law that the physical possession of the land should be taken only after compensation has been paid and resettlement arrangements have been completed. Usually, the acquisition authority takes formal possession of land after the completion of acquisition process and hands it over to the project authority without ensuring payment of compensation and preparation of resettlement requirements. The latter part of the work proceeds slowly since the acquisition authority is primarily interested in the acquisition of land. The resettlement of displaced persons and their rehabilitation, if any, is considered peripheral work and does not command matching attention. Quite often, force is used to take possession of the land under pressure from the project authority even though compensation has not been paid and resettlement site is not developed for people to shift. The affected persons in such cases are left with no resources to earn a livelihood and no proper place to shift immediately. The law should make a provision that possession shall be taken over only after the entire compensation is paid and resettlement site is fully developed for people to shift.

Utilization of Land: The project agencies express a great deal of urgency in obtaining the land for proceeding with their construction plan and, therefore, the process of acquisition land is accomplished expeditiously. The same degree of urgency is not reflected in utilization of the acquired land after it has been transferred to them. In many cases, the acquired land remains unutilized or partially utilized for years. This highlights the lack of concern and responsibility in accurately estimating the land required for the project leading to waste of public resources as well as unnecessary suffering to the affected persons. In most instances, unutilized land remains unprotected and is encroached upon by squatters or land grabbers. There is no provision in law for setting a time limit for utilization of land and, failing which, for return of the land to the government and, thereafter, to the land losers. This deficiency needs to meet.

Mode of Acquisition: The Land Acquisition Act provides for a single mode of acquisition of land. Under it, the competent authority, wishing to acquire land for a specific purpose, provides the affected citizens/ landholders opportunity to place their views, first, on the intent to acquire land and, later, having decided to acquire it, on the award for compensation. Once the processes are completed, the acquired land vests in the 'appropriate government' free from all encumbrances. In this mode, the landholder has no freedom to decline parting with his/her land and has no scope of retaining any interest in the land after its acquisition by the state. It is a permanent break with the attachment to land. The law does not envisage transfer of land for a specific period and its reversion to the landholder thereafter or leasing out land on specified conditions without transfer of ownership rights. This inflexibility (in the pattern of transfer of land and mode of acquisition) fails to make a distinction between different purposes for which land may be acquired and the terms and conditions accompanying such categories of transfers and different types of agencies for whom land may be acquired. All acquisition by the government is lumped under the term 'public purpose' and carries uniform obligations towards the land losers. The result is iniquitous in terms of the impact it creates on different sections of society faced with differing social situations.

Quantum of Land: No parameters have been laid down in the law nor a scrutiny by experts envisaged on the quantum of land proposed to be acquired for a given project. This has resulted in such distortions as widely varying quantum of land acquired for a similar project in different locations and the quantum of land acquired much in excess of the need in foreseeable future. The apex court has also imposed no such obligation on the state to provide the rationale for it. The assessment made by the requiring agency is accepted as valid for this purpose. The acquisition authority has no means or expertise to verify the rationality of demand coming from different project agencies and to compare whether the land asked for is on par with similar projects elsewhere. The profligacy shown in assessing requirement of land is also on account of low rates of compensation paid to the land losers and to take advantage of the appreciation of its value in future. This has led to excessive land being acquired which cannot be utilized for the project and is, therefore, diverted for purposes other the one declared in the proposal. There is no provision in the law for return of such land to the 'appropriate government' and, thereafter, to the land losers or imposition of a penalty on agencies for having inflated their requirement and for diverting it for a purpose other than the one declared or for using it extravagantly. There has been a strong demand for expert scrutiny of land acquisition proposals and return of the unutilized land to the erstwhile landowners.

Change of 'Public Purpose': Under this law, land is acquired for a 'public purpose'. The nature of public purpose is usually declared in the preliminary notification issued for acquisition. But after the land has been acquired, the government utilizes the land not for the public purpose so declared but for some other 'public purpose'. On the face of it, this change should invalidate the acquisition proceedings because the specific purpose for which land was acquired was not being accomplished. But the courts have not found any violation of law in this arrangement and have upheld utilization of land for some other 'public purpose'. There is no provision in the law that change of 'public purpose' would require reprocessing the case of acquisition and declaration of a fresh award. This deficiency not only permits declaration of 'public purpose' arbitrary but also deprives affected persons the opportunity to raise objections and seek a fresh award on the basis of prevailing 'market value' of land. The insertion of a provision for reprocessing the proposal of acquisition and declaration of a fresh award would put a check on indiscriminate use of 'public purpose' for acquisition of land or rushing with acquisition without ensuring that it would be utilized for the declared purpose.

Consultation with Experts: The laws on acquisition of land fix the entire responsibility on the acquisition authority (i.e. the 'Collector' in the case of Land Acquisition Act, 1894) to scrutinize the proposal for acquisition and take decisions on the subsequent processes of acquisition, from the publication of notification to the taking possession of the land and implementation of the reference court's orders. This agency does not have the requisite expertise to take sound decisions on many matters due to which the affected persons suffer. No effort is made, for example, to involve experts in estimating the quantum of land required, scrutinizing the rationale for location and least displacing non-displacing alternatives, carrying out the environmental and social impact of the project, surveying the extent of displacement, preparing development plan, working out the cost of resettlement and rehabilitation and examining the correctness of the cost-benefit analysis presented in the report, and assessment of market value for declaration of award etc. The acquisition authority and the staff available to it for assistance are not professionally equipped to handle this scrutiny. The price for this lack of competence is paid by the affected persons. The association of experts for different processes has, therefore, been considered necessary to generate less hostility against the acquisition of land and more credibility in respect of the decisions taken by it.

Compensation

(a) **Entitled Categories:** At present, compensation is paid to the interested persons who raise their claim before the competent authority. As the land records are not updated prior to initiation of acquisition proceedings, this entitlement remains confined to owners of land and such tenants, if any, whose names have been incorporated in the record of rights. For reasons stated earlier, a large number of tenants/sharecroppers who operate on informal basis or with oral contracts are not taken cognizance of for payment of compensation. The current practice also leaves out from the benefit of compensation other displaced persons who lose their livelihood or access to life supporting products provided by CPRs on account of acquisition. This renders them destitute since they have nothing to fall back upon to carry on their lives. There is no provision for compensating these categories of persons which intensifies their hostility to the acquisition proposal. There is a wide consensus that compensation and rehabilitation to all displaced persons should be admissible to all such persons.

(b) **Ambit:** Forced displacement causes multifaceted vulnerability—social, cultural, economic, emotional and psychological. Compensation merely takes into account loss of land as a productive asset. It neither covers loss of livelihood and deprivation of other life supporting material derived from CPRs which sustains quality of life and social environment nor it incorporate a commitment that alternative resources required for obtaining these benefits would be provided. The ambit of compensation also excludes assured access to social amenities such as drinking water, education, health, sanitation, and availability of physical infrastructure such as roads, market, water bodies, grazing land etc. The mechanism of assessment of compensation, in any case, ignores the disruption of social networks, cultural institutions, emotional vacuum created by uncertainty of future, and traumatic conditions created by loss of nerve. It is not even equipped to assess such losses. The ambit of compensation is too narrow and inadequate and needs to be broadened to cover all such losses along with the mechanism for assessing them.

(c) **Mode:** The mode of compensation is usually in the form of cash disbursement and is insensitive to the incapacity of certain communities such as tribes to handle money transactions and to utilize cash for pursuing income generating activities. The quantum of cash compensation, in any case, is inadequate to procure alternative land of equivalent productivity to carry on their traditional vocation. The tribal communities (certain other fragile groups may also be similarly placed) are ill-equipped to shift to other livelihood options. They are also extremely vulnerable to various machinations of vested interests to squeeze their cash compensation from them through devious ways. The law does not provide for giving alternative land as compensation though it does not bar the 'appropriate government' from resorting to this option. But this option has generally been avoided. This approach has, therefore, been particularly harsh to the vulnerable communities. The provision of law for land, at least in the case of the affected tribals, has therefore been strongly advocated.

Tribes and the Scheduled Areas: The law of acquisition does not make any reference to much less incorporate existing protective provisions regarding Scheduled Tribes and scheduled areas relating to laws on prohibition of alienation of tribal land to the non-tribals and PESA, 1996 on empowerment of tribal communities in management of natural resources. This is understandable as the colonial law of Land Acquisition Act, 1894 pre-dates them. But the amendments carried out in it as also other land acquisition laws enacted since independence have also failed to do so. Worse, the practice of acquisition has virtually ignored them though the acquisition authority i.e. 'Collector' is also responsible for enforcement of other laws on the statute book. In fact, in many cases, the acquisition authorities resort to land acquisition for taking over tribal land to bypass the land alienation laws. Further, the mandatory consultation of Gram Sabha in the cases of land acquisition as required in PESA, 1996 is either not effected or maneuvered unethically. The Supreme Court judgment in Samtha case which places restrictions on use of land in scheduled areas for mining, industry etc. is also ignored. The acquisition proceedings do not even contain a reference to these provisions and reasons, if any, for not complying with their provisions. This causes immense damage to the interests of tribes who have been the largest victims of displacement as a social group, negate the utility of protective provisions made for them and threaten their dignified survival. The law on acquisition of land therefore should be in harmony with the protective legislations and constitutional provisions for the tribes.

Collateral Damage: In several cases, the project activities carried out on the acquired land damage the productive potential of land in the vicinity which has not been acquired. There is no provision in the laws of acquisition either to compulsorily acquire such land or to adequately compensate the landholder for the irreparable damage caused to his/her life supporting asset. This is particularly so in the case of industrial and mining projects. The laws are silent on the subject as the damage is caused long after the land has been acquired. The project authorities do not assume any responsibility for this loss. The state too has failed to intervene in such cases to protect the interests of the victims. They deserve to be adequately compensated whether under the law or acquisition of land or some other statute.

Acquisition in the Case of Urgency: The law provides for speedy acquisition of land without going through the processes provided for normal acquisition where the 'public purpose' is of an urgent nature and cannot wait for their completion. The land can be acquired simply by issuing a notice allowing fifteen days time before taking its possession, skipping the process of seeking the views of the affected persons and declaration of award. In the cases of unforeseen emergency, the land can be taken possession of immediately after publication of notice and, in the case of a building, after 48 hours of the notice. In all cases of urgent acquisition, 80 per cent of compensation is to be paid to the land loser. This provision does not define parameters of urgency which leads to its misuse. Rigorous conditions need to be laid down to restrict its application. Further, no distinction has been made between temporary acquisition and permanent acquisition. There is no obligation on the part of the 'appropriate government' to arrange for alternative accommodation where the occupation is temporary and no provision for immediate resettlement where the acquisition is permanent. The return of land in the case of temporary acquisition is also not envisaged. No consideration is given to the subsistence needs of the land loser in wake of sudden acquisition of his land which leaves him/her little time to seek alternative livelihood. These dimensions need to be accommodated in the law.

Rehabilitation: The laws of acquisition entirely focus on the authority of the government and processes to be followed for acquisition of the privately owned land without a corresponding obligation to provide the affected persons' alternative land, or employment, residential house, infrastructure facilities and social amenities through proper resettlement and rehabilitation. Whatever action is taken by way of resettlement is through *ad hoc* executive decisions which cannot be enforced. In the absence of a legal obligation, the government has not seriously pursued the task of rehabilitation of displaced persons who are left to face the uncertain and bleak future. Even *ad hoc* assurances, limited as they have been, have been diluted or reversed. For example, state's earlier commitment to provide employment in industrial public sector projects to one member of the family losing a specified area of land was subsequently abandoned. The provision of land in the command area of an irrigation project by acquiring a part of land from the landholders benefiting from the project has never been seriously pursued. Though some States have enacted laws that provide assurance of rehabilitation in specific projects, the commitment has not been matched by performance. Government feels that payment of compensation and provision of a small house and a minimum of infrastructure at the resettlement site is sufficient to discharge its legal obligation. The task of relocation carried out under the government policy in vogue at different times has been most unsatisfactory in terms of coverage, content, suitability and effectiveness. This lack of concern for resettlement and rehabilitation is the main reason for hostility of the displaced persons to acquisition of their land, particularly the STs who have faced the brunt of land acquisition. This deficiency is also responsible for increasing hostility of the affected persons to acquisition proposals. A legally enforceable commitment to provide comprehensive rehabilitation and resettlement to all displaced persons has been sought to remedy the existing deficiency of provisions.

Participation: The Land Acquisition Act does not provide for mandatory participation of affected persons in the acquisition proceedings. It assumes that since the proceedings are quasi-judicial where the Collector/ land acquisition authority functions as a court, they are transparent and open for perusal to the interested persons or their lawyers. The issue of notices both for seeking views on acquisition proposal and later for declaration of award of compensation ensures adequate consultation with the affected persons and fulfils the requirement of participation. Both these assumptions are flawed because persons affected by acquisition have neither access to information required nor have capacity and resources to 'participate' in the proceedings through these modes. Besides, formal records of acquisition proceedings carry very limited information on decisions taken and even less on factors influencing the process of making these decisions. The attitudinal orientation of acquisition officials is most conservative in sharing information and providing genuine space for consultation with the affected persons. This results in a highly bureaucratized manner of acquisition. There is little effort to reach out to the affected persons and create an enabling environment for them to express their views much less to accommodate their concerns in decision making. There are no institutional arrangements for such participation to take place. The elements of participation relevant to the people are therefore missing from the law. A participatory acquisition process along with mechanism for effecting participation has been demanded so that the acquisition of land does not alienate the people affected by it and causes less friction.

Implementational Practices

Insensitive Execution: The acquisition process as it unfolds in practice makes the trauma of people even worse than what is conveyed by the law. This is on account of lack of sensitivity in officials handling it which aggravates the suffering of affected persons already traumatized by the prospect of acquisition of their land and loss of livelihood. This insensitively is reflected in lack of transparency, reluctance to share information with the affected persons, indifference towards their grievances and unwillingness to create institutional arrangements for their prompt redressal, and an enabling environment for their participation to minimize their trauma. It is also pronouncedly observed in the failure to undertake initial screening of the project proposal with a view to identifying adverse externalities on the environment and have them addressed before the acquisition process is initiated. The insensitivity is also evident in lack of action to update land records at the stage of initial screening of the project. This includes decision on the claims of occupants of public land/forest land who were entitled to regularization/validation in terms of prevailing government policy but have remained pending due to the negligence of revenue/forest officials, documentation of customary uses of CRPs benefits to be lost by the affected local community as a result of acquisition and interests of the non-landowners (agricultural labourers, tenants, sharecroppers, fishermen, artisans etc.) dependent upon land to be acquired for their inclusion in the award. The lack of concern for the affected persons, particularly the marginalized groups, is also observed in ignoring the legal requirement of consultation with the Gram Sabha as provided under PESA (1996). Where this consultation is effected, the process is carried out unethically and consent obtained under coercion or even manipulated showing utter intolerance to the expression of dissent of the participating members. The callousness is also observed in not making the payment of compensation before taking possession of the acquired land. No precautionary steps are taken against induced transfers of land prior to initiation of acquisition process and squeezing of cash from the tribals after the compensation has been distributed through numerous ways of cheating them resorted by the non-tribals. These are only a few examples of how the poor among the affected persons suffer due to insensitive handling of land acquisition process. The practice of acquiring land is designed to take possession of land at least cost and in the least time. This makes the process of acquisition virtually expropriatory. Further, there is a perceptible bias in the conduct of officials in favour of the project and a virtually adversarial attitude towards the affected persons. Far from protecting the interests of the affected persons by taking up their concerns with the project agency, they do not even function as a neutral agency and harmonize the interests of people with the needs of the project. They are particularly hostile if the affected persons oppose acquisition of their land. There is no hesitation to use force to take possession of the land and shift affected persons to the relocation site without ensuring that compensation is paid and resettlement arrangements have been completed. The practice of acquisition also highlights lack of accountability of officials engaged in acquisition. No institutional arrangement for this accountability is provided in the law except in respect of the compliance of the prescribed processes. The practice of acquisition also ignores the need for setting it up. The unresponsiveness of officials towards grievances of affected persons is, therefore, widespread.

Government Response

The critique of the law and operation of acquisition process is not new. People's organizations, social activists, NGOs, and sensitive professionals have raised these issues before and even submitted proposals for amendments in the Land Acquisition Act, 1894. But the Government did not show any inclination to amend it. The amendments undertaken in the past have been largely oriented to reducing time taken for acquisition proceedings and raising the quantum of cash compensation. Responding to the civil society's critique of the law, an attempt was made to amend the Act vide the proposed Land Acquisition (Amendment) Bill, 2000 (MORD: 2004). It failed to materialize because the proposals were strongly criticized by the civil society groups for ignoring major issues. They were also rejected by the top bureaucracy on the ground that the changes would stall all acquisition of land leading to time and cost over-run of projects (MORD: 2004). Faced with continued protests of people against acquisition of their land and the resulting delay in the execution of various projects, the Central Government introduced the Land Acquisition (Amendment) Bill, 2007 in the Lok Sabha (MORD: 2007a). This proposal has gone farther than the aborted Bill of 2000. It takes into account a number of issues raised in the preceding pages. The following are the main features of this Bill:

Exclusion of Acquisition for Companies

The Bill has removed the words "and for companies" from the title and the Preamble of the Act. In pursuance of it, Part VIII of the Land Acquisition Act, 1894 relating to "Acquisition of Land for Companies" covering Sections 38 to 44B have also been omitted. This implies that the government would not acquire land for companies under this Act and shall confine compulsory acquisition of land required for its own development activities.

Commitment to Rehabilitation of Displaced

The Government has notified the National Rehabilitation and Resettlement Policy, 2007 (MORD: 2007c) and has introduced the Rehabilitation and Resettlement Bill, 2007 (MORD: 2007b) to give a statutory backing to it in the Lok Sabha. The provision for rehabilitation and resettlement contained in the two instruments are applicable to those affected by the acquisition of land under any law in force or involuntary displacement due to any other reason. This commitment to rehabilitate and resettle the persons affected by involuntary displacement is also incorporated in the Land Acquisition (Amendment) Bill, 2007. This makes rehabilitation and resettlement of the displaced persons integral to the process of acquisition of land and also enforceable.

Social Impact Assessment

There is also an obligation to carry out social impact assessment in respect of land acquisition carried out under the Act where physical displacement of 400 families en masse in plain and 200 families in hilly/desert/scheduled areas is involved for the purpose of preparing the rehabilitation and resettlement scheme and tribal development plan within it.

Redefinition of Expressions

The expression "person interested" has been expanded to include tribals and other traditional forest dwellers who have lost any traditional rights recognised under the Scheduled Tribes and Other Traditional Forest Dwellers (Recognition of Forest Rights) Act, 2006. Those having tenancy rights under the relevant state laws have also been included in this definition. These two categories of persons have been enabled by this provision to claim compensation for the loss of rights in land.

'Public Purpose' Redefined

The expression "public purpose" is being revised to include provision of land for strategic purposes relating to defense forces or any other work vital to the state, infrastructure projects which benefit the general public and any other purpose useful to the general public for which 70 per cent of the land has been procured privately but the remaining 30 per cent is required. The expression "infrastructure projects" has been elaborated to include activities relating to electricity generation, transmission and supply, construction of roads, bridges, airports, ports, rail, mining, water supply, sanitation, irrigation besides any other public facility which may be notified by the Central Government.

Protective Measures for Landholders

To prevent motivated transfers of land involved in acquisition, transactions in respect of land proposed to be acquired or creation of any encumbrance on such land have been disallowed from the date of the publication of notification under Section 4(1) of the Act until the final declaration under Section 6 or award made or paid under Section 16, whichever is earlier.

The Collector is required to undertake and complete the updation of land records including the classification of land and its tenure, survey and standardization of land and its property values in respect of the land under acquisition within the period available between issue of preliminary notification under Section 4 and before declaration under Section 6 of the Act.

Transparency in Proceedings

To achieve greater transparency in land acquisition proceedings, the Collector is required to keep open to the public and display a summary of the entire proceedings relating to the acquisition of land including compensation award to each individual.

Compensation

Three norms for assessing and determining the market value of land have been specified. Of the three, whichever brings out a higher value of land is to be relied upon. These norms are: the minimum land value, if any specified in the Indian Stamp Act, 1899 for registration of sale deeds in the area, the average of sale price for a similar type of land situated in the vicinity ascertained from not less than 50 per cent of such sale deeds during the proceeding three years, the average of sale price paid or agreed to be paid for not less than 50 per cent of land already purchased for the project where a higher price has been paid in respect of provision of land for development under any scheme of the government which has been subsequently disposed of. Where these conditions are not applicable, the state government has been empowered to specify the floor price per unit of the said land based on the average of higher prices for similar land situated in the adjoining area to be ascertained from not less than 50 per cent of sale deeds registered during the preceding three years where a higher price is paid. In addition, the market value of land shall be assessed after ascertaining the intended land use category of such land for which purpose the value of land of the intended category in the adjoining areas or vicinity shall be taken into account. The latter provision brings within the ambit of market value appreciation in the value of land resulting from its proposed use. For determining the market value of immovable property, trees and plants and standing crops, consultation with experts in respective disciplines has to be made.

Compensation for Damages

The Compensation would also be payable for damages caused to land, which is excluded from acquisition proceedings, during survey and measurement of land following preliminary notification within a period of six months from the completion of such work. These norms for determination of compensation may also be applied to cases pending for award before the coming into force of the amended Act.

Enhanced Compensation for Urgent Acquisition

In cases of acquisition of land under urgency, an additional compensation of 75 per cent of the market value would be payable in addition to the existing entitlement.

Enhancement of Solatium

The package of cash compensation is being enhanced by increasing the solatium amount to 60 per cent of the market value as against the existing 30 per cent.

Payment in Shares

The mode of payment of compensation is being diversified with an option that part payment in the form of shares, debentures up to 20 per cent of the amount, stretchable to 50 per cent, where the land requiring agency is authorised to issue them may be made.

Time Bound Compensation Payment

The entire compensation amount would now have to be paid within a period of 60 days from the date of the award. The physical possession of land will also be taken within this period.

Speedy Acquisition of Land

For ensuring timely acquisition of land, it has been proposed that the proceedings shall lapse and no fresh notification can be made for a period of one year if the declaration to acquire land within the specified period following issue of preliminary notification is not made. In the case of a lapse for the second time, fresh proceedings cannot be started for five years.

Time Limit for Award Reduced

The period of making awards relating to compensation is being reduced to one year for quicker completion of acquisition proceedings. The failure to adhere to this time schedule would lead to the lapse of acquisition proceedings.

Financial Obligation of the Project

The agency requiring land is now being made liable to pay the cost of acquisition which includes expenditure incurred on infrastructure and amenities on the resettlement site, administrative cost of acquisition of land and planning and implementation of the rehabilitation and resettlement scheme besides compensation payable to the land losers and demurrage for damages in the process of acquisition, etc. This may help in expediting payment of compensation and execution of rehabilitation and resettlement operations.

Reference for Compensation Settlement

The provision of a reference court in the existing Act is being replaced by the Land Acquisition Compensation Disputes Settlement Authority to be set up each at the centre and state level for disposal of such cases within a period of six months. Detailed requirements have been laid down regarding its composition, and tenure, powers and functions of its members.

Expeditious Reference Process

The Collector would have to make a reference to the Authority for adjudication of compensation disputes within 15 days from the receipt of such an application. The failure to do so would entitle the applicant to directly approach the Authority requesting that such a reference be made within a period of 30 days. The period for entertaining a reference petition has been extended to a further period of one year if the Collector is satisfied about the reasons for not filing the application within the specified period.

Utilization of Acquired Land

Restrictions have been introduced in respect of the transfer of acquired land by the requiring agencies. No such transfer of the acquired land can be made except for a public purpose and after obtaining the prior approval of the government. If the acquired land remains unutilised for a period of five years from the date of taking over possession, it shall revert to the government. Where acquired land is transferred by the government for a consideration, 80 per cent of the difference in the acquisition cost and the consideration received shall be distributed among those from whom land has been acquired in proportion to the value at which the land was acquired. A separate fund shall be created for this purpose to be administered by the Collector.

Reference to the Standing Committee

The Bills (2007 a and b) were referred to the Standing Committee of the Ministry of Rural Development for scrutiny before its discussion in the Lok Sabha. The Standing Committee considered the Bills and submitted its reports (Lok Sabha: 2008 a and b). The report on Bill (2007 a) contains the following major recommendations:

Repeal of the Act

'The Land Acquisition Act, 1894' should be repealed and a new comprehensive legislation may be brought before the Parliament.

Unsettled Cases

The benefits of solatium and Rehabilitation and Resettlement provided under the Rehabilitation and Resettlement Bill, 2007 should also be made applicable to cases where award is pending or remains to be settled under the existing Act retrospectively.

Public Purpose

The definition of 'Public Purpose' in the Principal Act is more detailed and should be retained. The proposed amendments are not agreed to.

Notice of Acquisition

The notice of acquisition should be published in at least three newspapers of which two should be in vernacular.

Protection of Agricultural Land

With a view to protecting agricultural land, where agricultural land is to be acquired, a certificate should be obtained from the Collector that wasteland, degraded/ barren land was not available for a particular project. Where there is no escape from acquiring agricultural land, an equal area of waste/degraded/barren land should be developed and expenditure on it should form part of project cost. Where such acquisition of agricultural land is meant for setting up SEZ, equivalent or double the area of wasteland would have to be developed. The prime agricultural land should be protected by enacting a law on the pattern of the Forest Conservation Act, 1980. Land use Policy should be finalized to minimize conversion of agricultural land for non-agricultural purposes. Data should be maintained about conversion of waste land into agricultural land for taking initiatives for food security of the country.

Enhanced Period for Filing Objections

The time allowed for filing objections to the preliminary notification should be increased from 30 to 60 days.

Acquisition for Companies

Part VII of the Principal Act (Clauses 38A to 44B) which lays down conditionalities for acquisition of land for companies should also be retained. Fixing percentage of the land to be acquired by a private agency and 'appropriate government' in the ratio of 70:30 is contradictory and impractical and is not agreed to.

Social Impact Assessment

Where the number of the families to be displaced is lower than the stipulated criteria for carrying out social impact assessment, the discretion should be provided to the Standing Committee of Rehabilitation and Resettlement proposed in the Rehabilitation and Resettlement Bill, 2007 to decide whether it should also be carried out in respect of such projects. The Panchyati Raj Institutions, particularly the Gram Sabha, should be involved while undertaking social impact assessment.

Compensation

The assessment of 'market value' for determining quantum of compensation should be based on the highest price indicated in the sale deeds of the last three years plus 50 per cent of such highest price. In tribal areas, the highest price paid among sale deeds of the last three years in the adjoining non-tribal block plus 50 per cent of that amount should constitute the market value.

Extra monetary benefits should be provided to the affected person/ family as the value of the land appreciates after acquisition due to change in land use.

The issue of shares and debentures should not be considered as part of the compensation but over and above the admissible amount.

In case of multiple displacement, the benefits to be provided to the displaces under laws of land acquisition and rehabilitation and resettlement should be doubled.

Acquisition in Emergency

The power of acquisition in emergency should be restricted to a minimum area of land acquired by the Central Government for the purpose of defense or national security. It should be used in the case of natural calamities only if the displacement is permanent and subject to the institutional safeguards that may be prescribed to protect the interests of the affected persons.

Unutilized Land

The return of unutilized acquired land to its original owners is not practical. The 'appropriate government' should make a clear list of lands which were acquired but have remained unutilized. No future acquisition of land should be taken up without the government certifying that no such unutilized land was available in the area for the project. Different departments of the Central and State Governments should maintain data of unused acquired land and place it on the website.

Service of Notice

The service of notice relating to land acquisition to the affected person, where the person named cannot be found, may be made to any adult member of the family and not necessarily to an 'adult male member'.

Project Scrutiny

A committee of experts with inclusion of local elected representatives should decide on the optimum size of the land for the project before notification for acquisition is issued.

Rights of Tribals

The rights of the tribals should be recognized and settled before taking up any acquisition of land. All their rights and entitlements should be preserved in perpetuity in the area of rehabilitation and the area should be declared as scheduled area. Gram Sabha should be fully consulted before making acquisition in scheduled area. The overriding effect should be given to PESA.

Action Taken on the Report

Taking into account the report of the Committee, the Government introduced the revised Land Acquisition (Amendment) Bill, 2009 (MORD: 2009 a) in the Lok Sabha which was passed. But the Bill could not be passed by the Rajya Sabha before the Lok Sabha was dissolved for elections. As a result, the Bill would have to be reintroduced in the Parliament.

Revised Bill

The revised Bill incorporates the recommendations of the Committee on protection of agricultural land, constitution of Expert Committee for scrutiny of projects, publication of preliminary notification, enhanced period for filing objections, acquisition of land in the case of urgency, removal of gender bias in the service of notice. Even in respect of these amendments, the recommendations of the Committee have not been included in their entirety. Some additional changes have been incorporated by the government on its own. The most important among them is the provision that an Expert Committee to scrutinize Land Acquisition projects shall ensure that only minimum area of land would be acquired for the project and shall recommend a site for the project which entails minimum displacement of people and minimum acquisition of land after considering all possible sites. The period for payment of compensation has been increased from sixty days to ninety days. The Land Acquisition Compensation Disputes Settlement Authority has also been entrusted with the settlement of disputes relating to rehabilitation and resettlement of the affected families under the Rehabilitation and Resettlement Bill, 2009 (MORD: 2009 b). The provision for payment of enhanced compensation in unsettled cases has been deleted.

Appraisal of the Revised Bill

In the context of the five fold classification of issues relating to the structure of Land Acquisition Act, 1894, it is evident that the changes proposed in 2007 and, further, in 2009 do not go very far. While some of the relatively minor issues have been addressed, the more fundamental concerns have either been ignored or disregarded. On the conceptual framework, from the viewpoint of the affected persons, the most serious flaw in the Bill (2009) is that the government has shown no willingness to give up or dilute its absolute power to acquire land without the consent of the landowners using the notion of Eminent Domain. The Act, therefore, continues to retain its undemocratic character. There is little inclination to seek a less coercive and more consensual mechanism to acquire land for a 'public purpose'. While the Bill does take care of the anxiety to protect agricultural land from acquisition by incorporating some safeguards, there is no assurance that this would necessarily happen as the acquisition of agricultural land has not been barred altogether. But even this degree of concern has not been shown to protect tribal land from acquisition even though there are state laws to prohibit alienation of tribal land to the non-tribals and the Fifth Schedule also protects tribal land from such transfers. The Bill also ignores the recommendation of the Committee to insert provision giving overriding effect to PESA, 1996. Even the provision in the Rehabilitation and Resettlement Bill, 2009 that the requirements of PESA should be adhered to has not been inserted in the Land Acquisition (Amendment Bill, 2009). The refusal to even harmonize the two laws on this issue points to the excessive zeal in the government to smoothen the process of acquisition even if it involves riding roughshed over the interests of the tribals. Similarly, no sensitivity has been shown to the incidence of multiple displacements by disregarding Committee's recommendations.

In respect of the definitional limitations, the expression 'land' in the existing law which in official usage is interpreted to mean privately owned land remains untouched. No change has been made to enlarge its meaning by incorporating other categories of land. The narrow meaning given to 'land' in the practice of acquisition would, therefore, continue to exclude government/public/panchayat/forest land and other community property resources. The rights subsisting on such lands in the case of its acquisition would remain unrecognized for payment of compensation. Some concession has been made to the Scheduled Tribes and other traditional forest dwellers in this regard by recognizing their rights in relation to forest under the recently enacted The Scheduled Tribes and Other Traditional Forest Dwellers (Recognition of Forest Rights) Act, 2006 for computation of compensation. But this recognition of forest rights does not extend to other lands mentioned above. The lands managed by community outside the forest area which carries no individual rights of ownership but only of use with the sanction of community would continue to be excluded from compensation entitlement. Many areas containing such lands have not been surveyed particularly on higher slopes of hilly terrain. No records of user practices with regard to such areas may exist as the un-surveyed land would have been recorded as 'government land'. Such users would be disregarded for entitlement to compensation. The 'primitive' tribal groups in shifting cultivation areas whose lands have not been surveyed or wrongly recorded as 'government land' would be the worst victims of this omission.

The proposed amendments do expand the definition of 'person interested' so as to enlarge the circle of those who are entitled to claim compensation for loss of their rights in land. The tribals and other traditional forest dwellers who enjoy rights in forest land would now be recognized as 'persons interested' for this purpose in view of the recently enacted law mentioned above as also those having tenancy rights or interest in easement affecting the land under the relevant state laws. But occupants of government/panchayat land would continue to be excluded from its purview even though a sizable section of them are entitled to validation of their occupancy in terms of the existing policies of state governments. Besides, the definition would, as before, exclude agricultural labourers, artisans, fishermen, unrecorded tenants and sharecroppers etc. who may be using the acquired land for their livelihood. The Committee's recommendations to devise mechanism to cover informal tenancies has been ignored.

The revised definition of 'public purpose' far from restricting the scope of acquisition which the people had wished, increases it further by not only the very wide meaning given to 'infrastructure projects' but also inclusion, in its ambit, of 'any purpose useful to the general public'. The exclusion of acquisition of land for companies is also cosmetic because the requirement of land for companies, by and large, would be covered under 'infrastructure projects' where private investment is most concentrated. The requirement of land for companies can also be accommodated under another component of 'public purpose' which extends to 'any activities useful to the general public'. In such cases, land for companies to the extent of 30 per cent of their total demand would be acquired by the government if the remaining 70 per cent has been procured by companies directly. These two provisions camouflage the requirement of land for companies and use the provision of compulsory acquisition for them through the backdoor. Besides, leaving companies to directly procure land from the landowners through market transactions would be of no help to the poorer landowners particularly the tribals. This externalization of acquisition would have pronouncedly negative implications for them. This is because the scale of displacement would not get reduced, only the mode would become different. The poorer landowners would have no strength to resist private procurement of land, despite their unwillingness to sell it, due to coercion exerted by companies, given their resources, clout and propensity to use muscle power. This mode of acquisition would, in fact, be worse for them. With little bargaining capacity, they would be pitted against unequal buyers. Such transactions would lack transparency and accountability. Still worse, the displaces of private procurement would be deprived of the benefits of rehabilitation and resettlement which would be available only to those whose lands are acquired by the government. Tenants and sharecroppers would suffer most in this mode of acquisition as the owners would not associate them in the deal. The recommendation of the committee to delete this provision has been disregarded despite the strong reasoning given to support of it.

In respect of Processual matters the draft of the revised law makes no improvement in creating conditions for effective participation of the affected persons in acquisition proceedings. A majority of people in rural areas are illiterate. Although the provision of publication of notification has included three newspapers instead of two in the existing Act one of which would be vernacular, the newspaper circulation does not extend beyond the district and sub-divisional towns. The gazette notifications are not accessible even to the educated persons let alone the ordinary inhabitants of rural areas. The posting of information on the website of the 'appropriate government' would not help the people who are not even literate let alone computer friendly and internet users.

The placing of information for inspection in the Gram Panchayat is a welcome addition to the existing mode of its dissemination. But the poorer people who are engrossed in their livelihood activities are not in the habit of visiting Panchayat office unless specifically called for some meeting. The Bill should have cast responsibility on the Panchayat to disseminate the information in the Gram Sabha. Besides, the information contained in preliminary notification is too inadequate to enable them to express their views on the proposal. There is no provision to specify what information should be contained in the notification to serve this purpose. The revised provision therefore, continues to be inadequate.

The welcome additional change incorporated in the Bill (2009) is the provision of a committee to scrutinize acquisition proposals before the issue of notification under section 4. The committee has been entrusted with the responsibility of ensuring acquisition of minimum land, avoidance of agricultural land for acquisition, selection of suitable site for resettlement, updation of land records, clearance for conversion of land use and obtaining social impact assessment. The association of experts with this committee creates some hope that the scrutiny of acquisition proposals would not be a cosmetic affair, although they would be in a minority. Its limitation is that there is no popular participation in this scrutiny. The recommendation of the committee to associate elected local representatives with this committee has been disregarded.

The increase in time limit for filing objectives to the proposal for acquisition is a positive change which should enable people to gather relevant information and have collective deliberation with a view to placing their considered views. Its effectiveness would depend upon the information supplied by the acquisition authority and its reach to the affected persons. Similarly positive is the restriction inserted in the Bill (2009) in respect of application of urgency clause to the minimum area required for defense and national security. But even the requirement of the defense needed to restrict to only a few activities which cannot brook any delay. Not all activities relating defense should qualify for this concession.

The utility of initiative for mandatory social impact assessment of all projects involving acquisition of land for the purpose of preparing a rehabilitation and resettlement plan has also been seriously compromised by restricting it to areas involving a specified size of physical displacement. This would end up excluding a large number of projects from this obligation. This is particularly so in hilly terrain where tribal settlements are usually small in size and spatially dispersed. While the District Rehabilitation and Resettlement Committee has been authorized to carry out Social Impact Assessment in such areas under the Rehabilitation and Resettlement Bill, 2009, this has not been inserted in the Land Acquisition (Amendment) Bill, 2009. This implies that government does not intend to hold up the process of land acquisition for Social Impact Assessment in such areas. The positive aspect of the provision is further minimized by excluding the requirement of land for national security involving emergency acquisition of a minimum area. Considering that the Defense establishment is the agency responsible for the largest displacement of tribals (MORD: 2004), the restrictive

Government to review the amendments to the Land Acquisition Act, 1894 contained in the Bill (2009). Meanwhile, National Advisory Council submitted to the Government their recommendations on the proposed changes in the existing law which were a refreshing departure from the thinking of the Government reflected in the earlier Bills (2007 and 2009). The NAC favoured an entirely new law on land acquisition-cum-rehabilitation and resettlement, ban on acquisition of multi-cropped land, substantially enhanced compensation, provision of employment to all displaced persons, mandatory provision of land to all displaced tribals, return of unutilized land to erstwhile owners, disallowance of part acquisition by the Government and part by the company, addition of annuity to the subsistence allowance and pre-acquisition scrutiny of the acquisition proposal along with the rationate of the 'Public Purpose', restricting the use of urgency clause to the rarest of rare cases. During this period, the State Governments of U.P., Haryana and Gujarat also brought out their own land acquisition policies in which they provided for annuity in addition to compensation, allotment of a small plot of developed land in the case of urban development projects with a view to placating the agitating farmers and smoothen the process of acquisition. The recent judgment of the Supreme Court striking down land acquisition in the cases of Noida and greater Noida on grounds of misuse of urgency clause, insensitivity to loss of livelihood etc. hastened the need for a radical review of the ambit of amendments to land acquisition law proposed thus far. In the context of these developments, the Government has abandoned the Bill (2009) altogether and has proposed the Draft National Land Acquisition and Rehabilitation and Resettlement Bill, 2011. It has sought comments of the civil society and the State governments before giving a final shape to it. The Bill is scheduled to be tabled in the ongoing monsoon session of the Parliament.

The draft Bill (2011) is an improvement over earlier Bills (2007 and 2009) in several respects. It is an entirely new Bill replacing the existing Land Acquisition Act, 1894 which would be repealed. It also integrates Rehabilitation and Resettlement with Land Acquisition and abandons the earlier proposal of a separate Bill on the former. There is an unequivocal commitment not to acquire multi-cropped land for 'Public Purpose'. Similarly, land for private companies for private purpose shall not be acquired. The use of urgency clause has been restricted to defense, security and natural calamities to be exercised in rarest of rare cases. The compensation amount has been enhanced to six times the market value. When acquired land is transferred to another agency during the 10 years after acquisition, 20 per cent of the difference in acquisition cost and consideration received when it is transferred shall be paid to the erstwhile owners. The land acquired but remaining unutilized for 5 years shall be returned to the landowners. A pre-acquisition scrutiny of land acquisition proposal would be undertaken to examine if public purpose is served and displacement is minimum. When a private company acquires 100 acres or more of land and approaches the government for partial acquisition for public purpose, the Rehabilitation and Resettlement package shall apply to the affected families. The Rehabilitation and Resettlement proposal include annuity of Rs. 2000 per family per year for 20 years in addition to subsistence allowance for landowners and livelihood losers, and mandatory employment for one member of the affected family, failing which payment of Rs. 2 lakh in lieu thereof, In the case of irrigation projects, the affected landowners, if after acquisition are reduced to the status of a marginal farmers or landless persons, shall get one acre of land per family in the command area. But all affected tribal families in any project would be entitled to one acre of land. Where land is acquired for urbanization, 20 per cent of the developed area would be distributed among the affected land losers. There shall be a separate award for each family both for compensation and rehabilitation and resettlement. The possession over acquired land shall be taken only after payment of compensation and completion of rehabilitation and resettlement. The mandatory provision of house sites shall apply to all affected families without any discrimination.

Notwithstanding the improvements in the package of compensation and rehabilitation offered to the displaced persons, there is little change in the conceptual framework of land acquisition and the major concerns of the persons likely to be affected. There is no dilution to the notion of 'Eminent Domain' and the absolute power of the government to forcibly acquire land from the peasants unwilling to part with it. The definition of 'Public Purpose' revenues so wide as to cover practically any activity for which the government wants to acquire land. The structure of pre-acquisition scrutiny does not inspire much confidence because it is entirely bureaucratic with no participation of the affected persons. The provision of compensation continues to remain property centric, admissible to only landowners and possibly a share from it to the recognized tenants. Even the enhanced compensation would fail to enable to land losers to purchase alternative, agricultural land in the vicinity of the project. The definitions of 'Land' and 'Persons Interested' do not undergo any change. The former continues to exclude common land while the latter fails to keep informal tenants and livelihood losers out of its ambit. The allotment of land is too meager to provide any subsistence to the tribals to save them from destitution. The provision of employment has been robbed of its utility by substituting financial amount for it which is too meager to even provide for an agricultural worker's wage over a period of working life. The annuity per month is too little to provide any social security. The ban on multi-cropped agricultural land, though welcome, is an inadequate response to the enormity of diversion of agricultural land for non-agricultural purposes. It should have been extended to all cultivable land. There are reports that even the proposed ban on multi-cropped land may be diluted under pressure for some state governments. Above all, the new law provides no respite to people who are faced with the prospect of increasing land acquisition. They are not interested in higher compensation but in retaining control over their land. The tribals would continue to suffer the brunt of land acquisition as the new draft law fails to harmonise with the protective laws prohibiting alteration of tribal land to non-tribals. The Bill (2011) when passed by the Parliament would provide no comfort to the agitating peasants and others dependent on land. The approach to the amendments of Land Acquisition Act, 1894 all along is driven by the desire to smoothen the process of land acquisition rather than to meet the concerns of struggling farmers, to do justice to the displaced persons and to protect agricultural land for ensuring food security of the country.

Conclusion

A land policy conducive to accelerated agricultural growth would, first of all, aim at protecting agricultural land from diversion for non-agricultural purposes and its degradation. On both these counts, the land policy that has been followed has miserably failed. The extent of diversion of agricultural land would be evident from the fact that, in 2003 (59th round) the area operated by rural households has reduced to 107.5 million ha from 125.10 million ha in 1991-92 (48th round) and, in fact, has come down from 133.48 m ha in 1960-61 (17th round) (MoF: 2007). Not only this, huge areas of land have been/are being acquired for development projects. We have now reached a situation when we are scouting for productive agricultural land outside the country to meet food security in future. A more depressing picture is with regard to the extent and the level of degradation of land. Some estimates place the degradation between two-thirds to 50 per cent of the total land area. (Reddy: 2003) The most serious cause of land degradation is soil erosion affecting

45 per cent of the geographical area (around 130 m ha of land). Other factors responsible for degradation include water logging (2-46 m ha), Surface water stagnation (3-4 m ha) (MoF: 2007). Industrial pollution has also caused severe damage whose estimate has not been mapped out. The area of wasteland estimates by different agencies range between 175 m ha to 75.50 m ha (MoA: 2006). These estimates are outdated. The extent of wasteland would have increased as the degradation is increasing over time and at an increasing rate. The requisite concern in this respect and effective policies to stop further diversion and degradation of land are lacking despite the issue raised by civil society groups and experts.

In this context, the incorporation of provisions to protect multi-cropped irrigated agricultural land from diversion for non-agricultural purposes through compulsory acquisition in the Bill, 2011 is a belated attempt to arrest this trend. But this proposal is already facing opposition from some states and, even if it gets the legislative approval in its present from and sincerely enforced, would be insufficient to achieve the objective. It would not check acquisition of agricultural land by private parties through direct market transactions which is proceeding at a fast pace. On the land degradation front, the environmental laws are ineffective and are poorly enforced due to connivance of the concerned state agencies and the primacy given to economic growth. The prospects for preserving valuable agricultural land, therefore, are bleak. This calls for a radical change in land use policy and the growth paradigm.

From the viewpoint of peasantry, the Bill (2011) provides virtually no relief. There is going to be no let up in acquisition of land whether through compulsory acquisition by the government or through negotiated procurement by companies. Rather, it is likely to increase notwithstanding the provision relating to minimum acquisition and displacement inserted in the Bill, 2011. There is little hope that the level of deprivation of people of their land and consequent displacement would get appreciably reduced. A little more cash promised in the Bill 2011 by way of compensation and substitute for promised employment is least likely to enthuse the peasantry to part with its land willingly. The country may witness intensification of local resistance with its implication for governance and democracy.

References

Campaign for People's control over the Land Acquisition, Rehabilitation and Natural Resources (2001): *Resettlement Bill, 2000* (Draft by voluntary organizations). *Cities* V. *Chandrachud* on Eminent Domain.

Lok Sabha Secretariat (2008a): '*The Land Acquisition, (Amendment) Bill, 2007*' Thirty-Ninth Report of the Standing Committee on Rural Development (2007-8) October, 2008

Lok Sabha Secretariat (2008b): '*The Rehabilitation and Resettlement Bill, 2007*' Fortieth Report of the Standing Committee on Rural Development (2007-8), October 2008.

Ministry of Agriculture, National *Serving Farmers and Saving Farming,* Commission of Farmers (2006): Fourth Report (April).

Ministry of Finance (2007): *Report of the Expert Group on Agricultural Indebtedness*, Department of Economic Affairs, Banking Division, July.

Ministry of Rural Development (2004): *Report of the Expert Group* on *Prevention of Alienation of Tribal Land and its Restoration.*

Ministry of Rural Development (2007a): *The Land Acquisition (Amendment) Bill, 2007.*

Ministry of Rural Development (2007b): *The Rehabilitation and Resettlement Bill, 2007.*

Ministry of Rural Development (2007c): *The National Rehabilitation and Resettlement Policy, 2007.*

Ministry of Rural Development (2009a): *The Land Acquisition, (Amendment) Bill, 2009*, Bill No. 97-C of 2007 as passed by Lok Saba on February 25, 2009.

Ministry of Rural Development (2009 b): '*The Rehabilitation and Resettlement Bill 2009*' – Bill No. 98-C of 2007 as passed by Lok Sabha on February 25, 2009.

Ministry of Rural Development (2011): Government of India, *The Draft National Land Acquisition and Rehabilitation and Resettlement Bill,* 2011, July 27.

Reddy, Ratna V. (2003): 'Land Degradation in India, Extent, Costs and Determinants', *Economic and Political Weekly*, 43 (10): 4700-13.

Upadhyay, Sanjay and Raman, Bhavani (1988): *Land Acquisition and Public Purpose,* Other Media, New Delhi.

19

Land Acquisition: Compensation and Inclusive Development

P.V. Indiresan

Introduction

The proposed Land Acquisition Bill 2011, for compensation for land acquired for various purposes is an improvement on the original 1894 Act. As an economic solution it is generally acceptable. At the same time, the Bill does not include the issue of Corporate Social Responsibility which has become an important factor these days. As compensation for land is essentially an issue in rural development, both the government and the businessmen can discharge their true Corporate Responsibility only by making rural areas as good as cities are.

The issue can be improved in four different ways. One, though it is largely in consonance with the current policy of Inclusive Growth, land compensation can, and should, offer Inclusive Development too. Two, though it does compensate a few more other than landowners, it does not include all villagers; that omission needs to be rectified. Three, it does not specifically counter the fact that fertile land is acquired quite needlessly and against the wishes of the local people. That waste should be avoided as far as possible. Finally, ecologists often halt necessary projects. The Bill should offer protection against such obstruction.

In addition, landowners have not one but three kinds of assets: one, inflation-protected wealth in their land; two, inflation-protected income from the output of their land and three, prestige of landownership. Money, which is the main compensation that is offered in the Bill is not inflation protected nor is the income derived from it. Money may offer instant prestige but may also be wasted fast. These factors too need to be corrected.

Inclusive Growth, Inclusive Development

It has been the tradition among commentators, administrators and other policy-makers to treat Inclusive Growth and Inclusive Development as being one and the same. Even dictionaries make a similar confusion. However, growth is a physical phenomenon; it can increase and it can decrease. It is reversible. Essentially, it pertains to the lowest Maslow Need of Physical Needs and depends on income. A wise individual may save a part of the income and convert it into development – a chemical change that is not reversible and which pertains to the four higher Maslow Needs of Security, Status, Autonomy and Self-actualisation.

Inclusive Development is more social than individual or economic. An individual gets security with ownership of a residence, acquisition of employable skills, healthcare and the like. Acquisition of a residence may be individual but the other two – employable skills and healthcare – can be obtained only through social institutions. Likewise, status is for the individual but only in relation to others: it depends on education, on the professional status as well as the address where the person resides. Autonomy comes from the freedom or ability to choose – choose where one lives, what profession one pursues and what one does in one's own life. That too requires socio-economic development. Finally, self-actualisation depends on the way one spends one's leisure – that might be in reading, in entertainment, in sports, in travel or whatever. Self-actualisation depends on a person's culture. Thus, all higher Maslow Needs require social inputs as much as increased income.

A person with higher income may save some of it and use those savings to enjoy the higher Maslow Needs or may squander it or may not even be able to obtain higher Maslow Needs only because there are no accessible social inputs. For instance, the son of a peon in Delhi may become a high officer and enjoy the higher Maslow Needs easier than the scion of a *zamindar* in Orissa.

The issue is what the objective of compensation for land acquired from villagers should be. Is it merely to pay them well enough so that they do not create a ruckus and din? Or, is it the goal to provide them a quality of life that urban dwellers take for granted? Is it the aim to reduce the prevailing increase in rural-urban disparity?

Some More, or All Villagers

The Land Acquisition Bill has extended the kinds of people who will get compensation – from actual owners of the land acquired to those who work on it to make a living. That is good. Unfortunately, in Singur, the Tatas found an additional problem. There were many casual labourers who used to earn some income by working occasionally on the land but had no regular contract with the owners. They too wanted to be compensated. Mr Ratan Tata was willing to accommodate them but the state government would not let him. A similar problem exists with the new land compensation bill too. It has no provision for such casual labour; it offers no help even to landowners whose lands are not acquired.

Should we take land acquisition as a contract only between the government and landowners with an extension to contracted cultivators or should it include all villagers, particularly casual labour and other villagers? One can understand the reluctance of the government to make such an extension: that may lead to unsolvable disputes. In this situation, a good solution appears to be to offer social development which will not exclude anybody – through the installation of schools of high quality, training programmes for employable skills, secondary healthcare and mass transport at affordable prices to access such services as well as markets, sanitation, energy and telecommunications.

The question that has not been raised is: Is not land acquisition a good opportunity to develop rural areas and cut down the rural-urban divide?

Where Do We Acquire Land

There are occasions where there is no choice about acquiring land. That happens in the case of mines, hydro projects and even roads. At the same time, there is also a temptation to acquire large tracts of land to build new townships or industrial estates – even when it, quite needlessly, involves converting valuable cultivable land into non-agricultural use. The Bill has no commitment about conserving green fields. That is indeed a lacuna.

Can we have a policy that, except when it is absolutely impractical, only low value land should be acquired even when it will be in bits and pieces? An area, as small as ten hectares, is good enough to develop either an industrial estate or a residential habitat. If that principle is accepted, the villagers could be asked where and at what price they would part with their land and the cheapest option can be chosen. That will introduce healthy competition which will cut down prices; it will certainly protect our agriculture too.

Environmental Issues

Social activists decide much of policy these days; they are particularly active in issues of environment. They often decide what projects may be approved and what should be discarded. It is not unknown for their opposition to cause losses of thousands of crores of rupees. While they do have valuable inputs to offer, it is equally important to have a rational policy on what projects should be or should not be undertaken. The difficulty lies in the fact that the objectors need only raise objections; they do not have to offer effective alternatives.

It is important to decide how much energy we need for our economic growth and for the support of our population; what materials like steel, aluminium, plastics, fertilisers or pesticides we need to serve or feed our population. Many of these industries do pollute. Can we do without them? Is it not better to locate such industries in rural areas rather than in proximity to cities? The Bhopal disaster would probably have never been one if only the factory had been in a rural area and not close to a crowded, highly populated city. These issues may not be considered relevant to the law on compensation for land acquisition but they are important and somebody has to address them. Just as activists cannot be ignored, the issue of locating polluting industries too cannot be brushed away.

Recommendations

Normally, whenever a new Bill is introduced or major amendments are proposed to an earlier Act, critics try to dot the "i"s and cross the "t"s. I suggest here that the amendment of the Land Acquisition Bill offers an opportunity that should be taken to have a re-look into the entire process of development, particularly rural development. That is the only way either the government or business houses can discharge their true responsibility.

For that reason, it has been proposed here that in place of a policy of Inclusive Growth, we should attempt Inclusive Development. In other words, the aim should not be mere financial compensation for the displaced but the development of the entire population. Otherwise, both rural-urban and intra-village social and economic disparities will increase.

Incidentally, in several states, famers have taken exception to the habit of state governments selling land to private entrepreneurs at much higher prices than the compensation paid to them. Either the farmers should get virtually all the payments made by the ultimate buyer or the land should be taken on long lease with a share of the rent obtained from the land.

Inclusive Development is the provision of the four higher Maslow Needs in such a manner that they are available to the entire village. Then, the compensation should be more than money but in the introduction of high quality secondary schools, secondary hospitals, markets and places for recreation. As villages are too small to support such activities, it would be best to have them serve a cluster of villages whose total population is at least 30-50 thousand.

A modern habitat is equally important. At the time of Independence, New Delhi had a Floor Area Ratio (FAR) of 0.5 only. Now it has been increased in some places several times. That is particularly the case within, what are known as "Lal Dora" areas, areas where the original villages were situated. The increase in FAR has been disastrous; it should be held below 0.5 both in new developments and in the old village areas.

Whenever land is acquired, current practise is to build "gated communities". There is, in some states, a rule that shelter be provided for the Economically Weaker Sections also. That is usually inadequate. According to the 80:20 rule, the richer half of a population occupies 80 per cent of the area and the poorer half only the remaining 20 per cent. Hence, the rule should be that 20 per cent of the total residential space should be reserved for the poor and the number of dwellings for the poor should be at least equal to the number of residences for the well-to-do. It should also be the rule that the richer half pays for the cost of infrastructure for the poorer half also. Some of the poor may be accommodated in "servants' quarters"; the remaining poor may then be charged only the marginal costs. Then, the developer does not incur losses but does not either exploit the poor. That is true Corporate Social Responsibility; not making a token contribution out of declared porofits.

It is inevitable that whenever a rural area is developed, highly trained specialists are imported. In all justice, whenever an outsider is brought in, a scholarship should be provided to a local student to get a similar training. It is not necessary that those students should serve within the community; a scholarship is enough compensation for the local population. In fact, social mobility is a useful good that helps to integrate the national population.

Much of current education is esoteric; it does not lead to employment. Hence, it is important that employers be required to pay for employability training. That is actually a matter of self-interest. Employers should also indicate what kinds of skills they need and institutions should be developed that will provide such training.

These views are probably not in the mainstream of current thinking in India. Nevertheless, in my view, Inclusive Development is such an important and so badly neglected aspect of our policy that every opportunity should be taken to promote it.

Index